The
Bioterrorism
Sourcebook

All that is written in books is worth much less than the experience of a wise doctor.

Abu bakr Muhammed ibn-Zakariya al Razi
Persian Physician 860–c.923

In this respect our townsfolk were like everybody else, wrapped up in themselves; in other words they were humanists; they disbelieved in pestilences. A pestilence isn't a thing made to man's measure; therefore we tell ourselves that pestilence is a mere bogy of the mind, a bad dream that will pass away. But it doesn't always pass away and, from one bad dream to another, it is men who pass away, and the humanists first of all, because they haven't taken their precautions. Our townsfolk were not more to blame than others, they forgot to be modest—that was all—and thought that everything was still possible for them; which presupposed that pestilences were impossible. They went on doing business, arranged for journeys and formed views. How should they have given a thought to anything like plague, which rules out any future, cancels journeys, silences the exchange of views. They fancied themselves free, and no one will ever be free so long as there are pestilences.

Albert Camus, *The Plague* (p. 36)

Sixteenth-century woodcut of the Plague of Boils (Smallpox).
Courtesy of Pitt Theological Library Digital Archive Collection

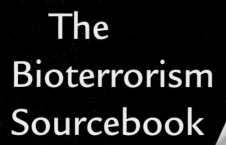

The Bioterrorism Sourcebook

MICHAEL R. GREY, M.D., M.P.H.

Professor of Medicine
Tufts University School of Medicine
Division of General Medicine and Geriatrics
Department of Medicine
Baystate Health Center
Springfield, Massachusetts
Division of Occupational/Environmental Medicine
Department of Medicine
University of Connecticut School of Medicine
Farmington, Connecticut

KENNETH R. SPAETH, M.D., M.P.H.

Saint Vincent Catholic Medical Centers
Department of Medicine
Department of Community Medicine
New York Medical College
New York, New York

McGraw-Hill
Medical Publishing Division

New York Chicago San Francisco Lisbon London Madrid
Mexico City Milan New Delhi San Juan Seoul Singapore Sydney
Toronto

The **McGraw-Hill** Companies

The Bioterrorism Sourcebook

1 2 3 4 5 6 7 8 9 0   DOC/DOC   0 9 8 7 6

ISBN 0-07-144086-0

Notice

Medicine is an ever-changing science. As new research and clinical experience broaden our knowledge, changes in treatment and drug therapy are required. The authors and the publisher of this work have checked with sources believed to be reliable in their efforts to provide information that is complete and generally in accord with the standards accepted at the time of publication. However, in view of the possibility of human error or changes in medical sciences, neither the authors nor the publisher nor any other party who has been involved in the preparation or publication of this work warrants that the information contained herein is in every respect accurate or complete, and they disclaim all responsibility for any errors or omissions or for the results obtained from use of the information contained in this work. Readers are encouraged to confirm the information contained herein with other sources. For example and in particular, readers are advised to check the product information sheet included in the package of each drug they plan to administer to be certain that the information contained in this work is accurate and that changes have not been made in the recommended dose or in the contraindications for administration. This recommendation is of particular importance in connection with new or infrequently used drugs.

This book was set in Berkeley by GTS/TechBooks.
The editors were Marc Strauss and Marsha Loeb.
The production supervisor was Sherri Souffrance.
Project management was provided by TechBooks.
The cover designer was Aimee Nordin.
The index was prepared by Bernice Eisen.
RR Donnelley was printer and binder.

This book is printed on acid-free paper.

Library of Congress Cataloging-in-Publication Data

Grey, Michael R.
  The bioterrorism sourcebook: a medical and public health primer / Michael R. Grey, Kenneth R. Spaeth.
    p. ; cm.
  Includes bibliographical references.
  ISBN 0-07-144086-0
    1. Terrorismx—Health aspects. I. Spaeth, Kenneth R. II. Title.
  [DNLM: 1. Bioterrorism. 2. Bacterial Infections. 3. Chemical Warfare Agents. 4. Disaster Planning. 5. Public Health. WA295 G844b 2005]
  RC88.9.T47G746 2005 2006
  303.6'25—do22   2005041665

For Barbara: blessed with the genius of faith and believer in the best of all possible worlds.

For our children—Nicholas, David, Kelly, and Jordan— that they will face a safer future.
MRG

To the hope for a world in which wisdom guides our heads and compassion our hearts.
KRS

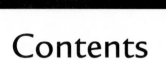

Contents

Preface

In the event of a bioterrorist attack anywhere, clinicians everywhere will assume significant roles in the public health and medical response in their communities. In 2002, Secretary of the Department of Health and Human Services Tommy Thompson stated that "Physicians today need to be ready to recognize and respond to unusual symptoms that might signal a bioterror attack. Primary care doctors might be the first to spot the danger signs, and their knowledge and rapid action could be crucial for the nation." We agree with this assessment and find it a cause of concern that so little attention has been given to the bioterrorism preparedness needs of community-based physicians. Although emergency response efforts warrant the investments of money and resources that are currently underway, failing to look beyond emergency issues means preparation is incomplete. Since community practitioners form the point of first contact with the health care system for most Americans and are key sentinels in the nation's public health system, they have significant responsibilities in meeting the challenge of bioterrorism preparedness and response.

The Bioterrorism Sourcebook was written first and foremost by two general internists for our clinical and public health colleagues. One of the authors (MRG) was involved in two of the more significant public health events in recent memory relating to bioterrorism—the 2001 U.S. Postal Service anthrax attacks—as well as a number of local and state-based bioterrorism preparedness efforts, including the smallpox vaccination program that began in Connecticut on January 24, 2002. Such experiences brought home the realization that any clinician may find him or herself drawn into the issue of bioterrorism. If this could occur in a small state like Connecticut—where

Ottilie Lundgren, a 94-year-old resident of a rural town died of inhalational anthrax two months after the September 11th attacks—it could occur anywhere. Our goal in writing *The Bioterrorism Sourcebook* has been to synthesize and integrate information into a substantive yet practical resource for community-based practitioners that is concise, lucid, and accessible.

We faced several challenges along the way. Bioterrorism is a dynamic area of medical and public health science. It is complex, highly technical, and intrinsically multidisciplinary, operating at the nexus of applied and basic research and encompassing fields as diverse as toxicology, pharmacology, engineering, public health, and health policy. Recent events in the United States and elsewhere in the world have accelerated interest in the field and a rapidly expanding reservoir of information on bioterrorism has resulted, including Websites, journal articles, and online and distance learning courses.

There are several problems with this growing reservoir of information, however. First, much of it focuses on emergency triage or the acute management of biologic, chemical, or nuclear events. For reasons alluded to already and discussed elsewhere in the text, such focus is shortsighted. Further, a large majority of these resources provide information on the so-called "Big Six" biologic agents: smallpox, anthrax, tularemia, viral hemorrhagic fevers, plague, and botulism. A host of other biologic agents as well as chemical and radiologic agents are thus given relatively short shrift. Another concern is that although a great deal of information is accessible electronically to any resourceful individual, clinicians are typically hard-pressed to find the time to efficiently comb through such disparate sources, whose quality and comprehensiveness vary greatly, just to find what they really need. Notwithstanding the many advantages that today's information technology brings to the practice of medicine, in certain circumstances books retain a familiarity, efficiency, and accessibility that newer technologies simply cannot match.

The primary purpose of this book is to fill a void that exists for frontline clinicians and public health workers who will be called on to respond quickly, knowledgeably, and sensibly should a bioterrorist event occur in their community. So much emphasis to date has been on both emergency response and acute care facilities—and not without justification—that we sought to focus instead on educating and preparing community-based physicians, health care workers, and local public health officials for the responsibilities they will assume in the event of a bioterrorist attack as a complement to the care of these victims in hospitals and emergency rooms.

Creating a singular resource for practicing physicians, medical students, residents, and public health officials who work in the outpatient setting that is detailed, practical, and easy to use. To date, no such resource exists despite the enormous amounts of information available on the Internet and medical literature addressing bioterrorism. The vast amount of information available is of varying scope and credibility, but again, no single resource exists that is comprehensive, practical, and meaningful to clinicians. As noted earlier, the Centers for Disease Control and Prevention (CDC) website crashed after the anthrax cases in 2001, underscoring the need for built-in redundancy in our information technology systems, surge capacity, and the value of a hands-on manual such as *The Bioterrorism Sourcebook.*

Organization of the Book

The Bioterrorism Sourcebook is divided into four main sections, the first setting out basic principles and providing background for addressing the class-specific sections that follow. Section I emphasizes relevant clinical and public health issues as they relate to biologic, chemical, and nuclear (BCN) terrorism and provides the foundation for approaching Sections II through IV. We considered separating the more public health-oriented chapters from the more clinical-oriented chapters; however, the threat of bioterrorism requires clinicians to know the clinical aspects of bioterrorism as well as their role in the nation's public health response. For this reason, we hope that our readers will view the cross-disciplinary approach of the introductory section as a foundation from which to best approach the agent-specific chapters that form the bulk of the book.

The introduction proper begins with a brief discussion of terrorism, including derivations and definitions of the term as well as a short overview of the history of terrorism in the ancient and modern world. Terrorism evokes different views depending on the perspective with which one views it, and at the risk of sounding trite, it is reasonable to make the point that one person's terrorist may be another person's freedom fighter or religious martyr. Our short review of terrorism writ large makes the simple point that acts of planned violence against individuals, groups, organizations, or governments for political, religious, or ideologic purposes—what some have offered as a reasonable definition of terrorism—is not a modern phenomenon.

Chapter 2 moves directly into a discussion of the clinical approach to bioterrorism. Recent experiences with public health emergencies, such as the SARS epidemic or Hurricane Katrina, demonstrate that practicing clinicians will be drawn into medical or public health emergencies and yet they remain largely ill-equipped to do so. Practitioners must be prepared to assess patients presenting with potential bioterrorist syndromes, be capable of taking a bioterror history and be comfortable at least initially with triage, treatment, and prophylaxis. This chapter provides one approach to clinical diagnosis in this area. Among the points stressed are that clinicians need to include certain questions traditionally relegated to the social history—if asked at all—to find those clues that might indicate a potential bioterrorism diagnosis. An obvious example of what is meant here is that obtaining a detailed travel or work history may prove critical to the early identification of a bioterrorist event. In addition to providing one approach to the targeted bioterror history, Chapter 2 summarizes likely responsibilities that will devolve to clinicians in the event of a bioterrorist attack. This includes familiarity with initial triage of patients, antibiotic prophylaxis and medical surveillance of primary and secondary contacts, collecting samples, infection control procedures, and the importance of recognizing psychologic sequelae to bioterror events in victims, near victims, and those who were not exposed but nonetheless present for medical evaluation and require knowledgeable reassurance. It bears restating that cumulative and fairly recent experience with bioterrorism and other significant public health issues suggests that for every exposed individual somewhere between 15 and 200 nonexposed individuals will present for medical evaluation. Clinicians will assume a central role in distinguishing between individuals requiring medical triage, treatment, prophylaxis, or expectant surveillance and those requiring only reassurance. Finally, clinicians also must be aware that they have a duty to notify key public health officials in their community in the event they are entertaining or have made a diagnosis consistent with a bioterrorist event.

As noted previously, the introductory section also addresses the public health responsibilities of community-based physicians in the event of a BCN event or, for that matter, any public health emergency. Chapter 3 summarizes clinicians' public health responsibilities, beginning with an overview of existing public health infrastructure as it relates to BCN terrorism and disaster management. This includes the specific roles played by federal and state health and law enforcement agencies, and strategic components of the national bioterrorism response, such as communication strategies, pharmaceutical stockpiling, and the laboratory response network (LRN). This chapter introduces the concept of the incident command system—an organizational approach to disaster management broadened to include BCN specific concerns. The chapter also reviews applicable principles of occupational health and safety and the role of personal protective equipment as they relate to BCN terrorism. If this information seems somewhat far afield from bioterrorism, it is worth pointing out that multiple reviews and analyses of the response to the September 11th attacks found a critical need to develop standardized protocols to maximize safety while enabling rescue workers, first responders, health care workers, and workers responsible for "cleaning up" to do their jobs.

The historical schism between medicine and public health is unacceptable in today's global village. Infectious diseases or environmental and occupational health threats do not respect national boundaries and can therefore intrude—often with sudden and massively disruptive consequences—into ordinary medical practice. One need only mention SARS, Chernobyl, or anthrax to be reminded of how significant these international public health issues can become for practicing physicians. And yet, few clinicians have had substantive medical training or experience in the field of public health. Indeed, one of the lessons driven home time and again whenever a major public health emergency looms is that most public health units in this country are understaffed when it comes to individuals with substantive clinical experience and, more importantly, that this dearth has negative consequences for a well-crafted and effective public health response. This point finds expression in several places throughout the book.

Chapter 4 picks up on the public health theme of the previous chapter by discussing potential environmentally based terrorist threats. The chapter begins with a discussion of the potential risk to the nation's food manufacturing and agricultural sectors and from there moves on to consider how indoor air environments—specifically ventilation systems—offer opportunities for terrorists to disseminate a variety of BCN agents within buildings, as well as options for protecting buildings from such threats. The final subsection of this chapter includes a discussion of water-borne terrorist threats, a possibility that the nation's water industry takes very seriously. Additional details relating to specific water-borne threats are included in agent-specific chapters, particularly the chapter on Category B agents.

Several chapters in the introductory section focus on areas where clinical medicine and public health come face to face in meeting the threat of bioterrorism. Chapter 5 addresses BCN surveillance systems. The term refers variously to approaches encompassing environmental, clinical, or laboratory-based reporting systems designed for early identification of potential bioterrorist events and rapid communication to key individuals or organizations forming the basis of the nation's emergency public health response system. Syndromic surveillance is one of the most rapidly developing areas in the field of bioterrorism. During the 2004 Republican National Convention, for example, state health departments throughout the northeast collaborated with regional hospitals, urgent care centers, and laboratory facilities to create a system by which emergency room encounters and hospital admissions diagnoses were entered into a clinical database programmed to identify selected clinical symptoms or diagnoses as a means of identifying potential BCN events. The goal of surveillance is to be able to facilitate the early identification of emerging public health threats—whether influenza or anthrax, for example— and to promulgate that information quickly to the medical and public health community so that steps may be taken to contain its public health impact. The chapter considers the purposes and capabilities that well-designed surveillance systems would entail. Many surveillance models are being investigated and field tested, including those that use primarily clinical data (such as noted earlier with the 2004 political conventions) to those that use fixed site environmental monitoring, such as exists for radiation worldwide (see Section IV: Radiation Syndromes). The current limitations of existing surveillance strategies and the generalizable value of improved surveillance models for non-BCN health issues are also discussed.

Chapter 5 also presents several of the clinical tools developed or promoted as a means for clinicians to differentiate between naturally occurring and intentional outbreaks. The chapter includes discussion of clinical syndromes—sometimes referred to as toxidromes—as well as clinical algorithms that may aid clinicians in making the diagnosis of a bioterror event. One of the greatest challenges to the community physician is diagnosing an index bioterror case following a small-scale attack with a BCN agent. Infectious diseases take days or even weeks to evolve, but even nuclear or chemical attacks can be subtle in their presentation unless the attack is large enough to generate a sudden cluster of sick individuals presenting geographically and/or temporally with similar clinical syndromes. Further, as noted previously, in the overview of the toxicology chapter, the stereotypical pathophysiologic response of organs and tissues to a variety of insults means that signs and symptoms can be frustratingly nonspecific.

Chapter 5 suggests several clinical tools to aid clinicians in the difficult task of identifying a BCN diagnosis. The first tool is the concept of the sentinel health event or sentinel health data. In the context of BCN terrorism, sentinel data are any information that by its very presence demands consideration of a BCN exposure in the differential diagnosis.

An example might best serve to illustrate this point. A person presenting with a black eschar on their arm might not elicit consideration of cutaneous anthrax, but the identical clinical finding in a person who works in a mailroom of the Pentagon, Senate Office Building, or a U.S. Postal Office should—in today's world—prompt consideration of anthrax. This chapter also considers epidemiologic clues that can raise or lower one's index of suspicion for a BCN event such as travel, atypical presentations of common diseases, or "at-risk" jobs, buildings, and organizations.

The second half of Chapter 5 builds on the clinical utility of sentinel data by suggesting strategies by which clinicians can begin to distinguish sporadic illness caused by indigenous or naturally occurring infectious diseases from planned terrorist attacks.

For example, naturally occurring plague is not unheard of in the Western United States so that an occasional diagnosis would not be unexpected. In contrast, a spate of plague diagnoses in this same region would have very different implications, whereas even a single case of plague in a nonendemic area would have equally serious implications. Weaponization of biologic and chemical agents as a means of amplifying their potential for causing harm is also considered in the section on distinguishing natural from deliberate attacks. For infectious diseases, weaponization usually involves genetic manipulation to add virulence, infectivity, or biopersistence. The anthrax used in the U.S. Post Office bioterrorist attacks was weaponized by greatly increasing its ability to remain airborne, a property that enabled the microbe to be widely distributed in contaminated buildings and lowered the infective dose required to cause disease. Analogous strategies can be applied to both chemical and nuclear agents. For example, adding solvents to chemical agents may increase lipid solubility and therefore dermal absorption, which in turn enhances toxicity.

The next set of chapters covers areas that are important to clinicians and public health workers in preparing for bioterrorism. These include clinically-oriented chapters covering topics such as vulnerable populations (Chapter 6), the psychologic consequences of bioterrorism (Chapter 7), toxicology (Chapter 8), as well as a selected overview of medicolegal and ethical concerns (Chapter 9).

The concept of vulnerable populations may be somewhat unfamiliar to some readers. It refers to the fact that by virtue of preexisting medical, psychologic, or social factors, certain groups are uniquely vulnerable to the adverse effects bioterrorism. Although the individual agent-specific chapters also include small subsections summarizing vulnerable groups as they related to that particular agent or class of agents, Chapter 6 sketches out the general principles behind this concept, one that has much wider applicability in the health and welfare of our society.

There has been a great deal of emphasis on the acute medical issues surrounding bioterrorist events in the medical literature as well as in the media and other information sources. Although this focus is certainly appropriate—particularly for the emergency response system, hospitals, and local, state, and federal public health agencies—community-based clinicians may be expected to address several areas that, to date, have received somewhat less attention. Two areas given only passing attention in most existing resources are the chronic effects of the biologic, chemical, and nuclear agents and whether some individuals are by virtue of medical, psychologic, or social factors uniquely vulnerable to various bioterrorist agents. There is much less literature on which to draw to answer these two questions, but they are questions to which practicing physicians will want answers or at least some guidance. After all, many exposed and nonexposed individuals will continue to seek out the opinion and care of their own physicians following a bioterrorist event and what, if any, possible long-term health consequences are reasonable concerns. Although the chronic health effects of bioterrorist agents and vulnerable groups are dealt with in each agent-specific chapter, we dedicated Chapter 6 to a longer discussion of the medical, psychologic, and socioeconomic factors that may conspire to make some individuals more vulnerable to bioterrorist agents.

Chapter 7 addresses psychologic aspects of bioterrorism and offers clinicians strategies by which they can begin to understand and manage its psychologic consequences. The emotional impact of disasters is an area that is too often given less attention than is warranted. Numerous epidemiologic studies support the view that 70% of all mental health problems in the United States are managed by primary care doctors. Nearly three-fourths of all physician visits are motivated by substantial or wholly psychosocial factors. In the aftermath of the September 11 terrorist attacks, clinicians reported a surge in terrorism-related psychosocial complaints in their patients. This statistic was even higher the closer the practice came to the epicenter of the attacks. There is little doubt that individuals and communities experience profound emotional and psychologic consequences following any disaster, be it natural or deliberate, and that the psychologic fallout usually far exceeds the medical consequences. Clinicians working in communities with large Russian émigré populations will no doubt recall how some of their patients remember vividly the impact of the Chernobyl Nuclear Power Plant disaster not only on individuals living in the region but also on all of Russian society. Many of these individuals were not themselves directly exposed to nuclear material, and yet long after they have left their native country, many still experience profound worry or frank anxiety, depression, or posttraumatic stress. Similar psychologic outcomes will result from any sizeable bioterrorist event. In any given situation, community practitioners will be confronted with a surge in psychosocial complaints in their patients, regardless of their proximity to the actual event or degree of exposure.

Chapter 8 introduces the reader to some of the more relevant principles of toxicology. This chapter is of particular relevance to understanding the health consequences of chemical agents. It is a fundamental precept of toxicology that most organs or tissues respond in only a limited number of ways to toxic insults. This stereotypic tissue response explains why it is not always possible to distinguish pathophysiologically the root cause of any given end organ damage. For example, ARDS can result from numerous specific respiratory insults, ranging from infectious causes (e.g., influenza) to occupational or environmental exposures (chlorine gas) to pulmonary agents, such as phosgene. Elucidating the cause requires careful inquiry as to preexisting conditions, occupational or environmental exposures, or in the case of BCN terrorism, where the person works, has traveled, or with whom they have been in contact, to name a few. Clinicians' understanding of toxicology is generally related to basic pharmacology. In fact, pharmacology and toxicology share many general principles in terms of kinetics, target organ effects, metabolism, and excretion. Although a detailed review of toxicology is well beyond the scope of this book, Chapter 8 offers a quick overview of basic toxicology, including how physical properties, routes of exposure, toxicokinetics, biotransformation, and detoxification influence the toxicity of any given chemical. Understanding these issues will facilitate readers' awareness of the pathways by which chemical agents enter the body as well as the body's response to these particular toxins. Because toxicology has broad applicability to other areas of medicine—for example, poisonings and occupational health—we hope that this "mountain top" approach will enable clinicians to apply this knowledge to other areas of their practice.

The final chapter in Section I focuses selectively on the legal and ethical issues raised by medical and public health preparedness and response. Chapter 9 includes brief discussions on the historical, legal, and regulatory framework in which bioterrorism preparedness and response operates and some of the ethical dilemmas posed by such efforts.

Taken as a whole we hope that the introductory chapters in Section I of *The Bioterrorism Sourcebook* provide a concise foundation by which to understand subsequent chapters focused on the clinical syndromes caused by various BCN agents.

Sections II through IV form the bulk of *The Bioterrorism Sourcebook* and provides detailed information on biologic, chemical, and nuclear agents, respectively. Each major section begins with a brief historical summary of the class's use as warfare or terrorist agents, an overview of basic terminology, and consideration as to why this particular class of agent is a viable threat within the context of bioterrorism preparedness. Readers are encouraged to expand their knowledge by going directly to the sources cited for each chapter and are included in the bibliography. Our aim is to summarize succinctly from authoritative sources the core topics that must form the foundation of our understanding of biologic, chemical, and nuclear agents.

One of the initial decisions was to determine how to categorize the various agents to provide a sensible and coherent structure to the text. Two options were considered.

The first—and the one eventually used—was the approach promulgated most commonly by the CDC and the U.S. military. In this categorization, infectious agents are classified into Category A, B, and C agents, and chemical agents by their dominant clinical presentation, for example, blistering agents, nerve agents, and the like. Alternatively, a syndromic approach across the BCN categories could have been used. A syndromic approach might include a "rash-fever" syndrome such as Category A agents smallpox and viral hemorrhagic fevers, as well as the Category B agents Psittacosis and Typhus fever. Similarly, an "ARDS and fever" syndrome might incorporate not only Category A agents (e.g., anthrax and plague) and Category B biotoxins (e.g., ricin and staphylococcal enterotoxin B), but chemical agents that cause serious pulmonary injury, such as chlorine gas or phosgene gas.

There are compelling advantages and disadvantages to either approach. In the end, the Categories A, B, and C approach to infectious agents is emphasized, chemical agents are classified into four designations used by the CDC and the U.S. military (nerve agents, pulmonary agents, blistering agents [or vesicants], and blood agents), and nuclear agents were left to stand alone in their own section. However, because there is clinical and public health utility for clinicians to appreciate the value of syndrome recognition, Chapter 5 in Section I is dedicated to syndromic surveillance, and whenever appropriate, whichever syndrome a particular agent would fall is noted for each agent or class-specific chapter. In this way, the most common nosologic approach to bioterrorism is emphasized—of value to readers as they look at other sources of information, both text-based and electronic—and a complementary and clinically useful method of making biologic, chemical, and nuclear diagnoses is provided.

Section II is the first of three large sections addressing the major categories of bioterrorism agents. Perhaps because the United State's first experience with bioterror in the post–September 11 era came in the form of a biologic attack with anthrax, the scientific and medical literature, media attention, and much of the national preparedness effort emphasizes the threat of biologic agents. The agent specific sections in the *The Bioterrorism Sourcebook* begins with biologic agents for several reasons, including that quite a bit more is known about these agents, experience with them is much more recent, and they are of particular concern to the public. Section II includes 10 chapters: five on the Category A infectious agents smallpox (Chapter 12), Viral Hemorrhagic Fevers (Chapter 13), Anthrax (Chapter 14), Plague (Chapter 15), Tularemia (Chapter 16), a long chapter on the numerous Category B and C agents (Chapter 19), and finally several chapters on biotoxins, in particular the Category A agent botulinum toxin and ricin.

Category A agents—smallpox, viral hemorrhagic fevers, plague, tularemia, and anthrax—are considered by most authorities to pose the most significant public health threat if used in a terrorist attack. These agents—along with botulinum, technically a biotoxin and not a microbe—are classified by the CDC as Category A agents. This term was adopted in 1999 through a consensus conference held under the auspices of the CDC. The conference drew together public health and infectious disease experts, government representatives, as well as civilian and military intelligence experts and law enforcement officials from around the country. The group identified three categories reflecting graded public health importance: Category A, B, and C agents.

Category A agents have the greatest potential for serious public health impact for three distinct reasons. First, these agents are capable of causing mass casualties, have a contagiousness that enables wide dissemination, and are likely to instill widespread public fear and anxiety. Second, preparedness training and planning for each of these agents necessitates broad multistaged public health approaches. For example, identification and management of an epidemic caused by any of the Category A agents require effective disease and syndrome surveillance systems, diagnostic laboratories trained and equipped for rapid and accurate identification, stockpiling and planning to distribute antidotes, antibiotics and other equipment needed to manage an outbreak, and improved systems of communication among the nation's public health, law enforcement, and health care communities.

Like Category A agents, Category B agents are disseminated easily, but they cause moderate morbidity and only infrequently cause death. Consequently, Category B agents are less likely to cause widespread public panic and require less investment in terms of preparedness. The last category, Category C agents, are emerging pathogens whose absolute risk remains uncertain at this point, but they are not viewed as a high public health threat based on current public health risk assessment. An example of the latter is the Hanta virus. Categories B and C agents are included in their own chapter within the biologic agent section.

Section III addresses the chemical agents that pose the greatest threat as weapons of terror. Once again, this section begins with brief chapters covering the class of agents and their historical use as weapons. The four major categories of chemical agents—pulmonary agents, vesicant or blistering agents, nerve and incapacitating agents, and blood agents—each include sections on the pathophysiology of various chemical agents, laboratory and radiologic findings, the clinical presentation, recommended management, existing sources, vulnerable populations, and the consideration of long-term or chronic effects from exposure.

Section IV begins with a brief historical chapter on the Atomic Age and nuclear issues, followed by a chapter that addresses basic definitions and issues relating to nuclear physics. The subsequent chapter (Chapter 28) on nuclear terrorism begins with a review of the likely scenarios for nuclear terrorism, the pathophysiology of nuclear injury, laboratory, and radiologic findings, the clinical presentation, recommended management, vulnerable groups, and consideration of long-term or chronic effects from exposure.

What follows next for each class of agent is the expected discussion of epidemiology, pathophysiology, clinical manifestations, laboratory and radiographic findings, and treatment. These topics are certainly covered in great depth in the extant literature and on many websites, such as that of the CDC.

Finally, the book includes a series of appendices and a bibliography of useful sources and resources. The first appendix provides detailed dosing regimens based on the latest consensus recommendations for management of BCN diagnoses. The second appendix contains a glossary of key terms, abbreviations, and concepts. The final appendix includes the Quick Reference Guides from the agent-specific chapters. The intent of this duplication is to enable readers' quick access to information that is already contained in various formats within the main text in its most concise format as quickly as possible without having to return to reading through individual chapters. The book closes with a selected bibliography of citations and websites used in the writing of this book, and a list of useful information sources, including key phone numbers and websites.

Readers will noticed that within each chapter are several sections that are not commonly included in standard references and sources but that we feel have particular relevance to practicing physicians. This includes a discussion of the sources of the agents and "at-risk" occupations where exposure to these agents is possible even without a deliberate terrorist attack. Our purpose in providing this information is to reinforce the point that clinicians should have a rough inventory of what facilities or industries in their community could be targets for terrorists or, perhaps even more relevant, where individuals might be exposed to the agent in the absence of a deliberate biologic, chemical, or nuclear attack. The latter adds emphasis to the importance of considering environmental and occupational factors in approaching not only BCN terrorism but also our routine clinical practice. Whenever possible we also summarize what is known about the long-term or chronic effects of these agents. Information on sources of exposure, at-risk occupations, vulnerable groups, and chronic effects also affords clinicians and public health officials with information that might be of use in their roles both during an emergency and in its aftermath.

After all, survivors of any BCN event—be they directly or indirectly exposed—will be confronted with a host of potential physical, psychologic, and social sequelae that cannot be given justice simply by summarizing the acute clinical manifestations, diagnostic approaches, management strategies, and preventive issues.

Creating an off-the-shelf resource for practicing physicians and public health practitioners was the primary goal in writing this book. Therefore, within each chapter are tables providing critical information for each agent in a readily accessible format. These tables address critical medical and public health information for that agent, syndrome-based diagnostic strategies, and public health responsibilities respectively. The Quick Reference Guide, Making the Diagnosis, and Critical First Steps each attempt to live up to their name—the salient points organized for quick access and ease of use.

Final Thoughts

Readers will likely notice aspects of organization, style, or content in this book that differ from many, but not all, medical textbooks. For example, a strategy to make more apparent the relevance of BCN terrorism to community practitioners was adopted by including clinical or public health vignettes at the beginning of each chapter that have—it is confessed—somewhat "headline grabbing" attributes. Some of these vignettes focus on a particular agent as a terrorist threat, whereas others present naturally occurring, occupational, or accidental case histories, the latter purposefully intended to underscore that episodic and occupational/environmental exposures can also lead to disease that mimics bioterrorism diagnoses. Although some of the vignettes are fictionalized accounts of potential clinical encounters, whenever possible actual or newsworthy events underscore the point that bioterrorism is a fact of modern life and requires vigilant awareness on the part of the medical and public health community, as well as of the general public. These vignettes were crafted to illustrate key clinical or public health aspects of that particular agent or class of agents, but they are also intended to stimulate the interest and attention of the reader. For similar purposes, readers will also note the liberal use of historical vignettes, literary excerpts, and quotes throughout the text. As the French novelist Albert Camus in his 1947 postwar novel *The Plague* wrote: "The plague had killed all colors, vetoed all pleasures." These excerpts are not mere leavening—although again it is hoped that they enliven the readers' experience—but are chosen because they speak to the text itself, illuminating evocatively various scientific, social, or psychologic themes as only historians or novelists can do.

The Bioterrorism Sourcebook does its best to provide state-of-the-art information, but it is also fair to point out that bioterrorism—not unlike the HIV/AIDS literature of 10 years ago—is a moving target. Further, knowledge of the health effects of most of the biologic, chemical, and nuclear weapons derives largely from historical experience and therefore firm evidence-based recommendations are few. Extrapolations from other sources, including toxicology studies and military research programs, accidents, and deliberate uses of these weapons do not necessarily provide all that we would like to know about the health impact of these agents. Given this situation, it is not hyperbole to suggest that clinical acumen and early diagnosis might mean the difference between early, controlled interventions and mass casualties.

Two caveats relating to terminology are in order, both of which were adopted largely in the interests of brevity and convenience. First, rather than repeat the phrase "biologic, chemical, nuclear," or combine the first letters of these words into one of several commonly used acronyms (e.g., BCN, NBC, or CNBR), we either use the term of "bioterrorism" to mean all three, or BCN. When referring specifically to one class of bioterrorist agent, we will use terms such as *biologic weapons, chemical agents,* or *nuclear events.* Second, although our primary target audience is practicing clinicians—a group we feel has been underrepresented in current preparedness efforts—the book's format, accessibility, and approach should prove equally valuable to individuals from the many disciplines that comprise the nation's health care and public health infrastructure and that are essential components in our preparedness efforts, in particular, first responders, law enforcement officials, local public health officials, and, of course, learners in the medical, allied health, and public health educational system. We ask readers to forebear with our admittedly arbitrary use of "clinicians," "physicians," or "community-based practitioners," rather than our full intended audience, for the simple reason of avoiding repetition of a string of professions within the text. We acknowledge here that our colleagues from many disciplines and professions share responsibility for the nation's bioterrorism preparedness efforts that is certainly equal to that of physicians and other health care providers.

In making a commitment to write *The Bioterrorism Sourcebook,* we asked ourselves two compelling questions. First, does biologic, chemical, and nuclear terrorism pose a threat to our society? Second, do practicing clinicians have unavoidable and unique responsibilities in meeting these threats? We believe the answer to both questions is an unqualified yes. The threat of bioterrorism is real. Just ask congressional staffers in the offices of Senators Bill Frist or Tom Daschle, the victims and families of the anthrax attacks, or the thousands of U.S. postal workers placed on prophylactic antibiotics as a result of the anthrax attacks. The impact on our country of even these few bioterrorist attacks has been enormous and lasting.

Today, the nation's health care system is a critical component of our national security apparatus, a new and historically unprecedented development whose ultimate impact remains to be seen. It is imperative that the nation's medical community increase its familiarity with the clinical and public health aspects of bioterrorism and it needs accurate, accessible, and concise information to better prepare for addressing the concerns of patients and the public health needs of our communities. *The Bioterrorism Sourcebook* represents our contribution to the educational needs of these medical and public health professionals, our colleagues in the vital effort to improve our nation's preparedness for the next—and sadly, probably inevitable—bioterrorist attack.

Michael R. Grey, M.D., M.P.H.
Springfield, Massachusetts

Kenneth R. Spaeth, M.D., M.P.H.
New York, New York

Acknowledgments

First and foremost, I acknowledge the efforts of my co-author, Dr. Kenneth R. Spaeth, without whom this book would not have come to publication. Our collaboration began while Ken was still a medical student, and I have had the unusual and gratifying experience of seeing a student and mentee emerge as a colleague in the fullest and best sense of that word. That much of the research and nearly all of the writing was accomplished while Ken was still an internal medicine resident is a testimony to his dedication, energy, and enthusiasm.

Many individuals and organizations contributed to the cumulative activities out of which grew the concept and content of this book. The staff of the College of Continuing Studies at the University of Connecticut, especially Martha McKerley; the staff of the Connecticut Directors of Health, especially Jennifer Kertanis, Dr. Ralph Arcari, and Hongie Wang, and the research staff of the Lyman Library at the University of Connecticut School of Medicine provided timely and detailed literature searches on the chronic health effects and vulnerable groups. The dedication, good sense, and good humor of staff physicians and nurses at the Connecticut State Health Department made them colleagues in the truest sense of the word. I would like to single out James Hadler, M.D., M.P.H.; Matthew Cartter, M.D., M.P.H.; Richard Melchreit, M.D., M.P.H.; Debra Rosen, R.N.; Monica Rak, R.N.; and Kristen Sullivan, R.N. Funding for many of the educational programs, from which material was culled for the book, is gratefully acknowledged—including the Department of Public Health, the Connecticut Association of Directors of Health, and the College for Continuing Education at the University of Connecticut Storrs campus. Many colleagues and staff of the University of Connecticut Health Center, in particular the Division of

Occupational/Environmental Medicine, provided hands-on support, thoughtful guidance, and back-up in innumerable ways since our division first became involved in the issue of bioterrorism. I would like to single out for particular praise the following colleagues: Dr. Marcia Trape-Cardoso, Dr. Doug D'Andrea, Dr. Marc Croteau, Dr. Lisa Benaise, and Dr. Franklyn Farrell. Special thanks are owed to Anne Bracker, C.I.H., M.P.H., and Timothy Morse, Ph.D., who researched, supervised, and participated in many educational activities taken on by our division over the last few years. They both provided helpful comments and materials on chapters addressing public health aspects of clinicians' responsibilities. We gratefully acknowledge the work of Dr. Stephan Kales of Cambridge Health Systems, Barbara Blechner of the University of Connecticut School of Medicine, and Dr. Zygmunt Dembek of the Connecticut State Health Department for their insights on the chapters on chemical toxidromes, legal and ethical issues, and syndromic surveillance, respectively. Colleagues from the Departments of Medicine and Emergency Medicine deserve mention as well, including Richard Garibaldi and Robert Fuller. Cheryl Steciak, Melissa Woodword, Bill Cook, and Kathy McDermott provided administrative and IT support for over two years. Their creativity, steadfastness, and good humor were always welcome. I want to recognize Nella Field's immeasurable assistance in the final push toward manuscript preparation. Additional administrative and library support was provided by Sherry Kinsey and the librarians at Baystate Health. The early enthusiasm that Marc Strauss showed for this project was responsible for us putting our faith in McGraw-Hill and we owe a particular debt to the careful and very necessary attention that Karen Edmonson gave to the book in months preceding its publication. Thanks as well to Stephanie Lentz and the staff of TechBooks who shepherded the book through all phases of publication and for their forbearance on many issues. Last, we relied heavily on information available in the public domain for much of the material synthesized in *The Bioterrorism Sourcebook*, in particular the enormous reservoir of information maintained by the Centers for Disease Control on their excellent website (www.cdc.gov).

A number of quotes, stories, and graphics in this book were found by my ever resourceful wife, Barbara Lynn Russell Grey. She knows best of all what it took to bring this project to its conclusion and how many of our mornings together were necessarily shortened in order to do so. Her insight, unfailing friendship, and love without wane makes all burdens lighter.

MRG

Michael Grey has, through the past decade, been my teacher, my mentor, and with this, my colleague. More than merely a great co-author, though, he is a great physician and a great friend. Thanks to Drs. Holger Hansen and Gerard Kerins at the University of Connecticut School of Medicine for their input when this project was in its nascency. Doctors Margaret Smith, Dennis Greenbaum, Carmen Ramis, Jacquelyn Perez and John McAdam at Saint Vincent Catholic Medical Center all generously gave their support and understanding without which the writing of this book and a residency could never have co-existed. Much thanks to Neil Gladstone for suggestions, support, assistance, and a lifetime of friendship. Thanks, too, to Dr. Rose Goldman, at Cambridge Hospitals, for encouraging me to turn a thesis into a book. So many at McGraw-Hill and TechBooks need thanking—in particular John Williams, Marsha Loeb, and Stephanie Lentz, who were both incredibly helpful and incredibly patient. Much gratitude to Marc Strauss for believing in this project. Karen Edmonson deserves special thanks for being the driving force in getting this project on track and completed. Very special thanks to Maya Lopez for her help. My thanks and love and so much more to Melissa Ehrenreich the gifts of for her love, understanding, and support: Amidst the chaos and darkness, she is my light and my joy. Finally, thanks to all who believed, supported, tolerated, endured, waited, forgave, and understood.

KRS

Bioterrorism Timeline

Radiation hazard

Biohazard

Chemical hazard

An historical overview of those events that are felt to have contributed in important ways to the current global threats of biological, chemical, and radiological terrorism.

 1500–1200 BC During epidemics, the Hittites of Anatolia send infected animals and people into enemy realms in order to cause outbreaks.

1200 BC Poison-tipped arrows are used during the Trojan War (described by Homer in 800 BC).

600 BC Solon of Athens puts hellebore roots in the drinking water of Kirrha.

1200–500 BC The Byzantines master the use of "Greek Fire" and direct a naptha mixture through long siphons at enemy navies. It is described as "the ultimate weapon of its time."

500 BC Sun Tzu writes the *Art of War* where he discusses strategies for using smoke and fire against enemies.

420 BC An Athenian stronghold is overtaken by Spartan forces with the use of irrative fumes created by the burning of sulfur, coals, and tars during the Peloponnesian War.

200 BC Hannibal hurls clay pots filled with poisonous snakes at the ships of King Eumenes of Pergamum.

65 BC Mithridates use poisoned honey to incapacitate Pompey's army who indulgently eats the combs. These combs are probably contaminated with botulinum.

90 AD Claims abound that insurgents in and around Rome are speading plague via contaminated pins that they use to prick others.

1346 A Tartar attack on the city of Kaffa in which the warriors catapult the corpses of plague victims into the walled city successfully causes an epidemic.

1495 The Spanish contaminate wine with the blood of lepers and gives it to the French.

1650 The Polish military general, Siemienowicz, reportedly puts saliva from rabid dogs into artillery shells and fires them at his enemies.

1763 Sir Jeffrey Amherst, the Commander-in-Chief of the British, orders smallpox contaminated blankets to be distributed to Native American tribes supporting the British during the French and Indian Wars.

1895 Wilhelm Conrad Roentgen discovers X-rays.

1899 Pierre and Marie Curie discover radiation.

1899 The Hague agreement is reached internationally to prohibit the use of projectiles filled with chemical weapons.

1905 Albert Einstein, who is working full-time as a patent clerk in Bern, Germany, conceives and derives what is perhaps the most recognizable equation in history: $E = MC^2$.

1915 In Ypres, Belgium, the German army uses chlorine gas to attack British and Canadian troops.

1916 The Germans develop phosgene.

1917 The Germans develop mustard gas.

1925 The League of Nations develops "Geneva Protocol for the Prohibition of the Use of Asphyxiating, Poisonous or Other Gases, and Bacteriological Methods of Warfare" in response to the horrors of chemical weapons seen in WWI.

 1932–1944 The Japanese-maintained Unit 731 performs experimentation with biological weapons on POWs including: anthrax, botulism, brucellosis, cholera, dysentery, gas gangrene, meningococcal infection, and plague.

 1935–1936 Italy uses mustard gas in its invasion of Ethiopia.

1938 Nazi scientists discover sarin gas.

 1938 German chemists, Otto Hahn and Fritz Strassmann, are able to split the uranium atom into two roughly equal parts by bombarding it with neutrons.

 1938–1944 Nazis use Zyklon B to systematically murder millions of Jews, Gypsies, POWs, and the mentally ill. The Japanese army uses lewsite, mustard, and possibly other chemical agents against the Chinese.

 1941 The Japanese airforce bombs Pearl Harbor, and the United States enters WWII.

1942 The Manhattan Project is fully coordinated.

 1942 A civilian agency, the War Reserve Service, is established to develop a biological weapons arsenal. The program is headed by George Merck, head of Merck Pharmaceutical. The research and development is done at Fort Detrick, Maryland. Similar efforts develope in numerous other countries.

1943 The British test anthrax bombs off the coast of Scotland for use against the Nazis.

1943–1948 T-2 mycotoxin contamination of grain stores kills at least 100,000 Russian people.

 1944 Palestinian political leader and Nazi collaborator, Haj Amin el-Husseini, organizes an attack on the water supply of a Jewish area in British-governed Palestine. The British are able to apprehend the attackers who are carrying enough Nazi-supplied toxin to kill 25,000 people.

 1945 Atomic bomb tests are first conducted in New Mexico on July 16th.

1945 The first use of the Atomic bomb occurs. "Little Boy" is dropped from the Enola Gay on August 6th. Three days later "Fat Man" is dropped on Nagasaki.

1949 The Soviet Union successfully tests its first nuclear bomb.

1950s The Cold War and the associated Arms Race are underway.

1962 The Soviets place ICBMs in Cuba and the Cuban Missile crisis results.

1964 China performs its first test of a nuclear weapon.

 1969 President Nixon converts the War Reserve Service to USAMRIID, whose sole purpose is maintaining biodefense rather than developing weapons.

1970s The Strategic Arms Limitation Talks (SALT) treaties are initiated.

 1972 The Convention on the Prohibition of the Development, Production, and Stockpiling of Bacteriological and Toxin Weapons and Their Destruction, otherwise referred to as the 1972 Biological and Toxin Weapons Convention (BTWC) is established. It is signed by the United States, the USSR, and 144 other countries.

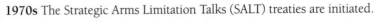

1974 India performs its first test of a nuclear weapon.

 1976 Ebola virus is identified in the Congo in Zaire.

1978 A Bulgarian defector is assassinated in London by the KGB who fires a ricin pellet into his leg from a umbrella gun. He dies 4 days later.

 1979 A core meltdown in a main reactor at the Three Mile Island Nuclear facility leads to the release of a small amount of radioactivity into the environment on March 28th. No injuries or deaths result.

 1980s The Soviets reportedly give mycotoxins to the communist governments of Vietnam and Laos for use against CIA supported-resistance movements. There are similar reports from Cambodia and Afghanistan.

 1980s Iraq, under the leadership Saddam Hussein, uses mustard and nerve agents against Iranian forces in the Iran–Iraq war, killing and injuring roughly 10,000 Iranian soldiers.

1983 The Soviets shoot down commercial airliner Korean Air Line 007 that crosses into Soviet airspace; a United States congressman is aboard. Later that year, the United States deploys ICBMs throughout Europe.

1985 Perestroika policies implemented in the Soviet Union by Mikhail Gorbachev leads to de-escalation of the Arms Race.

1986 The Chernobyl Nuclear Facility accident occurs near Belarus, Ukraine on April 26th. Human and mechanical errors plus major flaws in the safety features result in the release of radioactive material. In the aftermath of the accident, some 100,000 people are evacuated and the surrounding area is contaminated with radioactive fallout. Thousands are injured, thousands die, and the death toll continues to climb.

1987 An abandoned medical source of cesium-137 exposes 249 Brazilians to ionizing radiation and results in 52 hospitalizations, 4 deaths, and thousands of post-event medical visits, most of which are "worried well."

1987–1988 Iraq uses chemical agents against ethnic Kurdish citizens residing in northern Iraq killing thousands and injuring tens of thousands.

1991 United States coalition-led forces uncover 50,000 shells and bombs containing mustard gas, sarin, and cyclohexyl sarin.

 1992 Russian police confiscate 1.5 kilograms of highly enriched uranium at a train station in Podolsk, Russia in August.

1993 Two Russian naval servicemen are arrested in July. They were attempting to smuggle highly enriched uranium out of the country that they stole from a naval base.

 1993 The Convention on the Prohibition of Development, Production, Stockpiling, and Use of Chemical Weapons and on Their Destruction (Chemical Weapons Convention) is signed by 130 countries in Paris. The Geneva Protocol, from 1925, had only banned use of such weapons.

1994 Aum Shinrikyo, a Japanese cult, release sarin gas in the Japanese city of Matsumoto killing 7 and injuring about 600 people in March.

 1994 A Slovakian man is arrested in June for attempting to sell highly enriched uranium to undercover agents in Germany. The materials are traced back to a Russian nuclear facility.

 1995 Aum Shinrikyo releases sarin gas into the Tokyo subway system killing 12 and injuring 3,800 people.

  1998 Pakistan, Iran, and North Korea test nuclear devices.

1998 Turkish police arrest 8 people (4 from Turkey, 2 from Kazakhstan, and 2 from Azerbaijan) confiscating uranium and plutonium. The smugglers were selling the material to undercover police.

 1999 Nipah virus is identified as the causative agent in an encephalitis outbreak in Malaysia and Singapore sickening over 250 people and killing over 100 people.

2000 Four residents of the former Soviet state of Georgia are arrested in April for possession of highly enriched uranium.

2001 Three men are arrested in France in June for smuggling enriched uranium. It is believed they were attempting to sell it on the black market.

 2001 In October, one month after the attacks on the WTC and the Pentagon, anthrax attacks occur in the United States. Spores are sent through the mail and consequently causes 22 infections with 5 deaths.

 2002 American citizen, Abdullah Al Muhajir, is arrested in June along with two Pakistani immigrants for planning to detonate a "dirty bomb."

2002 The G7 nations in June give Russia 20 billion dollars to protect and dismantle its nuclear arsenal.

2002 Two men are arrested in Turkey carrying weapons-grade uranium worth 5 million dollars on the black market. It is believed they were heading to Syria or Iraq.

2003 A manager at a state-run Russian nuclear facility is arrested in August trying to sell uranium and other radioactive materials. He was accumulating these materials in his garage.

 2003 In January, 7 Algerians are arrested in London, England for possessing small amounts of ricin. The Algerians also had the means to produce more ricin.

2003 Two letters are found containing ricin in October and November. One is addressed to the Department of Transportation, and the other to the White House.

 2004 Authorities in Johannesburg, South Africa arrest a man in September for illegally distributing 11 shipping containers of uranium-enrichment materials. The man has been linked to prior efforts in helping Libya acquire such materials.

SECTION I

Clinical Principles and Practices

CHAPTER 1

Introduction to Terrorism and Bioterrorism

The modern understanding of the word "terror" dates from the Reign of Terror following the French Revolution of 1789. One of the key figures in the Revolutionary Council—the governing body responsible for thousands of deaths by guillotine—was Maximilien Robespierre (1758–1794). Robespierre's observation that "Terror is nothing more than justice, prompt, severe, inflexible; it is therefore an emanation of virtue" rings familiar to anyone who has read or listened to the righteous rhetoric of terrorist organizations, be they Islamic militant groups, the Protestant and Catholic factions in Northern Ireland's civil strife, or even in the words of homegrown terrorists like Timothy McVeigh. There are numerous examples that fit this definition of terrorism throughout both modern and ancient history.

The United States has had its share of actions that meet a modern view of terrorism. Whether one is speaking of the assassination of President Abraham Lincoln in 1866, or the 1998 bombing of the federal building in Oklahoma City, terrorism may reflect homegrown conflicting worldviews as well as imported ones. As the 20th century drew to a close, religious justification for terrorism emerged across the globe and in widely different societies and cultures—from Northern Ireland to Africa, from Indonesia to the Middle East. Throughout human history, terrorism has been employed by religious, political, and ideologically driven individuals, organizations, or governments to communicate their agenda, their resolve, and their capacity to disrupt life as usual.

One definition of terrorism is the deliberate, planned violent act whose purpose is to cause injury or sow fear to achieve political, religious, or ideological objectives. According to the *New York Times*, "terrorism seeks to hurt a few people and to scare a

3

lot of people in order to make a point" in contrast to traditional warfare whose purpose is to conquer territory and capture cities (January 6, 2000). Terrorism may take the form of individual or state-sponsored terrorism but in all instances maintains these characteristics. When we talk about terrorism in the modern sense, we are usually referring to acts that target individuals, organizations, or governments.

In 1992, Francis Fukuyama, then Deputy Director of Policy Planning for the State Department, declared that the fall of the Soviet Empire was the final event marking the coronation of Western liberal democracy. The world, he argued, had fully and finally been delivered. The title of his book on the subject summed up this view: *The End of History*. At the time, it seemed a possibility: our enemy had been defeated and the new world order was at hand. On September 11, 2001, this comforting view shattered.

Instead, a new world enemy has been ordained, a different type of enemy: different worldview, different culture, different means. This enemy is unfamiliar, elusive, without institutions or uniforms or clear structure, and it wages its violence in unconventional ways; among these are thought to be biological, chemical, and nuclear (BCN) weapons. The existence and threat of such weapons became etched into the public consciousness beginning with the index case of the post-September 11 anthrax attacks in October 2001 and became injected indelibly into it with the first administrations of smallpox vaccine that began with four physicians at the University of Connecticut Health Center, Farmington, Connecticut, in January 2003 (Fig. 1–1).

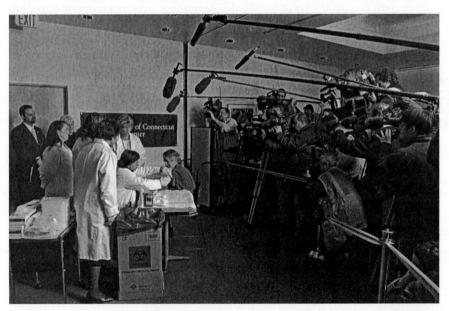

FIGURE 1–1   The short-lived National Smallpox Vaccination Program being initiated with vaccine administration at the University of Connecticut Health Center, Farmington, Connecticut in January of 2003.
Courtesy of Jim Walter, M.D.

Although the intensity of our nation's focus on and concern with terrorism waxes and wanes—driven by media stories or changes in the alert level issued by the Homeland Security Agency from yellow and back again—underneath these oscillations is a vigilance that would have been inconceivable only a few short years ago.

Experts typically summarize three general categories of weapons of terror: weapons of mass destruction (WMD), weapons of mass casualty, and weapons of mass disruption. WMDs are devices—such as missiles or bombs—that are capable of inflicting serious damage to infrastructure, for example, buildings, bridges, or dams. In contrast, biologic, chemical, and radiological agents are viewed as weapons of mass casualty because they may result in widespread disease, debility, or death. Weapons of mass disruption are devices or strategies intended to inflict social, political, or economic injury on a targeted group, organization, or society. Examples of the latter are poisoning or contamination of agricultural supplies or food manufacturing systems (agroterrorism), or cyberterrorism, where vital systems are disrupted or critical records are destroyed by computer hackers. These three categories are by no means mutually exclusive as the September 11 attacks on the World Trade Center (WTC) and Pentagon made all too clear. In each instance, the terrorists destroyed nationally prominent buildings, killed over 3,000 persons, and inflicted profound and lingering social, political, and economic disruption of American society from which it is still recovering (Fig. 1–2).

Much can be and has been written about the methods by which these deliberate attacks occur. Nevertheless, experts tell us that the terrorist's overarching desire

FIGURE 1-2 Clean-up at the remains of the World Trade Center.
Courtesy of FEMA/Andrea Booher.

is to injure or kill, destroy objects or infrastructure, or otherwise disrupt the normal routine of society, whereas the mechanism of achieving these outcomes is less important. An obvious corollary of this is that terrorists are free to use unconventional or unexpected means intended to surprise their chosen target. Terrorists can often use their numeric, logistic, and technologic inferiority to achieve a tactical or strategic advantage leading some experts to refer to terrorism as "asymmetric warfare." That is, a confrontation between an "enemy" whose resources are unequal to their opponent and thus use apparently random acts of violence in order to communicate political, religious, or ideological objectives to the rest of the world. This point was made quite explicit in then Secretary of Defense William Cohen's Annual Defense Report to the President and the Congress in 1998. In it, Cohen acknowledged that asymmetric warfare offered strategic advantages for terrorists in their efforts to gain a tactical advantage. "Those who oppose the United States," wrote Cohen, "will increasingly rely on unconventional strategies and tactics to offset U.S. superiority, such as biological or chemical weapons."

✯Biological, Chemical, and Nuclear Terrorism: Present Realities and Concerns

Even before September 11, 2001, the federal government was preparing with relatively little fanfare for the possibility of BCN terrorism within the nation's borders. As part of this preparation, Congress passed legislation requiring the Justice Department's Federal Bureau of Investigation (FBI) to conduct field tests of our crisis response systems. One of these field tests, referred to as TOPOFF 1, took place in the city of Denver, Colorado, on May 17, 2000, at a cost of approximately $3 million. A brief review of this real-time simulation of our preparedness for bioterrorism is both instructive and sobering.

Denver was considered an optimal location to run the TOPOFF experiment because this particular region of the country had received significant funding and training in disaster management over the years due, in part, to the substantial military presence in the state. The biological agent chosen was *Yersinia pestis*, the causative agent for plague. The combined resources of local, state, and federal health and law enforcement officials as well as local law enforcement agencies were brought to bear following the mock release of plague. Although some members of Congress were debriefed on the TOPOFF 1 trial, for reasons that will become clear, details were kept from the media and the public for quite some time. While many aspects of the test are worth commenting on, a few will serve to highlight lessons learned as a result of the TOPOFF test and exemplify the larger issues that *The Bioterrorism Sourcebook* is intended to address.

Perhaps the most serious problem was that the locus of decision making was unclear even to participants. For example, conference calls used to make critical decisions often involved 50 to 100 individuals. Not surprisingly, within a very short period of time, communication began to break down as key personnel were besieged by multiple competing demands coming from different authorities. A reasoned decision to provide antibiotic prophylaxis to the public could not be implemented as there were insufficient resources. For example, existing clinical personnel could not keep up with the demands of providing 140 individuals per hour with antibiotics against the plague. Within a short period of time, antibiotic stores, hospital bed and emergency room capacity, and the energies of participants were exhausted. As the simulation proceeded, the epidemic breached the original "containment" zone established within the city limits and a call for the implementation of quarantine went out. Notwithstanding the enormous social, political, and economic repercussions of such a decision—which we discuss further in Chapter 9—it turned out that the resources needed to enforce a quarantine simply did not exist.

The TOPOFF I simulation taught other lessons as well. For our purposes, one of the most important was that, although there was sufficient public health talent available, it was of the wrong kind to respond to this level of public health emergency. What was needed during the acute period were not just epidemiologists, policymakers, or biostatisticians, but individuals with clinical expertise who could provide triage, treat affected individuals, assess the need and potential risks of antibiotic prophylaxis in potentially exposed individuals, and communicate with the public. The bald fact of the matter is that relatively few health departments have individuals with significant clinical experience working for them, nor do these organizations have ready access to individuals who can take care of people in a crisis of this magnitude. The largest qualified pool of individuals who could serve in such a capacity are community-based clinicians, particularly primary care doctors, a group who, in our estimation, has been underutilized, if not marginalized in bioterrorism preparation efforts. A second TOPOFF simulation using a "dirty bomb" or radiation dispersion device was conducted in 2003 in Seattle, but as of this writing, information from the second TOPOFF remains closely held by the FBI. A third TOPOFF stimulation was conducted in the spring of 2005.

Some might argue that TOPOFF—being a simulation of a bioterrorist attack—is not necessarily representative of what would happen in a real crisis. Very similar issues, however, arose in 2003 during the severe acute respiratory syndrome (SARS) epidemic in Toronto, Canada. The physical, emotional, and professional demands the situation placed on provincial and city health officials were exhausting, and the public health system itself was taxed severely. Further, one of the primary difficulties they recognized as the crisis evolved was the barrier to effective and timely communication with community-based physicians, an issue that has not yet been adequately addressed in the context of bioterrorism. The same experience occurred in the United States during the 2001 anthrax attacks, and in subsequent anthrax hoaxes. In the weeks following the anthrax attacks, the Centers for Disease Control and Prevention (CDC) assigned over a thousand employees to a situation that, when all was said and done, involved only 22 cases. The CDC's excellent website, containing extensive information on BCN agents and bioterrorism preparedness, was inundated during the anthrax attacks and at one point crashed. The serious implications of these few facts are obvious. Even a small epidemic caused by a not highly transmissible disease nearly paralyzed the nation's public health community. In the case of the TOPOFF trial, a biological attack in a single city could not be contained. But these were not the first indications that the nation was woefully ill-prepared for a bioterrorist attack.

☣The Dark Winter Exercise

In June 2001—three months before the WTC and Pentagon attacks and four months before the anthrax attacks—a simulated smallpox attack on U.S. soil was conducted at Andrews Airforce Base. Dark Winter, as it was called, took place over a two-day period and involved four organizations involved in bioterrorism preparedness.[1] Former and current government officials and military officers were asked to role-play the parts of the National Security Council, with former Senator Sam Nunn of Georgia as the president. Observers to the exercise included members of the National Governor's Association, FBI, CDC, National War College, and others. The results of Dark Winter were shared with senior administration officials, eighty members of Congress, and the international diplomatic community in the United States and abroad. Three "weekly" meetings of the faux National Safety Council (NSC) were held, and at each, those running the simulation gave information to the "president" and the NSC regarding the outbreak. The premise of Dark Winter was an aerosol release of weaponized smallpox in the ventilation systems of three buildings in Oklahoma City, Philadelphia, and Atlanta.

The initial facts were these: on December 9th, the NSC was informed of an attack exposing some 2,000 people; 12 individuals were subsequently diagnosed with smallpox. In response to this information, the president called for the implementation of ring vaccination to protect primary and secondary contacts. Efforts to inform the public were taken, including a Presidential Address. The president ordered the distribution of 1 million of the existing 12 million smallpox vaccine doses held in reserve for the Department of Defense. A media frenzy and public panic began.

At its next meeting held the following "week," the NSC was informed that ring vaccination had failed. The epidemic had now spread to fifteen states, killed 300 individuals, and involved another 2,000 diagnosed cases. Tens of thousands of health care workers were not showing up for work, violence had broken out around vaccination clinics in Philadelphia, and neighboring countries had closed their borders with the United States.

---

[1] Center for Strategic and International Studies, Johns Hopkins Center for Civilian Biodefense Strategies, ANSER Institute for Homeland Security, and Memorial Institute for the Prevention of Terrorism.

The president ordered the National Guard to quell the violence, and both national quarantine and Marshall law were considered. The final meeting took place on "December 22nd," 14 days after the initial attacks. By this point, 1,000 people had died of smallpox and 16,000 cases had been diagnosed in twenty-five states and ten countries. Vaccine stores were depleted, hospitals and emergency rooms were filled, and economic activity had come to a standstill. In short, a state of near chaos existed across the country. Dark Winter earned its name as the hypothetical epidemic propagated. Given the 17-day incubation period for Variola major, a second epidemic wave was predicted in early January with over 30,000 cases worldwide and 10,000 deaths. A third epidemic wave would involve some 300,000 infected individuals and 100,000 deaths. By the fourth epidemic wave, total cases worldwide would exceed 3 million with 1 million deaths from smallpox.

The Dark Winter exercise was a wake-up call for our nation's political, public health, and health care institutions. Whether one agrees entirely with its assumptions and the scenario, it conforms to other experiences both real and simulated. According to participants and postexercise analyses, a major bioterrorist attack in the United States has the potential to cause enormous loss of life, the disruption of essential institutions, civil unrest, and loss of public confidence in government, public health, and health care institutions. Additional conclusions were that the nation lacked adequate strategies, plans, and information systems to manage such a crisis, that national security needs had now placed public health preparedness at the forefront of strategic planning and funding, and that a crisis of this magnitude engendered social, ethical, legal, political, logistical, and economic consequences that we were not yet prepared to manage with any degree of confidence.

The Dark Winter and TOPOFF simulations, and the *real* world international SARS epidemic, demonstrate convincingly the dilemma the historical schism between medicine and public health poses in our efforts to respond to major public health threats in societies where mass communication, international travel, and strong traditions of civil liberties all come to bear. Bioterrorism clearly represents a critical convergence of the health care and public health systems whose traditions and perceived missions are not integrated as readily as will be necessary to meet this new and unfamiliar threat.

The traumatic events of September 2001, the U.S. Postal Service anthrax attacks, the nationwide smallpox immunization campaign, and the ongoing war on terrorism form the social and political substrate out of which this book emerged. Experts offer a wide variety of scenarios for how a bioterrorism event might unfold, but one aspect remains constant throughout: once an attack is identified, the focus of planning and preparation centers on managing the emergency response. This is a reasonable place to begin, of course. A bioterrorist attack could be quite large and cause great panic. Looked at differently, however, this focus on emergency response addresses only part of the planning that is necessary in order to prepare the health care and public health communities and the American public. Why so? The presenting signs and symptoms of nearly all the viable BCN weapons are often nonspecific, and indeed, individuals may not even know they have been affected as was seen early in the anthrax attacks. Furthermore, for terrorists, the logistical difficulties involved in a large-scale attack are such that smaller, more subtle attacks might be even more likely. As a result, it seems quite possible that after an attack the first cases could well be seen in an outpatient setting. Yet, a survey of the burgeoning body of bioterrorism literature indicates a failure to address this possibility. Medline searches reveal virtually nothing directed at responding to or managing bioterrorism in the outpatient setting. This oversight is further reflected in the emergency response emphasis seen at the federal and state levels, where, with the exception of the CDC, relatively little is offered in the way of training, preparing, or generally involving primary care and community clinicians in bioterrorist events. This lack of attention is troubling because formal and informal surveys indicate that the large majority of community practitioners feel terribly unprepared to respond to the issue of bioterrorism. One study found that only 22% of physicians felt that they or their hospitals, clinics, and offices were adequately prepared to respond to a biological attack. Studies addressing the issue of health care workers' willingness to care for patients during an epidemic are also troubling. In one survey of physicians in New York and Washington, DC, it was found that although 80% would care for patients in the event of a smallpox outbreak, this number dropped to 33% if the physicians had not yet been vaccinated. Further, 45% of respondents did not feel obligated to care for patients if doing so meant endangering their own life or that of their families. In light of these observations, it is not far fetched to suggest that knowledgeable and well-prepared community-based practitioners might well be the difference between a quickly contained, limited bioterrorist attack and a propagating epidemic.

Another important lesson from both genuine and simulated bioterrorist attacks is that medical care is local. With guidance from federal, state, and local public health officials, local facilities provided virtually all of the care to affected individuals, and local physicians made the majority of decisions regarding care and prophylaxis during the post office anthrax attacks. Accurate and timely communication among the CDC, state and local health departments, law enforcement officials, and community practitioners remains a potentially weak link in our response to public health emergencies. This concern remains today, although it has been lessened by the investment of significant resources over the last few years. A pertinent example occurred in New York City following the diagnosis of a case of cutaneous anthrax there. A public information hotline established in the wake of the September 11 attacks received over 15,000 calls in a single day. For every case, there were over 15 noncases, and this figure is probably an underestimate. By some estimates, for every individual actually exposed and in need of medical care, nearly 200 non-exposed individuals will seek medical attention following a bioterrorist event. This will put enormous pressure on the local health care system and physicians. It further underscores the need for adequate preparation for primary care doctors, as well as the need for ready access to timely information during an event. What's more, how the medical community responds not only will facilitate the efficient implementation of

FIGURE 1-3 President Bush observing an emergency response drill for a simulated chemical weapons attack.
Courtesy of Georgia Institute of Technology.

FIGURE 1-4 Repair and clean-up after the September 11 attack on the Pentagon.
Courtesy of FEMA/Jocelyn Augustino.

any acute *clinical* management and public health action, but also has the possibility of affecting social order. Local physicians will be critical in providing reassurance and calm to many who will be frightened and insecure in the aftermath of any such event, regardless of their proximity to the actual event. Indeed, one of the recurring observations following any terrorist attack—whether it be traditional terrorism such as the WTC, or a bioterrorist attack with ricin, such as in the Tokyo Subway—is that psychological sequelae and "the worried well" outnumber direct victims sometimes by an order of magnitude. This fact alone highlights the critical need for the availability of nonemergency response providers, if only to triage physically distressed victims requiring reassurance from those who require medical treatment, isolation, or surveillance.

Just as importantly, the medical community will be involved in addressing a wide range of issues in the aftermath of any acute event. Once the urgent issues are addressed—be they crisis or disaster management, triage, mass vaccination clinics—lingering medical, psychological, and social issues will remain to test the nation's health care system, most certainly requiring ongoing involvement on the part of community practitioners. A pertinent example is the WTC attack, a traditional terrorist attack whose scale, inventiveness, and results dwarfed our previous experiences with terrorism, whether homegrown or international in origin. The emergency system, medical, public health, and law enforcement mobilization, that occurred immediately was remarkable (Fig. 1–5). Few now realize that over 72,000 individuals, sometimes referred to as the "WTC" cohort are now enrolled in ongoing medical surveillance programs. This fact testifies to the necessity of approaching bioterrorism with a long-range perspective, as well as to the vital role that community-based practitioners will assume in the care of the myriad possible long-term health issues that follow such attacks.

In the post-September 11 world, clinicians will be asked to carry a great number of responsibilities in the context of bioterrorism: early identification and diagnosis of a possible bioterrorist event, responding to the consequences of an attack, managing patients medically in both preexposure and postexposure circumstances, surveying and interfacing with emergency response systems, educating patients and the community at large, and finally, allaying the legitimate concern and anxiety of patients, families, and communities in the face of uncertainty or imminent danger. These are daunting responsibilities. Our aim in this book is to begin to address the unique needs of practicing clinicians who, by circumstance and status, will be expected to meet these responsibilities.

Both the medical and public health communities are embarking on a challenging journey. Identifying, containing, responding to, and managing bioterrorism in all of its manifestations is a monumental task. It is clear from our experiences with anthrax, the smallpox immunization efforts, and scenarios such as the TOPOFF trials that there is still much work to do, laudible progress notwithstanding. Particularly troubling from a primary care perspective is how relatively little effort is being made to prepare those working in the community to respond to bioterrorism events. A bioterrorism event is likely to be similar to the

FIGURE 1-5 The cloud of dust and particulates that resulted from the collapse of the buildings at the World Trade Center.
Courtesy of the EPA.

post office anthrax attacks—small scale, subtle, and nonspecific—and will likely be seen in an outpatient setting. Further, depending on the biological agent's transmissibility, if such an attack was identified outside the hospital or emergency room setting, it could mean the difference between a self-limited epidemic and a propagating one. Think of the lessons provided by the Dark Winter exercise. Even in a large-scale attack, well-prepared physicians will become a national resource for clinical and public health efforts aimed at containing and managing the event. This fact is troubling because every study and experience with bioterrorism indicates that affected communities near and far will bombard their primary care givers seeking diagnosis, prophylaxis, answers, and comfort. This last point must be weighed against the fact that outpatient physicians feel unprepared to respond to bioterrorism.

Developing surveillance and emergency response plans is critical, but failing to include primary care doctors and other practicing clinicians in bioterrorism preparedness efforts is potentially a critical oversight. *The Bioterrorism Sourcebook* is a small step in helping those in the outpatient setting to be prepared from both a public health and clinical perspective. By covering areas such as toxicology, public health efforts, agroterrorism, syndromic surveillance, and other topics, we have tried to convey basic concepts and practices that shed light on bioterrorism from a broader perspective. Clinical chapters on the major biological, chemical, and radiological agents offer succinct details on historical, diagnostic, management, and preventive issues that we hope will serve as an informative and practical resource for clinicians.

Several events in recent years offer insight into the depth and breadth of the challenges posed by bioterrorism, from real world epidemics such as SARS, the U.S. Postal Service anthrax attacks, and September 11 attacks to simulated attacks such as the Dark Winter exercise and the TOPOFF trials. To different degrees and in an array of ways, each also offers insight into where weaknesses, even voids, still exist. The purpose of *The Bioterrorism Sourcebook* is to provide clinicians and the public health community the information and resources they need to prepare for and respond to future biological, chemical, or nuclear threats.

CHAPTER 2

A Clinical Approach to Biological, Chemical, and Nuclear Terrorism

The U.S. Postal Service anthrax attacks demonstrated that while diagnosing an index case of a BCN event is difficult, knowledgeable and vigilant clinicians can and do play vital roles in lessening the extent and severity of such attacks. Increased clinical vigilance results in earlier recognition and earlier intervention. Likewise, more vigilant public health efforts facilitate preventive interventions (e.g., antibiotic prophylaxis) and environmental decontamination. This in turn protects exposed workers and prevents further exposures. The clinical and public health experiences gained from the anthrax attacks serve as a valuable frame of reference for anticipating the clinical and public health needs generated by any future BCN attacks.

Certainly, an acute BCN event will activate immediately the machinery of the nation's public health infrastructure and alert clinicians to evaluate all patients in a different way. With index cases of bioterrorism, the challenges are much greater: clinicians need to recognize BCN exposure even when it is subtle and unheralded, as with the early cases of anthrax.

Clinicians need to be able to take an appropriate history and conduct a targeted physical examination not only to ensure an index case does not get missed following a BCN event, but also to evaluate all patients with a syndrome consistent with BCN exposure following a recognized attack. A second element of clinicians' responsibilities relates to infection control. Early and strict adherence to established infection control practices is essential to protecting health care workers and first responders, medical and ancillary staff, and secondary contacts, and to limit the spread of an epidemic. Finally, in addition to their bedside skills and

15

OF NOTE...

Hijackers in Florida

It is believed that one of the September 11 hijackers was seen by a Florida physician for what was initially diagnosed as a skin infection but was later (during the September 11 attack investigations) diagnosed as cutaneous anthrax. In the doctor's defense, such a diagnosis would have been extraordinarily rare, particularly when the United States had not yet recognized the dangers to come. Nonetheless, a proper diagnosis initially may have altered history—serving to highlight the importance of properly trained and vigilant clinicians.

awareness of infection control practices, clinicians must also be prepared to engage with both the public health and legal systems when responding to any real or potential BCN event. This chapter provides guidance to clinicians in the three essential responsibilities of clinical diagnosis, infection control, and public health intervention.

Taking a BCN History

Clinical vigilance in today's geopolitical climate has become a requirement for clinicians. Barring a sentinel terrorist event that changes the clinical approach radically, BCN possibilities should be ever present, albeit hovering low and distant on differential diagnoses of appropriate clinical pictures. Such a practice is rife with challenges, particularly because the signs and symptoms of the biological and chemical agents are typically nonspecific, especially early on. This is particularly true in an unrecognized attack as the first cases will likely present with syndromes indistinguishable from more common diseases or may be nested within a normative seasonal rise in the syndrome. For example, respiratory bioterrorism syndromes may be very difficult to identify in the midst of a typical winter-time influenza epidemic.

As with the establishment of any diagnosis, a meticulous history is one of the clinician's best tools. In the setting of biological, chemical, and to a lesser extent radiological events, history taking must be particularly thorough, even pointed at times. The purpose of the BCN history is to establish the probability of an exposure; to determine the nature of the exposure; to identify subsequent management

OF NOTE...

Case of *Bacillus anthracis*

On November 16, a 94-year-old socially isolated Connecticut woman presented to a small community hospital after several days of worsening cough, shortness of breath, and flu-like symptoms. She was febrile and mildly dehydrated, but it was her advanced age and social isolation that led doctors to admit her. Her WBC and CXR were both unremarkable. The next day, an alert and surprised infectious disease specialist recognized *Bacillus anthracis* in a gram stain of the patient's blood. The patient was begun on Ciprofloxicin and other broad spectrum antibiotics and questioned about potential exposures. The diagnosis of anthrax was confirmed subsequently by blood cultures. The patient had no known connection to government offices, postal facilities, or news outlets, which had been linked to nearly all of the earlier anthrax cases. On November 19, the Connecticut Department of Public Health (CDPH) was notified by the hospital of the positive blood culture results. Despite orders from the FBI to wait for test results from the CDC in Atlanta, Griffin Hospital officials decided to tell the staff about the anthrax case. On November 20, the isolate was identified as *B. anthracis* at the CDPH laboratory; the CDC confirmed the next day that the anthrax was indistinguishable from that found in the earlier cases. Even with aggressive and appropriate supportive and antibiotic treatment, the patient's respiratory status deteriorated and she died 5 days after admission. Postmortum examination found hemorrhagic mediastinitis with positive immunohistochemical staining for *B. anthracis* in the spleen and medistinal lymph nodes.

Over the ensuring three weeks, state and federal health officials expended nearly 1,500 person-hours on the investigation. Nevertheless, the source of the anthrax was never identified with complete certainty, although cross contamination from one of the contaminated mail-sorting facilities is probable.

decisions relating to treatment, decontamination, and infection control; and to dictate public health measures that may be indicated, such as postexposure prophylaxis, immunization, surveillance, or epidemiologic and possibly criminal investigations. Clearly the clinician provides a pivotal link in the recognition, management, and public health response to BCN terrorism.

The initial clinical assessment of patients with suspected exposure to BCN agents begins with a thorough history of the present illness (HPI). Soliciting details about the onset, timing, rate of progression, and nature of signs and symptoms will be essential to identifying the presenting toxidrome. Toxidromes refer to a constellation of symptoms and signs that taken together form a pattern of illness consistent with a biological or chemical event. (Chapter 5 discusses several of the most relevant and more readily recognizable toxidromes in more detail.) Recognition of a

TABLE 2–1 Components of a BCN History

History of present illness: Note time and location of onset, rate of progression, symptom clusters

Detailed past medical history: Medical conditions, medications, allergies

Detailed social history:

Memberships in organizations

Environmental exposures (animals, unusual foods, water sources)

Occupations (prominent or strategically important positions, agencies, or organizations)

Travel (international and domestic, tourist attractions, national monuments, or political rallies)

Clusters and syndromes of co-workers, family, or co-travelers with similar symptoms and time of onset, animal die-offs

---

toxidrome, or including a particular BCN agent in the differential diagnosis, must necessarily guide medical management, infection control, decontamination, and public health responses (Table 2–1).

Thinking along the lines of toxidromes should be familiar to clinicians. The art of differential diagnosis in many respects begins with recognition of a pattern of symptoms or signs that suggests a particular disease or set of diagnoses. Until recently, few BCN agents would have shown up in the differential diagnosis of specific clinical syndromes even in the most rigorous of training programs. After all, prior to September 11, how many of us would have added smallpox to the differential diagnosis of a toxic-appearing patient who presents with a fever and rash? The probabilities against such a diagnosis would have been enormous. As is often the case, however, a small shift in detail or circumstance changes these probabilities entirely. To continue our smallpox example: if it was 1965 in Somalia—over a decade before the "eradication" of smallpox by the World Health Organization—one would have been negligent not to include smallpox high on the list of possible causes for a patient presenting with fever and rash. If the time and place was not 1965 in Africa, but rather 2004 in Atlanta, one would once again think smallpox was a near impossibility. But what if the stricken individual was a CDC employed laboratorian? This minor change in detail alters in important ways the differential diagnosis, including, in our example, at least two Category A agents: smallpox and viral hemorrhagic fevers. Although arguably less weighty in the probabilistic approach to differential diagnosis, the circumstances of medical and public health practice have been substantially altered by the fact that terrorists have used and may continue to use these agents as weapons to pursue their political and social agenda.

Following the HPI, clinicians should conduct a past medical history (PMH), the objective of which is to identify significant co-morbidities, medications, and allergies. Any positives from this aspect of the clinical history have the potential to impact directly on clinical decision-making. An example from the recent national smallpox vaccination campaign will suffice to make this point more concrete. Individuals with atopic dermatitis, eczema, or significant skin compromise were medically excluded from receiving the vaccinia vaccine because of the increased risk of several serious adverse events, including autoinoculation, generalized vaccinia, and eczema vaccinatum. Should a case of smallpox be diagnosed somewhere in the world, mass vaccination clinics might well be implemented. The preexisting medical conditions that would have precluded vaccination in a preevent situation might not prevent a recommendation for vaccination in the setting of a known case of smallpox; however, the clinician still needs to know the individual's medical history in order to allow an informed decision to be made of the risks versus the benefits of receiving the vaccine.

The social history is an often-neglected component of the clinical history, but in the setting of BCN terror it assumes an unusually prominent role. The overarching objective of this component of the clinical history is to identify social, environmental, and occupational factors that may have a direct bearing on clarifying the likelihood of a BCN exposure. If an attack has occurred, then aspects of the social history have relevance from a public health perspective, identifying potential means by which an epidemic from a deliberate biological attack might progress. Some individuals may be in high-risk occupations or high-profile institutions. Government, research, military organizations, as well as mass media organizations and financial institutions come to mind. Identifying political affiliations, as well as attitudes toward certain ethnic, racial, religious, political, activist groups, organizations, or associations may also be important but should be handled delicately.

A detailed occupational history will determine if the person works in a high-risk occupation. Examples include government workers, such as post office employees, elected officials, or other governmental representatives or employees (Fig. 2–1). Military personnel; law enforcement workers, such as the FBI; and city, state, or federal employees of any capacity could be targets because they work in public facilities that are more probable targets. Certain individuals work in higher-risk locations in these industries as well; the most obvious example being mail room workers. Nongovernmental workers who may be at greater risk include individuals employed in high-profile positions in major corporations or media companies such as television, radio, or newspapers. Lower-profile positions at high-profile organizations—for example, Wall Street firms and major media outlets—could again include mail room workers. Occupations at greater risk would also include laboratory technicians, scientists and researchers, and of course, health care workers. In addition, all of these "at risk" occupations will have close contacts—spouses, children, parents, and the like—who themselves are at greater risk than the general public (Fig. 2–2).

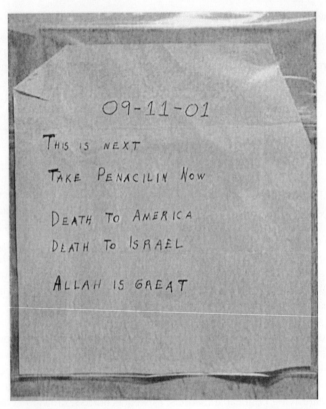

FIGURE 2-1   Anthrax letter sent through U.S. Post Office following September 11 attacks.
Courtesy of the Federal Bureau of Investigation.

Other components of the social history should include identifying potential exposure to animals, unusual packages or foods, recent travel, change in diet, water sources, and sick contacts. Inquiring as to whether exposure to animals—rodents, rabbits, exotic animals, or dead animals, including birds—is important since certain biological agents are zoonotic in their natural state. The person may have received strange or suspicious packages in the mail or from direct delivery sources. Imported gifts, and any correspondence or package with foreign addresses are again potential sources. Obviously exposure to sick contacts who may have a similar constellation of symptoms must be included. A dietary history should be taken to determine whether the person has ingested any unusual foods contained in cans or jars. Dietary history must also include types of food as well as food sources. Others with similar food exposures might themselves become ill or could transmit the agent to others, so determining the timing, location, and nature of eating patterns can be useful. Whether a person has been recently exposed

FIGURE 2-2   Vial containing small amount of ricin sent to the Department of Transportation in 2003.
Courtesy of the FBI.

to certain public water sources, such as in camping, swimming, and public drinking fountains, can be significant as well. All of these have relevance not only from a clinical perspective, but also from a public health perspective by identifying potential means by which an epidemic from a deliberate biological attack might progress.

A critical element in the BCN history is determining whether the person has had any recent travel. On the heels of the recent SARS epidemic, the CDC advised

TABLE 2-2   Summary of Clinician Responsibilities

Take a targeted history in order to assess potential for BCN event

Properly and safely collect samples for laboratory analysis

Be familiar with the BCN triage, treatment, and prophylaxis

Engage knowledgeably and reassuringly in patient and community education in preparedness or postevent scenarios

Follow up with patient, secondary contacts, and health care workers

Notify local and state public health and law enforcement officials all health care organizations and physicians' offices to post signs asking anyone with a respiratory syndrome and recent travel to Southeast Asia, China, or Canada to take precautions upon entering a health care facility. A thorough travel history includes questions about international travel to developing nations or locales where sporadic naturally occurring Category A, B, and C infections occur. Travel to such countries or regions, especially if coupled to contact with animals, health care facilities, or sick individuals, represents a risk factor that may influence the probability of a BCN diagnosis. For example, viral hemorrhagic fevers (VHFs) are endemic zoonoses in many central African countries. As an example of a BCN risk factor, travel to nations with strong anti-American sentiments or harboring people with such sentiments would obviously make such a possibility more likely. It is important to keep in mind that BCN terrorists run considerable risks themselves of inadvertent exposure, and it is not unreasonable to expect that the earliest warnings might ironically come as a result of disease or toxidromes in terrorists themselves or individuals who associate with them, as mentioned earlier regard to one of the September 11 hijackers. Suspicion should not be limited strictly to flagged locales of international travel. Domestic travel itself is another potential risk factor that requires consideration, especially if the travel took the individual to popular sites, such as tourist spots, national monuments, government buildings, political rallies, or major U.S. ports of entry, such as New York, Boston, San Francisco, or Los Angeles.

Clinicians need to synthesize the collected physical and historical data to determine whether there is a disease or syndrome cluster present. Identifying similar constellations of symptoms in close contacts—such as family members, co-workers, travel groups, or clustering of animal deaths in local area or traveled to areas—will be essential to raising the index of suspicion for a BCN event.

Clinical Algorithms

Clinical algorithms are familiar to clinicians. They have been developed and promulgated for a wide range of clinical conditions, ranging from preoperative assessment of cardiac risk for noncardiac surgery, to community-acquired pneumonia, to management of low-back pain. In clinical medicine, algorithms have been an active area of research and development for many years and have involved collaboration between federal agencies and universities or other academic organizations. For example, the Agency for Health Care Quality (AHCQ) has pioneered clinical algorithms that have found utility in diagnosing a number of important conditions.

Diagnosing a BCN syndrome is difficult, but efforts are underway to create workable tools that can assist clinicians. For example, in the months leading up to and during the early phase of the national smallpox vaccination program, the CDC developed a toolkit for clinicians that included a number of clinical algorithms designed to help clinicians diagnose smallpox and to manage adverse events from the vaccination itself. The CDC featured these clinical algorithms on its website and promulgated them in written materials as well. They proved useful to many organizations involved in the vaccination program.

To be useful, diagnostic algorithms must not only be able to accurately distinguish between those with and those without the condition, but should also be readily understood and followed. Complex algorithms will find little clinical utility if they cannot be applied quickly and with confidence, especially in an emergency or crisis situation. This is an area of intense research. Over the coming years such collaborative efforts are likely to provide clinicians with a set of bedside tools to assist them in the diagnostic management and prevention efforts in BCN preparedness and response.

Infection Control Practices

Clinicians should be far more familiar with infection control practices than is currently the case. Practicing physicians are famously unwilling to recognize the importance of infection control as a means of stemming the spread of infectious disease. Perhaps no other example demonstrates more aptly this fact than the story of puerperal sepsis and Ignaz Semmelweis (1818–1865) (Fig. 2–3). Sadly,

OF NOTE...

Ignaz Semmelweis (1818–1865)

Semmelweis was a Viennese-trained Hungarian obstetrician (Fig. 2–3). Decades before the Germ Theory of Disease was first promulgated by scientists such as Louis Pasteur and Robert Koch—general acceptance in the medical community came much later—Semmelweis proved conclusively that one of the most serious complications of pregnancy puerperal sepsis, was caused by physicians and medical students who moved from the autopsy room to the birth room without first washing their hands. Semmelweis observed that mortality rates between a ward staffed by students and physicians was far worse than one staffed by midwives, and further, that midwives carefully washed their hands when examining a woman and delivering the child, whereas the medical staff did not. Implementation of strict hand washing and hand disinfection with chlorine caused mortality rates to plummet from 18% to 2% in a brief span of time. He published his findings in 1861 in a work entitled *The Cause, Concept and Prophylaxis of Childbed Fever*. Semmelweis's observation that doctors were causing death and disease in their patients understandably threatened the medical establishment. Semmelweis's ethnicity and his volatile personality did not further his cause. Ignored and disdained by his medical colleagues, Semmelweis became increasingly unstable. He was committed to an insane asylum where, in 1865, he died under mysterious circumstances. Ironically that same year, the English surgeon Joseph Lister began using carbolic acid to achieve improved postoperative outcomes and ushering in the modern era of antisepsis with which his name is indeliably attached. "Without Semmelweis," Lister once wrote, "my achievements would be nothing."

FIGURE 2-3 Dr. Ignaz Semmelweis, a Viennese-trained Hungarian obstetrician observed that mortality rates improved with washed hands.
Courtesy of the National Library of Medicine.

numerous studies continue to show that health care providers, particularly physicians, are poorly compliant in such basic hygenic practices as hand washing. The statistics on nosocomial and iatrogenic infections are so abysmal, that it should not be any surprise that infection control strategies are little understood and irregularly enforced. In the context of BCN terrorism, however, infection control practices are central to protecting health care workers, first responders, and of course, the public.

Standard Precautions

Standard precautions aim to reduce the inadvertent transmission of pathogens from moist body surfaces. Although transmission of microbes is bidirectional, adherence to these simple strategies leaves much to be desired in many treatment settings. Components of standard infection control practices are rigorous hand washing or use of alcohol-based hand sanitizers, simple personel protective equipment (PPE) such as gloves, face (with eye) protection, and gowns. These PPEs should be worn if exposure to body fluids could occur through examination of the patient, removal or cleaning of secretions, body fluids, or contaminated objects. Standard precautions should be used with all patients even in the absence of a known infection. Standard precautions are recommended for three Category A agents: inhalational anthrax, botulinum, and tularemia, nearly all of the Category B and C agents, and, presuming adequate decontamination of patients, for both chemical agent and radiologic exposures as well.

Contact Precautions

The next level of infection control practices are contact precautions. Contact transmission is the most significant means of transmitting hospital-acquired infections. Contract precautions include standard precautions along with the placement of the patient in a private or semiprivate room. Individuals examining or coming into contact with any surfaces in a patient's room should wear gloves and must change those gloves after contact with infective material or potentially contaminated surfaces. A gown should be worn upon entry in the room if direct patient contact is anticipated, or if the patient has diarrhea, a colostomy, or a draining wound that is not covered by a dressing. Face shields and eye protection should be worn when procedures are done to patients that generate aerosols or droplets, including suctioning, bronchoscopy, aerosol nebulizers, or sputum induction. PPE should be removed after leaving the room, with respiratory protection coming off last. Hand washing before entering and after leaving and disposal of all PPE in appropriately marked trash receptacles are essential. Visitors should also use contact precautions. Patient movement should be limited as much as is practical. Finally, daily room hygiene by hospital maintenance personnel is needed to decrease the likelihood that

TABLE 2–3   Summary of Infection Control Recommendations for Category A Agents

Standard precautions	Anthrax, botulinum, tularemia
Droplet precautions	Plague
Airborne precautions	VHFs, smallpox

bedside equipment, frequently touched surfaces, and other patient care items will become sources of infection for those entering the room and for those who may reside in the room after the source patient is discharged. In selective circumstances, commonly used equipment (e.g., stethoscopes) should be kept for dedicated use in that room (Table 2–3).

Contact precautions are required for specific clinical situations. Patients with severe gastrointestinal, dermatologic, or wound infections that may be transmitted easily by touching the patient or by handling objects the patient has touched should have contact precaution signs on their doors. These procedures are especially critical if the patient's infection is caused by a drug-resistant organism. A mundane example where contact precautions would be used is methicillin-resistant *Staphylococcus aureus* (MRSA). Additionally, contact precautions are recommended when patients have substantial compromise of their skin, such as might occur with weeping skin infections, decubitus ulcers, impetigo, or scabies. Contact precautions are indicated as well for surgical-site infections and infectious diarrheal diseases. In the context of BCN agents, dermal or gastrointestinal anthax, generalized or progressive vaccinia, or eczema vaccinatum, as well as many Category B agents— enteric pathogens such as cholera and shigella—require implementation of contact precautions.

✵Droplet Nuclei Precautions

Droplet nuclei precautions represent the next level of infection control. Droplet transmission refers to airborne infectious droplets generated when infected persons cough, sneeze, or talk. Invasive procedures, such as bronchoscopy or nasopharyngeal suctioning, generate infective droplets capable of being inhaled by unprotected individuals as well. Droplet nuclei do not remain suspended for long periods, but settle out onto surfaces (e.g., mucus membranes, skin, or bedside commodes). Two corollaries of this fact are that special air handling requirements are not required but that careful attention to room hygiene and decontamination is needed for patients for whom contact precautions have been implemented. For example, noncritical patient care equipment (e.g., IV poles or stethoscopes) either should be used for a single patient or disinfected before being used with a different patient. A familiar example of droplet nuclei precautions would be in the care of individuals with vancomycin-resistant entetrococci (VRE), an increasingly common nosocomial pathogen. In the setting of bioterror agents, droplet nuclei precautions are indicated for documented or suspected plague meningitis.

Aerosol Precautions

Airborne precautions represent the highest level of infection control practice. In select circumstances, such as VHFs, these may be even further amplified. Airborne precautions are designed to protect health care workers and patients from exposure to airborne respiratory pathogens that may remain suspended for long periods of time in the air. Familiar examples where respiratory isolation is indicated include *Mycobacterium tuberculosis*, rubeola, and varicella. Patients' doors should remain closed and they should be restricted to their room, except when transportation is required for diagnostic testing or treatment. During transport, the patient must wear a mask. Health care workers and visitors entering the room should wear an approved high-efficiency particulate air (HEPA) respirator for which they have been fit tested and trained in proper use. The room should be under negative pressure, ideally with air being exhausted directly to the outside environment. Complete air exchange should occur a minimum of six times per hour, maximally twelve per hour. If recirculation is unavoidable, then all air exchanges should be through a HEPA filter system.

Worst case scenarios in the setting of bioterrorism would be multiple cases requiring airborne isolation, in which case the bed capacity of most hospitals would be overwhelmed quickly. The CDC is working with state health departments and the private sector to address the need for "surge capacity" to handle a large influx of cases in the event of a mass casualty situation, but this is far from being realized at the present time. Alternative strategies would be to require HEPA respirators for all individuals entering and leaving the patient's room and strict adherence to VHF-specific isolation precautions. These augmented isolation precautions are summarized in Table 2–4.

TABLE 2–4 VHF-Specific Infection Control Recommendations

Hand hygiene—strict hand washing before and after all patient contact and entering of room

PPE—double gloves, booties, impermeable gowns, leg and shoe coverings, face shield, eye protection, N-95 or PAPR respirators

Negative pressure room

Limit access to essential personnel only

Dedicated equipment (e.g., stethoscopes and IV poles)

Disinfection with 1:100 bleach solution or similar EPA-approved disinfectant

Maintain dedicated care unit for VHF-infected patients if multiple cases occur

Source: Bioterrorism Guidelines for Medical and Public Health Management, AMA Press, 2002, p. 210. Adapted from AMA

TABLE 2–5   **Bleach and Hypochlorite Conversions**

Household bleach is made of 5.25% sodium hypochlorite

A 10% bleach solution corresponds to one and a half cups of household bleach per gallon of water

0.5% hypochlorite solution = 1 part bleach to 9 parts water

Decontamination and Disinfection

Decontamination and disinfection are essential steps with which clinicians and health care facilities must be familiar with whenever BCN agents are present or suspected. A simple but effective decontamination strategy is soap and water. Soap and water works very well as a rule of thumb with most BCN agents. Household bleach is another inexpensive option that is readily available but has fallen out of general favor. More expensive decontamination solutions are available, but choices should be restricted to those approved by the Environmental Protection Agency (EPA) for this purpose. Patients and most surfaces can be disinfected safely using a 1:10 dilution of bleach (1part bleach, 9 parts water). This solution will kill most microbes and spores (Table 2–5).

Contaminated medical or other equipment (e.g., endoscopy equipment) should be decontaminated with full-strength bleach (5% sodium hypochlorite). Once decontaminated, patients should be rinsed off with copious amounts of water to limit the irritation caused by dermal bleach exposure. This can be used as a sporicidal and disinfecting agent for all surfaces.

The "Worried Well"

Since the post-September 11 anthrax attacks and ongoing military action in Afghanistan and Iraq, there is an underlying anxiety and vigilance that burns in the public's consciousness. This concern is stoked by intermittent reports of suspicious powders, discovery of smuggled BCN weapons, and foiled BCN attack plans. In many of the cases of suspicious powders, "patients" were brought for medical evaluation complaining of a range of symptoms. Prior to the assessments that these powders were hoaxes, these patients were legitimately afraid for their health. This is a reminder that the "worried well" patient is not to be written off as paranoid or neurotic but to be taken seriously and met with understanding. Their concerns can be well founded, and the state of the world makes such patients a legitimate part of the provider's responsibility. A more detailed discussion of the psychological issues involved in BCN preparedness and response is presented in Chapter 7. Clearly any serious medical issues will need to be ruled out and addressed to the physician's satisfaction, but such patients can be helped tremendously with education on bioterrorism, plausible versus implausible risks, specifics about the various agents, and sources for up-to-date information. The psychological sequelae from an acute BCN event will affect many of our patients, even those not exposed. Examples abound where anxiety, posttraumatic stress disorders (PTSDs), and depression affect the mental and physical health of communities following major disasters or public health threats regardless of proximity to an attack or source of exposure.

Chronicity

One of the recurring themes of this book is that so much in the way of attention, planning, and funding for the BCN threat has revolved around acute events, such as a discreet chemical agent attack. This is shortsighted. Should such an event occur, the medical and psychological sequelae could last for years, even decades. Recent studies of New York City school children indicate a disturbingly high percentage of psychological distress in the form of mood disorders, somatic complaints, behavioral changes, and insomnia. From a medical perspective, recrudescent infections—known to occur with a number of biological weapons, such as Glanders or Nipah virus—would have to be addressed. Long-term effects of chemical and radiological agents such as mutagenicity, teratogenicity, neurological, hematological, and other chronic health complaints will require monitoring. Another consideration is medical management, which includes the long-term treatments that many of the biological agents necessitate. Although there is little experience with BCN events, the WTC collapse serves as an example of the chronic issues that arise, in this case with the toxicity of debris exposure primarily inhaled or ingested by rescue and demolition workers. The WTC Registry currently has fifty thousand enrollees, many of whom have continuing signs and symptoms.

Another aspect of BCN sequelae that will need to be addressed is pharmaceuticals. The treatments themselves have toxicities that will need to be addressed, whether used prophylactically or as treatment. The anthrax attacks after September 11 are an example of this among postal workers who were at risk for possible exposure. Public health authorities and local health care professionals were called upon to address issues of informed consent; exposure risks, pros and cons of vaccines and antibiotics (e.g., prophylactic ciprofloxacin), and management of side effects, while dealing with underlying chronic illnesses or drug interactions. It was a major source of anxiety and conflict.

Many of these issues will become the domain of the patient's primary caregiver. An acute BCN event might last minutes or might last days, but the long-term medical and psychological effects will linger. There has been little acknowledgment of this fact in the medical or public health community. Part of the reason for the silence is that long-term consequences of these agents are poorly understood. Indeed, the acute consequences of these agents are not well understood either.

More fundamentally, there is simply less in the way of energy and resources dedicated to anticipating the potential long-term issues and subsequent long-term needs after a BCN event. Community physicians will be the ones managing these consequences. As BCN-related public health and preparations expand, planners must recognize that an attack does not end when the first responders are recalled or when the emergency room is finally cleared. Indeed, at this point, many medical, psychological, and social issues are just beginning.

Assessing a BCN-Exposed Patient

The CDC and the U.S. Army Medical Research Institute on Infectious Diseases (USAMRIID) have taken a prominent role in the nation's preparedness efforts. In the context of bioterrorism, U.S. military planners have developed what they refer to as the "Ten Commandments of Biological Weapons Attack." These commandments are designed largely for mass casualty scenarios and they serve as a useful, concise summary of what the medical and public health community should be aware of. Clinicians need to maintain an index of suspicion, must protect themselves, must appropriately assess their patient(s), and must decontaminate as the nature of the agent dictates. In addition, a rapid bedside clinical assessment must be made to determine which organ systems are injured; whether any allergies exist; medications; past medical illness; including immunization status; most recent meal; and the nature of the event or exposures. Prompt clinical treatment with appropriate infection control rounds out the immediate clinical responsibilities that should ensure that the patient is well cared for and that the health care team is protected adequately. The final recommendations relate to the broader public health responsibilities that devolve to those involved in the identification, management, and aftercare of BCN events. These include notifying the proper authorities, both public health and law enforcement; conducting follow-up epidemiologic investigations that might limit the extent of the epidemic; and identifying individuals in need of acute or long-term surveillance, postexposure prophylaxis, immunization, or education. The final commandment admonishes clinicians to remember what they have learned and to share their knowledge with colleagues and the public.

Physician Preparedness for BCN Terrorism

The public health issues that stem from preparedness and response to BCN terrorism are of paramount importance. It is crucial that clinicians and public health officials understand the important roles and responsibilities involved. Multiple scenarios exist in which these skills might be called into play. For example, physicians may be called on to participate in local disaster planning in anticipation of a potential terrorist attack (Fig. 2–4). If an actual terrorist attack occurs in a community, patients may present in the offices of primary care physicians requiring recognition, triage, and containment. Events in other areas would place local health care providers and public health officials in the position of offering knowledgeable and rational responses to the concerns of their patients and their communities, including familiarity

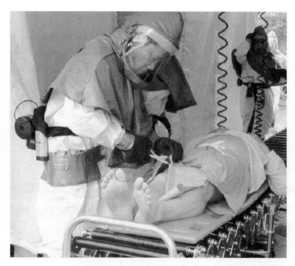

FIGURE 2–4   Bioterrorism mass casualty drill showing decontamination procedures.
Courtesy of OSHA.

with existing preparedness and emergency response plans locally, statewide, and nationally, such as those of the CDC, Federal Emergency Management Agency (FEMA), and the National Guard. In short, physicians must be prepared to seamlessly interact with whatever public health response is needed during a BCN event. Interfacing efficiently and effectively with the public health response and providing patients and communities with accurate information, directive guidance, or reassurance will in many instances involve local physicians.

Physicians need to understand the basics of effective emergency response and Incident Command Systems (ICS) that will be called into play include not only for BCN emergencies, but also for natural disasters or industrial accidents. The ICS will be responsible for coordinating the containment, decontamination, patient triage, surveillance, and immunization efforts during an actual emergency. Most practicing physicians—except those perhaps in rural areas—will not be asked to coordinate or organize their local ICS, but they may be asked to participate in it. It is important for health care workers to be aware of how an ICS operates and what role they might have in interacting with the ICS should an event occur (see Chapter 3 for more detail).

One of the purposes of this book and the proliferating training programs available to physicians and other health care providers is to provide them with sufficient pattern recognition to identify a sentinel case or sentinel exposure early enough so that the public health systems and ICS can be called into action to prevent further spread. Atypical presentations and unusual health events should prompt physicians to consider bioterror and to alert the public health authorities so that investigations can be initiated. Before outside assistance arrives (regional, state, interstate, and

FIGURE 2-5 Targets for sabotage: industrial plant where harmful and potentially dangerous chemicals may be used or stored.
Courtesy of ATSDR/Joseph Hughart.

federal), emergency medical teams will play an essential role during the initial response to an incident. Community physicians may be recruited to expand the pool of available providers. Outside assistance must be seamlessly integrated into the response effort. In anticipation of an event, clinicians should understand potential dangers in their community, as well as the existing contingency plans and key contacts in the local emergency responce or ICS. It is important to take an inventory of the potential sources of nonintentional BCN agents in your community because it is easier to attack an existing industrial or manufacturing source than to create and transport what might be used in a terrorist attack. Military and commercial defense and weapons sites are possible targets. Civil targets of terrorism could also include power plants, water treatment facilities; or infrastructure, transportation, and public and private buildings that have social, economic, or political meaning, such as

FIGURE 2-6 Targets of sabotage: rail lines, particularly those carrying chemical or other hazardous materials.
Courtesy of ATSDR/Joseph Hughart.

TABLE 2–6   Critical First Steps: Suspected BCN Event

Treat, isolate, or refer affected individual(s).
Contact local law enforcement and FBI.
Contact local and state health departments, or CDC.

centers of commerce and telecommunications. Industrial sabotage—whether it be an explosion at a local plant that uses chlorine gas, blowing up of a train carrying chemicals, attacking sites where chemical or nuclear materials are stored—is a possible scenario. Familiarity with potential exposures means better response and care following such an event (Figs. 2–5 and 2–6).

Once an incident occurs, community physicians may play several roles as the emergency response unfolds. First, they may be assigned to non-hospital treatment locations in order to triage, treat, or initiate prevention strategies. They may also be called to assist in attending homebound residents throughout the community. Community-based clinicians represent an auxiliary pool of trained providers to help staff local emergency rooms, hospitals, and other treatment facilities as part of each community's surge capacity. Last, they will be asked to participate in any needed postexposure surveillance and provide community immunizations to prevent, if possible, continuance of the epidemic or to limit the impact of future epidemics.

Health care providers should have a readily available list of all necessary phone and fax numbers for health and law enforcement officials at the local, state, and federal levels to be notified if BCN exposure is suspected. Local health officials should be sought for assistance in making a decision to activate the public health response, and for implementing action should it turn out to be an immediate attack. Given that a criminal activity is behind such an event, clinicians should expect that legal and law enforcement issues may supersede medical ones in an emergency or mass casualty setting where law enforcement agencies or local emergency planning agencies will have authority (Table 2–6).

Early identification, notification, diagnosis, and treatment of bioterrorist or potential bioterrorist events are vital for the patient, the community, and possibly the nation. Physicians must be capable of temporarily functioning in a BCN event until public health and law enforcement officials are available. Proper management includes not only medical management, infection control, and decontamination, but also administration of prophylaxis and activation and participation in the public health response.

Summary

Clinicians will almost certainly be on the forefront of any rational medical and public health response to a BCN event. They must be prepared to assess patients presenting with potential BCN syndromes by taking a detailed BCN history and familiarizing themselves with how to triage, treat, and prophylax. Additional clinical skills include knowing how to properly and safely collect samples for laboratory analyses and instituting appropriate PPE and infection control procedures. Clinicians have a central role to play if our society is to respond effectively to the medical, public health, and social impact of a BCN terrorist attack. This begins first with an honest acknowledgment that bioterrorism is a plausible threat and that clinicians have a responsibility to acquire a working knowledge of the medical aspects of BCN agents. Second, clinicians must begin to implement proper infection control practices in their offices and, when applicable, when working in the hospital setting. This implies a responsibility to their office colleagues and employees, as well as to the health care community more generally. Better awareness of the role of workplace health and safety would do much to improve the quality of care for the myriad of more common infectious disease risks that currently exist, from the common cold to blood-borne pathogens like HIV or infectious hepatitis. Fundamental to achieving this end is knowing and consistently implementing the use of PPE and infection control practices. These are generalizable principles and knowledge whose dividends would in all probability be seen in a decrease in nosocomial and iatrogenic infections and concomitant social and economic benefits. Regarding BCN terrorism specifically, clinicians who maintain a high index of suspicion and incorporate the concept of the sentinel health event (see p. 85, Chapter 5, Surveillance Systems and Bioterrorism), into their routine approach to clinical practice may very well help to identify early on a potential BCN event and to limit the impact of such an event on their patients, families, colleagues, and the public at large. Last, clinicians must be part of any postexposure management plans, including managing long-term sequelae in all its manifestations, and ultimately, interfacing fully with the public health response comprised of public health, medical, and law enforcement professionals.

CHAPTER 3

Clinicians' Roles in Public Health Preparedness and Emergency Response

An Overview of Homeland Security, Public Health, and Emergency Preparedness

In the post-September 11 world, clinicians need a basic understanding of the national homeland and public health security structure. Fundamentally there are two major areas of responsibility in the event of any emergency: crisis management and consequence management. In 1995, President Clinton signed Presidential Decision Directive 39 (PDD 39), the purpose of which was to define agency-specific responsibilities during an emergency. PDD 39 delegates crisis management authority—defined as overall control of the local, state, or national response to a bioterrorism event—to the law enforcement community. In the event of a terrorist attack, any and all criminal investigations would be controlled by the Department of Justice, specifically the FBI. PDD 39 assigns responsibility for consequence management—defined as the public health and safety issues resulting from a terrorist attack—to FEMA. In this organizational structure, health care and public health are the jurisdiction of FEMA.

The demands of law enforcement and the public health responsibilities are not mutually exclusive domains. That being said, it is important to acknowledge that historically these two communities have very different missions and culture. These differences cannot help but influence how each arm of the nation's security system will react in the event of an emergency. Recognizing these distinctions has relevance in the context of emergency preparedness and emergency response. The nation's preparedness efforts have included a great deal of cross-institutional community building in recognition of this fact.

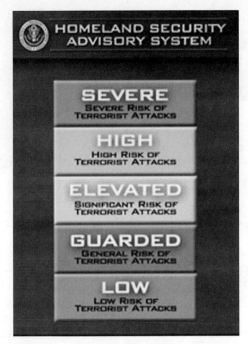

FIGURE 3-1   Homeland Security Advisory System.
Courtesy of the Department of Homeland Security.

An important example of an outgrowth of PDD 39 is the creation by the CDC of the Health Alert Network (HAN). The HAN is a dedicated CDC-sponsored website where state and local health directors may view securely posted documents, submit or collect data, obtain town- and district-specific aggregate data, enter planned absences, e-mail, or view a bulletin board. The overall goal of the HAN is to securely facilitate communication of critical health, epidemiological, and bioterrorism-related information on a 24/7 basis to local health departments, health organizations, clinicians, and other organizations with a vested interest in staying abreast of developments with potential public health implications. As a nationwide service, the HAN can disseminate late-breaking, updated, or new public health information. The HAN is intended to rapidly alert local health departments, public health officials, and the medical community to any issues that may impact the public's health. It also offers educational and training programs for public health and medical professionals. The HAN should improve communication between state and local health agencies, departments, and care providers for better coordination of knowledge, information, and practices in the event of an emergency of any variety. For example, it was widely and effectively used during the 2003 international SARS epidemic. The nation's public health response to SARS is a fine example of the "dual use" approach to bioterrorism preparedness and planning. Simply defined, broadening and strengthening the nation's public health infrastructure and resources—spurred at least in part by September 11th and the anthrax attacks—provides public health benefits well beyond bioterrorism. This system played a direct role in limiting the impact of a new global health threat in this country and should improve still further in the future. The SARS epidemic was one of the first truly national field tests of the nation's preparedness effort, and the response was widely considered a successful one. Currently, the HAN system provides public health information to over a million individuals across the country. According to the CDC, most states have HAN systems in place covering over 90% of their population. The HAN website is not encrypted, so that at the present time confidential data are not entered through this portal. It is an excellent resource for community physicians and other health care providers. Access to the HAN is through the website maintained by the CDC (http://www.phppo.cdc.gov/HAN/Index.asp).

In 1999, Congress gave the CDC formal responsibility to expand and improve the readiness of the nation's public health workforce and infrastructure to respond to a bioterrorist attack. CDC's Bioterrorism Preparedness and Response Office addresses the full breadth of preparedness issues, from biological to chemical to nuclear agents. This includes medical therapeutics stockpiling, enhanced communication, rapid laboratory diagnostic capacity, improved surveillance, and augmented state and national epidemiologic capacity and emergency planning. One of the important outcomes from such efforts is the risk stratification of the potential biological weapons into Categories A, B, or C. These categories have become the frame of reference for medical and even political discourse in regards to biological weapons. Needless to say, these categories play a central role in the organization and discussion of biological weapons in this text.

Since 2001, Congress has passed a series of legislative initiatives, all aimed at improving state and local emergency preparedness, public health infrastructure maintenance and expansion, public health workforce development, and outreach and education. Billions of dollars have been directed to this effort. Timetables and metrics were deliberately kept short in order to ramp up the nation's public health system rapidly. A major driver in state and local public health preparedness has been the CDC's Public Health and Preparedness Bioterrorism Cooperative Agreement. Funding through the CDC Cooperative Agreement has been directed to a wide range of activities. For example, states and local public health departments have 24/7 access to several rapid communications networks. One of the most accessible is the HAN. Other developments that have the potential to facilitate the nation's preparedness efforts are the Wide Area Notification System (WANS) and the MEDical SATellite (MEDSAT) program. The WANS is a telephonic system with autodialing and voice messaging capabilities. It allows 24/7 notification of key components of the emergency response, law enforcement, medical, and public health community via landline phones, cell phones, pagers, faxes, and e-mail. The goal of the MEDSAT program is to enhance the emergency communications capacity for departments of public health and their key public health emergency response partners. The MEDSAT program provides a highly reliable telephone connection to be used in response to a public health emergency or in the event of a localized or catastrophic failure of the conventional telephone network.

The Strategic National Stockpile (SNS), formerly known as the National Pharmaceutical Stockpile (NPS), is another national initiative coordinated by the CDC and one with which clinicians should be familiar. The purpose of the SNS is to meet the acute vaccination, prophylactic, and medical management needs in the event of a bioterrorist attack. The stockpile includes vaccines, antibiotics, medical supplies, and medical equipment. These resources are maintained at strategic locations throughout the country and are available for immediate delivery. The locales of the sites are known only by CDC officials. The SNS is structured such that it is perpetually in a state of readiness should a biological or chemical attack occur in the population. The supplies are packaged so that no specific request need be made; rather, a request for any particular item elicits a delivery of a prepackaged armamentarium of treatment and protection. The designated term is a "12-hour Push Package," reflecting the CDC's commitment to have them available in less than twelve hours following an event, and it "pushes" any possible supplies needed regardless of the specific agent involved. Existing plans call not only for the delivery of the "Push Packages," but also for a rapidly mobilizable a team of CDC advisors to help states and municipalities respond to a crisis. The CDC is currently distributing "Chem-Packs" in much the same manner. What began as a pilot project in 2003 is now a fully developed $30 million a year strategy to distribute preassembled packages of available chemical antidotes to states and municipalities for use in case of an attack involving chemical agents. Each Chem-Pack is thought to be capable of treating up to 1,000 people. Like the Push Packages, the CDC plan allows for Chem-Packs to arrive at an event site anywhere within the United States within a twelve-hour period.

☣Emergency Response Systems and Incident Command: Problems and Promise

The September 11 attacks on the WTC and Pentagon highlighted the inadequacies of the existing incident command systems in the United States. Hundreds of first responders and health care workers who came to Ground Zero worked long hours with wholly inadequate safety measures. Coordination was lacking, so that even when PPE was made accessible, knowledge of its availability and whereabouts too often did not translate down to those who could have distributed it to the men and women working in the smoking and pulverized debris.

Further, when PPE was donned, it was often used improperly and therefore not protective. Consequently, hundreds of firefighters, police, ironworkers, and other emergency responders found their unselfish efforts rewarded with medical problems that have, in many instances, continued to plague them. Indeed, thousands have since been enrolled in federally funded cohort studies to better define the health consequences of this unique national trauma. The largest and most prominent of these is the WTC Health Registry, designed as a prospective longitudinal cohort. A study of the health effects of individuals exposed to Ground Zero in the early hours and subsequent days following that trauma. Currently there are some 50,000 enrolled workers in the WTC Health Registry.

Sadly there were some at the time who espoused the opinion that anyone who had the temerity to raise health and safety issues during such a catastrophe was, at best, unpatriotic. Now that the dust has truly settled and the trauma is less immediate, it is evident that those who raised such concerns had the best interests of the heroic men and women at heart. Further, the experience of September 11 has provoked important attitudinal shifts within the emergency response community itself: emergency response and disaster management planning must integrate health and safety awareness into both training and preparedness efforts in order to protect those who are called on in such an event. Bravery should not require foolhardiness, or unnecessary or inappropriate risk taking, regardless of the circumstances. Further, emergency responders who become either injured or sick unnecessarily due to inadequate advance training, use of protective strategies, and the like are unable to do their jobs. In this instance, the emergency and public health response suffers along with that of the individual workers.

☣Incident Command Systems

The ICS model has been adopted to address the critical need for a formally organized and trained disaster management structure that can be called on in a crisis. The ICS is a planned system that provides for the development of a complete emergency management organization, whether comprising a single agency or optima agencies. The ICS can function as a bridge that connects resources outside of a community to those within a community in order to respond to temporary but overwhelming demands. Although the ICS provides a predictable chain of crisis management, it does so without sacrificing flexibility in terms of tasks and responsibilities and scale of effort. In addition, the ICS affords the opportunity to prioritize responses and accountability, and by streamlining and clarifying roles and responsibilities, it should prove to be a cost-effective management strategy. Clearly, advance planning and training of the ICS in any community or state or region is critical to providing the sort of experience needed in an emergency. Put bluntly, emergency planning with health care providers and public health systems needs to be accomplished prior to an emergency and practiced in order to have confidence that it will do what it was intended to do in an emergency. An effective ICS also allows for the overall coordination of containment, decontamination, patient triage, surveillance, and immunization efforts. These are addressed in more detail later in this chapter.

At the state level, there are fairly well-delineated crisis and consequence management teams in place. Training and advance preparation teams will respond to a crisis as planned, but neither field tests, nor table-top exercises are the same as the real thing. As noted in Chapter 1, the TOPOFF trials have highlighted the significant obstacles that even well-prepared crisis management and consequence management teams will find difficult to navigate. The reasons are many. First, each incident has its unique features and idiosyncrasies that complicate advanced planning and real-time response. Second, the emergency management response requires significant interactions with ordinary citizens whose diverse reactions will make an ordered response more difficult. Often communication systems are swamped by the sheer volume of information and traffic that they are being asked to carry, a circumstance that makes already confusing situations even more chaotic. Individuals trained to respond are themselves under enormous physical and emotional stress, and relationships that work well in a field exercise might deteriorate quickly during an acute event. The pace, intensity, and confusion of a true disaster often confuse lines of authority and hinder clear task assignment. Finally, as the nation observed during the first horrifying hours after the WTC collapse, safety practices and training may be inadequate to the task, thereby increasing the risk to those individuals who are putting their lives on the line.

The initial response to any emergency or disaster comes from local first response systems and hazardous materials (Hazmat) teams. The incident commander is, in fact, typically the local fire chief. Local agencies or the ICS commander notifies the state health department. State health departments have teams trained to respond quickly to any emergency and conduct onsite environmental monitoring in order to determine what, if any, chemical or radiation hazards exist. Regional response teams have been assembled throughout the country, allowing shared expertise. This is critical for bioterrorism as few communities have direct experience with BCN events. For example, these regional response teams have both state-based radiation experts and radiation safety experts from the Department of Energy to assist local and state officials in dealing with a radiation incident.

As noted earlier, FEMA is the overall lead nationally for managing the medical, public health, and logistical aspects following a bioterrorism event. President Bush has designed the Department of Homeland Security as the lead agency for coordinating the crisis and consequence management for all bioterrorism events. The Homeland Security Department supervises the FBI in its criminal investigations and FEMA in its public health management. These systems have yet to be fully tested in a true crisis, although table-top and field exercises have simulated disasters as a means of training and testing the preparedness systems at the local and state levels.

Surveillance is a key aspect of the ICS. This can take more than one form, including monitoring of water or air for chemicals, biologics, or radiation or surveillance for medical syndromes that could be a clue to a release. Identification of resources that would be needed to respond in a timely and appropriate way to the event is yet another task. This includes defining those individuals, institutions, or organizations that will participate in the emergency response and exactly what their roles will be. In the setting of a significant public health emergency, this includes local medical, public health, first response, and law enforcement resources. Choosing the proper response to the specific event is an important task of the ICS as well. During crises, timeliness and efficiency are crucial in order to limit as much as possible social, economic, and health consequences. Finally, the ICS is responsible

TABLE 3–1   Emergency Response and Incident Command Systems: Key Responsibilities

Identify sources of exposure and potential threats

Identify routes of spread and exposure from threat or event

Establish surveillance mechanisms

Identify resources needed to respond

Define the role of responders

Select the proper response

Establish training programs

in many communities for establishing training programs to prepare the community for such an event. These key responsibilities are summarized in Table 3–1.

Hospital Preparedness and Emergency Response

Hospitals are critical to the nation's preparedness efforts both in the response to disasters or bioterrorist events and as potential targets themselves. Emergency departments, intensive care units, security, and first responders will have crucial roles in any bioterrorism event. Each hospital should have a response plan in place that delineates emergency preparedness teams (EPTs). EPTs include clinical and support staff that are fully trained and whose roles are clearly defined. Hospitals should prepare institutional policies and procedures for the institutional response to issues such as mail-handling protocols; decontamination; chain of command; communication protocols; general safety and security of the institution, including locking of entrances at night and monitoring of visitors and staff; requiring all staff to have a visible identification badge with a photograph; fire and evacuation plans for the institution and communication with town and state emergency plans; discussions about radiation and biohazards safety and protocols to protect other patients and workers; mass casualty preparation and procedures; infection control institutional monitoring and reporting system to hospital epidemiologist, employee health service medical director, state department of public health, FBI, and CDC; and employee psychological support system through connections with Department of Psychiatry and Employee Assistance Program for debriefing and crisis management in the event of a bioterrorism incident.

Developing a resource binder or similar readily accessible resource where staff may review information available from the CDC, state and local health departments, the FBI, the EPA, or other sources is extremely important. A computer workstation with bookmarked key Internet sites for the most up-to-date information should be available to individuals involved in any organizational response. It is also useful to include directions for handling a suspicious package or letters, or someone bringing

TABLE 3–2   Critical First Steps: Suspicious Mail or Materials

Do not disturb, transport, or investigate package or envelope

Do not allow others to disturb or handle or investigate

If leakage has occurred, do not inspect, sniff, or touch

Close all doors, shut off ventilation, fans, and air conditioners

Alert coworkers, and everyone should leave the area

Wash hands with soap and water

Document individuals potentially exposed

Notify law enforcement, infection control officers, and local health department

suspicious materials into an emergency room or office (Table 3–2). These resources will enable EPTs and other interested parties to familiarize themselves with protocols and practices prior to an event and also serve as a valuable resource in the event of an attack. It is essential that the lines of communication be very clearly delineated: who should be called, when, and for what purpose. Specimens from patients suspected of having been exposed to a bioterrorism attack are a good example of the importance of communication. Determinations of biological, chemical, or radiation exposures may require special collection techniques, and the material collected may need special storage and shipment to the CDC or other specialized laboratories. Lines of communication with the state department of public health laboratory and the CDC will be necessary in order to ensure the collection, storage, and if necessary, transport and delivery, are all done with proper attention to chain of custody, health and safety, and, of course, the testing requirements needed to ensure accuracy.

Occupational and Employee Health Services

Many large institutions and all hospitals have occupational or employee health services (EHS). In most cases, occupational and employee health staff are likely to assume vital roles in advanced planning efforts and in an emergency itself. In the setting of hospitals, EHS staff may be asked to plan and prepare for mass onsite vaccination programs should a biological terrorist event occur locally or elsewhere. Effective response of staff in such a setting requires adequate training and drilling. Hospitals have run disaster drills for years, so that this approach is familiar, although it requires modifications to address the unique features of a bioterrorism event. In addition to training staff and educating the hospital community, EHS staff are often asked to develop protocols and procedures for emergent infectious disease problems within the institution, or to area employers with which the hospital's occupational or EHS may have contracts.

Occupational health services and EHS have important roles to play within their institutions as a filter for a wide variety sentinel health events (see Chapter 5). Early detection has implications for affected individuals, co-workers, the organization itself, and potentially for the community as a whole. Health care providers at EHS have a unique opportunity to practice primary prevention through the preplacement evaluation, review and update of immunizations, and education. When the EHS plays a role in delivering surveillance care for groups of workers, such as hazardous materials teams or firefighters, the medical services must be available to develop screening protocol for emergency responders. There is no better example of this than what occurred in the aftermath of the WTC attacks. Hundreds of workers with diverse skills were unknowingly exposed to the hazardous materials in the air following the attack. It was later determined that they were being exposed to asbestos, silica, organic compounds, lead, and a host of other chemicals. Contact with academic institutions or private organizations that can assist in developing a structured surveillance protocols is recommended. For example, the WTC registry mentioned earlier is a collaboration between the New York City Department of Health the Agency for Toxic Substance and Disease Registry (ATSDR), the New York City Fire Department, and Mount Sinai Medical Center (Fig. 3–2).

☣Safety Considerations

Safety is a collective responsibility for first responders at the scene of any disaster or emergency, be it bioterrorism or otherwise. There is a clear need for better planning in the ICS and for creation of workable standardized approaches that maximize safety while enabling first responders to do their work (Fig. 3–3). Arguably there is a need to develop a cadre of highly specialized disaster management managers to be chosen from within each agency and at every level: local, state, and national. These

FIGURE 3–2 OSHA officials assessing the potential risks at Ground Zero of the WTC.
Courtesy of OSHA/Shawn Moore.

Incident Command System Graphic:

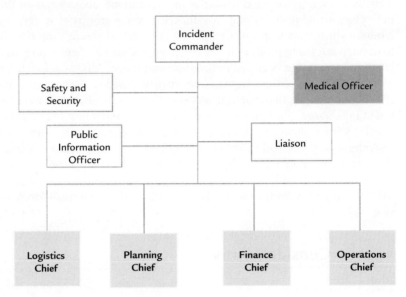

FIGURES 3–3   Incident Command System Schematics.
Adapted from Legasse, with permission.

individuals need to be able to integrate health and safety concerns into every training exercise so that the health of those working at the scene is not an afterthought. It is necessary to have a list of key personnel and skills needed to protect and advise others in regards to health and safety practices and training. This will include disciplines such as industrial hygiene, occupational–environmental medicine physicians, and medical toxicologists. The ultimate goal should be to standardize response training, hazard assessment, credentialing, equipment, and protocols.

The greater the degree of standardization, the more efficient the response will be from incident to incident and from state to state. This concept is embraced by the "all hazards" approach to bioterrorism preparedness. A useful and familiar analogy is blood-borne pathogens (BBP) training. It is now a standard of care in the health care field to assume that all patients are potentially infectious and to incorporate universal precautions. This is codified further in the Occupational Safety and Health Administration (OSHA) requirement for annual BBP training in health care workers and very recently in requirements for hospital workers to be given formal respirator fit testing to protect against airborne infections, such as tuberculosis. Routinization of training, use of safety equipment, and protocols for assessment, response, safety, and surveillance will be necessary in order to move the nation's emergency response and health care communities toward the chronic disease model of bioterrorism. By this we mean that a long-term sustainable model of bioterrorism preparedness and response is needed for this threat of bioterrorism is likely to be with us for years to come. Fortunately, much can be learned from the nation's responses and experiences with natural disasters such as floods, tornadoes, and earthquakes. These experiences can be generalized and incorporated into the public health response to bioterrorism.

Assessing Vulnerability

Public health preparedness is a multistep process. Planning ahead and anticipating what one might experience by conducting a vulnerability assessment at a community, state, or national level is crucial to developing appropriate response plans. Key questions to assess vulnerability include: are there industries or storage facilities in which hazardous materials are located or used? What are the nature and properties of these materials? Are current first responders properly trained and equipped to deal with these materials in a crisis? Useful resources to begin the process of vulnerability assessment are available, including local fire marshall offices and two excellent websites: www.Scorecard.org and the list of chemical hazard fact-sheets maintained by the State of New Jersey (www.state.nj.us/health/eoh/rtkweb/rtkhsfs.htm).

Identifying not only what risks or hazards are present in your community, but also the probability that it might be a target and how severe the consequences would be if it was attacked, allows for development of appropriate response plans. This probability–outcome analysis is useful for conceptualizing a given risk. For example, in a typical community, a bioterrorism event has a low probability of occurring; however, the severity of such an event should it transpire would be enormous. In short, it is a low probability–severe outcome event. In contrast, a fire has a relatively high probability of occurring in a community, but the severity in most instances is modest and certainly there is vast experience on how to handle such an occurrence. This would be a high probability–modest outcome event. A heat wave, in contrast, falls in between these two in terms of probability, but depending on local demographic and geographic considerations (i.e., a large number of homebound elderly) the impact could be moderately severe. Consequently, it would be considered a moderate probability–moderate outcome event. Second to anticipating what vulnerabilities exist are recognizing whether a bioterrorism event has occured. Much of this book is dedicated to that particular responsibility. It is important to assess in advance the risks to local first responders, health care workers, and the general public during and subsequent to a public health disaster. Doing so allows preparation in the way of adequate training, education, or acquiring necessary equipment tools.

❧Principles and Practices of Worker Health and Safety

The occupational health model of health and safety incorporates a hierarchy of controls to eliminate or at least limit the adverse health consequences of work. The hierarchy includes engineering out hazards, substituting a less hazardous material or process, administrative strategies, and PPE. This section summarizes the hierarchy of controls philosophy of occupational health and safety and emphasizes how this approach can be used to limit the health and safety consequences of BCN terrorism.

It is axiomatic that engineering out a problem is the best solution to a known or suspected industrial or occupational health threat. Engineering out a problem is a proactive approach whereby preventive strategies are implemented to reduce the likelihood of the anticipated danger. An example would be using a downdraft ventilation system at the site of a solvent tank. A properly functioning ventilation system draws solvent vapors away from the breathing zone of the worker and thereby minimizes the risk of inhaling these central nervous system toxins. In some cases it might be prohibitively expensive to engineer out a problem, in which case one might try to substitute a less hazardous material for the one being used. To continue our solvent example, a less toxic solvent such as acetone can be substituted for a more hazardous chlorinated solvent, such as 1,1,1,-trichlorethane. An example of an administrative strategy would be reassigning a worker who has developed a medical condition from a particular work process to a different job in the plant or, alternatively, limiting the number of hours they work at the process of concern. Although PPE is supposed to be the last control measure implemented, it is often the first one used because all too often it is the least expensive and most rapidly available control strategy. Unfortunately, PPE is often difficult to wear and in some instances—notably, respiratory protection—OSHA requires individuals to be properly fitted and trained in its use.

This hierarchy of controls model is generalizable to bioterrorism, but adapting it requires significant changes in the way our nation's industry operates. Engineering controls are also being used in preparedness efforts nationwide. Efforts to increase the number of HEPA and negative-pressure emergency rooms and hospital rooms are well underway. In many cases this requires retrofitting or building of new facilities, both expensive propositions. It is feasible to modify federal, state, and local building codes and inspection to formally required incorporation of safety consideration. (This is already done, for example, in many Western states where the risk of earthquakes is high and new building construction or renovations must incorporate earthquake-proofing methods.) Financial incentives could be used to entice hospitals to engineer into new facilities, or as part of planned plant upgrades, ventilation systems that will better protect patients and workers from bioterrorist attacks. Similar approaches to the design of any major new building, monument, museums, schools, and the like could also have benefits in terms of improved indoor air quality, lower rates of asthma, and irritant symptoms that are now common in many indoor air environments. These are not inexpensive to do, but again it is at least worth considering the dual benefits of incorporating some of these concepts in terms of protecting workers and the public.

Substitution is the second hierarchy of control in occupational health and safety. Wide-scale adoption of less toxic materials throughout industry would have the dual effect of protecting workers and lessening the risk to communities should those industries become the target of terrorists. A highjacked tanker truck of the solvent acetone, to use our earlier example, would pose less of a health threat than a truck filled with benzene. The concept of toxic use reduction has been embraced already by a number of industries in this country. How much more might be possible if, as part of our antiterrorism preparedness efforts, local, state, and federal governments offer incentives for industry to adopt these practices? Doing so will have immediate benefits to workers and secondary benefits to the public at large in terms of decreased environmental health hazards and possible bioterrorism terror.

Administrative strategies, too, are generalizable to BCN issues. A familiar example is the adoption of universal precautions as a standard infection control practice in health care and emergency response. Practiced diligently, universal precautions would do much to limit risks to health care workers and emergency workers during any disaster or in an unfolding epidemic. The overall goal is to decrease the risk of propagating an epidemic. In most instances, careful investigation of potential secondary contacts can and should be done by telephone an administrative strategy that works to limit unnecessary exposure to potentially infected individuals.

Administrative strategies can be viewed as work practices or rules for operating safely during an emergency. A less familiar example to most clinicians are the cold, warm, and hot zones established by the ICS during an emergency. Individuals operating in each of these zones must respect the purpose behind the differing classifications and use the health and safety precautions appropriate to each. Although it is unlikely that community practitioners will be working in hot or even warm zones, they should know the broad outlines of how these protective zones work and what PPE is needed in each zone. PPE needs to be available to emergency workers and individuals charged with organizing logistical support need to be able to track PPE when the ICS has been activated. Emergency and incident command planning and training needs are paramount, and local and state public health, health care, and law enforcement agencies are receiving considerable funding in this area. These issues are addressed in more detail below.

☣Health and Safety Issues During Acute Events

During a disaster or terrorism incident involving BCN agents, the ICS will be implemented, and decontamination areas will be established. Once victims have undergone decontamination, they can be seen and treated by public health and medical personnel. Because decontamination will have been done, health care providers should not need full protective equipment, and most clinicians will not be operating in hot or even warm zones. Universal precautions are always necessary and would be followed in any ICS-determined decontamination area as they would in any circumstance. However, other issues that are likely to require physicians, nurses, and other trained individuals' participation are preevent and postevent vaccinations and postexposure prophylaxis. For example, if a case of smallpox is diagnosed, mass vaccination of health care workers and first responders will be a first priority, followed very soon thereafter by mass public vaccination clinics that will be staffed by those health care workers previously trained and vaccinated. Fortunately, in the case of smallpox there is a 3- to 4-day window of opportunity following exposure to the virus when vaccination is still effective at limiting or preventing the disease. Mass vaccination is discussed in more detail in Chapter 12.

☣PPE

PPE is essential to safeguarding health with any BCN agent. Traditionally four levels of protection are recognized—Level A, B, C, or D—and each level affords progressively more protection and is recommended depending on the circumstances. Level A PPE affords the maximum degree of skin, eye, and lung protection. Level A is an encapsulated ensemble that includes a self-contained breathing apparatus and chemical protective suits, face shields, or eye-protecting glasses, gloves, boots, and hard hats. In some instances, Level A respiratory protection might include purified-air-powered respirators (PAPRs) or full-face respirators instead of self-contained breathing apparatus (SCBA) gear. Level A is the standard level of protection for entering contaminated, unsecured scenes. Once a Level A-equipped person has determined the scene to be secure and there is no significant risk of vapor contact or vapor risk, then lower levels of PPE can be employed. Level B protection provides maximal respiratory protection but is somewhat less protective of skin exposure. Level B gear includes full-face, PAPRs or SCBA gear, along with chemically resistant clothing. Compared to Level A gear, Level B does not protect the individual from exposures to vapors. Level C protection is recommended for use if both the agent and its concentration are known. Level C protection includes either negative-pressure or positive-pressure respirators (e.g., PAPR), chemically resistant clothing, and gloves, usually nitrile based. Level D protection is simply a work uniform. It provides minimal dermal protection and offers no protection to mucus membranes or the lungs (Table 3–3).

The CDC has issued recommendations for PPE based on the level of exposures likely in a given bioterrorism scenario. First responders operating in any environment in which significant uncertainties exist should use the maximal level of protection. These include situations in which the specific agent, the mechanism of dispersal, the duration of dissemination, or the concentration are unknown, or if there is ongoing dissemination of a potential bioterrorism agent. In these circumstances, workers should use a National Institute for Occupational Safety and Health (NIOSH)-approved pressure-demand SCBA along with a Level A protective suit. Workers may don Level B protective gear in circumstances where the agent is no longer being disseminated, or where splash hazards may exist, such as a tanker truck spill. When it can be determined with confidence that an aerosol dispersal device was not used, that concentrations are not inordinately high, or if a potentially hazardous item or object can be safely contained or bagged, workers may elect to use somewhat less constraining respirators,

TABLE 3–3   Personal Protective Equipment

Level	Degree of Protection	PPE	Environment
Level A	Maximum skin, eye, and lung protection	SCBA face shield, glasses, boots, gloves chemical protective suit	Contaminated, unsecured spaces
Level B	Maximum respiratory protection, moderate skin protection	PAPR, full face or SCBA, chemically resistant cloting	No protection vs. vapors
Level C	Moderate skin, lung protection	Negative or positve pressure respirator, chemically resistant clothing, nitrile gloves	If chemical and concentration is known
Level D	Minimal skin protection only	Work uniform	

such as a PAPR with HEPA filter or a full-face respirator with a P100 filter. Both types of respirators offer a fifty-fold reduction in inhalation exposure. Additional information on respiratory protection and dermal protection is provided next.

Respiratory Protection

Agents that can be aerosolized and inhaled represent the greatest potential risk in terms of BCN terrorism. For this reason, protection from airborne exposures is one of the essential goals of any public health preparedness effort. There are a variety of options for respiratory protection. NIOSH is responsible for testing and approving CNBR-approved PPE (chemical, nuclear, biological, and radiation). The review process is specific for each type of PPE and for the various bioterrorism agents; specifically, a particular respirator might work well for a droplet nuclei infectious risk, whereas a respirator designed to protect against a chemical mist is not necessarily protective for airborne microbes or aerosols. In nearly all circumstances, protection is limited because even the best respirators, garments, and gloves have laboratory-determined breakthrough times. Penetration of the protective barrier will occur given enough time in a contaminated zone or with sufficiently high levels of exposure.

Much is not known about the value of PPE in the setting of a bioterrorism event. Often wearing PPE for any length of time is uncomfortable, and depending on environmental conditions, raises concern with the risk of heat and cardiovascular stress. Respiratory protection standards help to determine what type of respirator is indicated for a given threat, what cartridges are approved for use, and the medical screening needed before someone is given a respirator to use. In all cases, fit testing is mandated. As part of fit testing, individuals must be assessed concerning a range of issues that might preclude approval for use of a respirator—medical conditions such as asthma, claustrophobia,or personal features, including the presence of facial hair or dysmorphic facial or skull features that could prevent a proper seal. For example, tight-fitting respirators are never permitted if there is facial hair. Depending on preexisting medical conditions or risk factors, additional testing (e.g., exercise stress testing or even full cardiopulmonary testing) might be warranted to determine whether one can safely use respiratory protection, especially full SCBA gear. Physicians should be familiar with the physical demands required to wear respiratory protection and may be asked to conduct preplacement examinations for first responders, fire fighters, police, health care workers, and public health workers relating to the use of PPE. An often neglected issue is that PPE is only good so long as it is properly maintained and, when needed, repaired. Durability during transportation of PPE, storage needs, and formal training programs for workers who will use the equipment, as well as preplacement evaluations and record keeping, are all aspects of preparedness that need attention and in some instances may well involve community-based clinicians.

An air-supply respirator (ASR) or PAPRs are versatile options for respiratory protection in many settings. APRs and PAPRs provide a steady stream of filtered air into the wearer's breathing zone, and some have integrated overhead hoods that provide a less confining but still protective breathing zone for the worker. They offer a high level of protection so long as the area is not heavily contaminated. They can be used in a non-oxygen-deficient environment (>19.5% oxygen), if the contaminant's concentration is known or is below published safety standards, and if dermal hazards are not present (if present this requires a chemical protective socalled "space" suit). PAPRs can be attached to belts and operate on battery power, thereby offering mobility as well as protection. SCBA gear provides workers with air from a cylinder. Combination equipment, where air is supplied by a compressor or cylinder distant from the user and a small portable cylinder is carried by the worker, is also available. A positive-pressure SCBA or combination SCBA and APRs are the only viable options for respiratory protection in any environment in which the oxygen concentration is below normal (<19.5% oxygen), when the type of contaminant is unknown, and when the concentration of the contaminant is either unknown or exceeds safety levels as set by one or more standard-setting agencies, such as NIOSH or the American Council of Government Hygienists (ACGIH).

Once the type of respiratory protection is chosen, it is necessary to use the proper cartridge or filter in order to provide selective protection against the agent of concern. Cartridges and filters are not the same thing. Cartridges are used to trap gases and vapors, whereas particles and airborne microbes are trapped by filters. Filters are designed for different types of particulates and are so designated. For example, in the N-95 filter familiar to many clinicians, the N refers to "not for oil," and the 95 refers to the efficiency of the filter for particles as small as <3 microns in diameter; meaning the filter will remove 95% of all airborne particulates, including microbes, when worn properly. For some chemical agents, standards have yet to be developed by NIOSH. In this instance, manufacturers will have to be contacted directly to determine whether a given cartridge will work for a particular agent.

Emergency management plans that include respiratory protection as a component must observe the OSHA Respiratory Protection Standard (29 CFR 1910.134). Respirators work only if they are properly fitted and worn. Improperly fitted, improperly worn, or poorly maintained respirators allow leakage into the wearer's breathing zone thereby defeating the purpose of the respirator. Bearded individuals may find it impossible to use a respirator whose seal is against the face. Instead, a PAPR that has a hood or helmet will be required. Chemical agents require using respirators with chemical cartridges capable of adsorbing the agent. Respirators with filters approved for CNBR have been developed. These respirators protect against twelve chemical agents but do not protect against all known potential chemical agents. Adequate respiratory protection for biological agents—such as tuberculosis, tularemia, or SARS—is provided by a N-95 type respirator. Although caregivers may be required to wear a higher-level respiratory protection—such as an N-95, N99, or APR—patients who are potentially infectious (e.g., coughing) are expected to use only surgical masks should they, for example, need to be transported out of their room for ancillary testing.

Dermal Protection

Most health care workers are familiar with dermal PPE. As part of good infection control and OSHA BBPs standard, hospitals, health care facilities, and many community physician offices have gowns, hoods, gloves, and booties that can be worn when exposure to BBPs is possible. Some of these are reusable but in most instances should be decontaminated and disposed of. Some chemical agents penetrate latex and other gloves readily, potentially augmenting dermal absorbtion due to the occlusion caused by the glove. That is why it is imperative that the glove—and indeed all PPEs—be chosen based on the properties of the agent or contaminant. Gowns, hoods, gloves, and booties should be removed and thrown away before taking off any respiratory protection because biological agents can be aerosolized in the process.

☤ Emergency Planning Procedure Manuals

Most communities and industries are required to have Emergency Planning Procedure manuals and to train first responders in their use. This document covers all aspects of what an individual should know during an emergency, such as exit routes, how to identify chemicals, maintenance and safeguarding, fire protection and extinguishing systems, and detection and alarm devices. Some of these devices—for example, carbon monoxide alarms or smoke detectors—are commercially available and are widely used in homes, businesses, and government offices.

Physicians should be familiar with OSHA standards that relate to the area of worker protection and hazardous material handling. Although these standards were developed to protect workers, they are pertinent to bioterrorism terrorism preparedness and have been widely promulgated in numerous CDC publications and other bioterrorism resources:

1. Emergency Responder Standard (OSHA 1910.120): addresses employer requirements in responding to hazardous material releases, environmental clean-up sites, including sites where BCN agents may have been used.
2. Hazardous Communication Standard (1910.1200): this standard summarizes the responsibilities of employers for evaluating and communicating the potential hazards of chemicals and needed PPE to employees.
3. Respiratory Protection Standard (1910.134): provides guidance to employers regarding respiratory protection against occupational diseases caused by breathing contaminated air.
4. Personal Protective Equipment (1910.132): outlines employer responsibilities for providing employees with PPE, for example, protective clothing, respiratory devices, and protective shields and barriers, when an environment with chemical hazards, radiological hazards, or mechanical irritants are present at levels capable of causing injury through absorption, inhalation, or physical contact.
5. Evacuation Plans and Processes (1910.37): addresses employer responsibilities to provide safe exit routes, procedures to ensure safe exiting by employees. This standard has potential applicability to hazardous industries located in communities where employees, emergency response teams, and the local populace may need to be evacuated.

Several websites offer more detailed overviews of each of these standards and should be consulted for additional information. They are also indexed under the Appendices (Table 3–4).

The Laboratory Response Network

Clinicians should be aware that the federal government, acting through the CDC and the FBI, created the Laboratory Response Network (LRN) in 1999. The LRN is a collaborative multilevel partnership that aims to more fully integrate laboratories with advanced testing capacity throughout the nation. Not all of these facilities have the same testing capabilities, but the network as a whole links laboratories with advanced diagnostic testing for clinical, military, agricultural, water, food, and veterinary

Table 3–4 Websites Offering Detailed Overviews of Relevant Standards

www.osha.gov
www.fema.gov
www.training.fema.gov
www.cdc.gov/niosh/topics/emres/responders.html

applications. The overall purpose of the LRN is to link the laboratory detection and surveillance capacity of the nation to a formal public health responsibility. The LRN was designed with bioterrorism in mind, but like many such efforts brings enhanced capacity to the nation's public health infrastructure that is generalizable to other public health threats. The LRN is comprised of some 140 government-supported public health laboratories or academic laboratories in all 50 states in the U.S., as well as laboratories in other nations, such as Canada, Australia, and Britain. These laboratories target biological threats, such as the Big Six Category A agents and emerging public health threats, such as SARS. There are an additional 62 laboratories in the LRN network that focus on chemical agents. LRN also operates the Biowatch program. Biowatch provides 24/7 air monitoring for biologic and chemical agents in dozens of cities throughout the U.S. The location of these sampling devices is a closely guarded secret.

There are three major classifications used by the LRN. Until recently, this classification system followed essentially the biosafety level designations familiar to many clinicians. In 2005, however, the CDC simplified the LRN structure to three categories: national, reference, and sentinel. The classification of any given laboratory depends on its capacity to handle certain biologic agents and how they function in regards to worker health and safety and public safety. "National laboratory" is the highest designation. These labs are trained and operated to handle the most dangerous biologic agents and to provide maximal protection to workers and the public. The next level of LRN is the "Reference laboratory." Such labs are capable of detecting or confirming

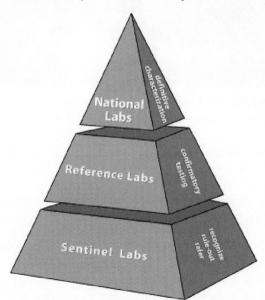

FIGURE 3–4 **The Laboratory Response Network is made up of laboratories across the country to recognize, confirm, and/or identify emerging infections and bioterrorist threats.**
Courtesy of the CDC.

FIGURE 3–5   Firefighters demonstrating decontamination procedures.
Courtesy of Edwards Air Force Base/Monte Congleton.

the presence of specific biologic agents. With the support of federal funding, many state public health laboratories and a number of academic laboratories have been upgraded to enable them to rapidly identify agents, a capacity that shortens considerably the response time for local and state emergency responders. Prior to the upgrading of these laboratories, it was often necessary to send samples to the CDC for identification and confirmation. Finally, the nation's hospital-based laboratories are designated as "sentinel laboratories." The sentinel designation is an acknowledgment that such laboratories receive and process hundreds of thousands of biologic specimens each day and that in the event of a covert bioterrorist attack, may be the first laboratory to come into contact with a biological agent. Sentinel labs are expected to send any suspicious specimens to one of the reference labs for identification and confirmation.

The LRN structure as it relates to chemical terrorism is similarly categorized into three distinct levels and functions. In descending order of safety, they are Level 1, 2, 3. Currently, there are five Level 1 laboratories. Such labs are capable of detecting a wide array of chemicals, including mustard, nerve, and other toxic agents. Some 41 laboratories have received the designation of Level 2 laboratories. These laboratories are prepared to detect a more narrow range of toxic exposures, but do include cyanide and toxic metals, such as mercury or arsenic. Last, there are the Level 3 laboratories. Their primary function is to be aware of the clinical presentation of various chemical agents in order to properly collect, store, or transport clinical specimens that may contain as yet unidentified chemical agents. Level 3 labs also are expected to interface with local and state emergency, hospital, and public health response systems in preparedness training, planning efforts for chemical events, and of course, actual incident response.

The LRN is a fine example of the "dual-use" concept advocated by many public health authorities. By clarifying the roles and responsibilities of the nation's laboratories, expanding staffing, implementing training initiatives, and adding advanced technologies, the nation's overall public health infrastructure is enhanced. This approach has the advantage of better preparing the medical and public health communities for bioterrorism-related events as well as other public health threats.

In addition to creating the LRN, significant efforts are underway to expand the nation's biosafety laboratory capacity tied explicitly to the perceived threat of bioterrorism. Currently, there are four BSL-4 level facilities in the United States: the CDC in Atlanta; USAMRIID at Fort Detrick in Frederick, MD; the Southwest Foundation for Biomedical Research in San Antonio, TX; and the University of Texas at Galveston. The National Institutes of Health (NIH) operates a small BSL-4 laboratory in Bethesda, MD, but at the present time it is conducting research on emerging infectious diseases at the BSL-3 level only. The reasons given for this increase in funding and planning are that current numbers of BSL-3 and BSL-4 laboratories would be unable to handle a large influx of specimens in the event of a bioterrorist attack or multiple attacks, whether a propagating epidemic ensues or not. In addition, proponents of this expansion insist that more BSL-3 and BSL-4 laboratories are needed to conduct research and testing of vaccines. These proposed new facilities are being sited at university or academic health centers and have encountered significant resistance in many of these communities.

❖Red, Yellow, and Green Zones

Clinicians should be familiar with designated zones used by hazmat experts to define levels of risk. These levels of risk correspond to level of contamination that is either confirmed or suspected, and of course, the level of protective gear must correspondingly follow from the zone in which work is being done. "Red Zone" refers to an environment in which contamination has been confirmed or is strongly suspected. Such was the case in 2001 when the U.S. Senate Office Building and several U.S. Postal Service facilities were found to be contaminated with weaponized anthrax. The Yellow Zone designation refers to environments where possible contamination exists but remains unconfirmed. This was the situation in a number of mail facilities that had received mail from the contaminated facilities. The final designation is Green Zone. Green Zones are environments in which contamination is unlikely.

These designations are useful in a preevent circumstance as well. For example, given the recent experience where weaponized anthrax spores were distributed through the U.S. mail, postal facilities should be viewed as potential targets for terrorists. It follows that all mail-sorting facilities, mail rooms, or post offices are "at-risk" work environments. Even without known contamination, however, facilities that receive mail from previously contaminated facilities, working next to high-speed mail-sorting machines, or within physical proximity of a nonmail facility near a mail facility would make these areas Yellow Zones. Similarly, the absence of identified bioterror agents in the majority of American workplaces would by definition imply that these facilities are Green Zones. This does not, however, imply that there are not significant risks associated with these industries. Documented security deficiencies in industrial facilities using toxic materials leaves open the possibility that terrorists could use sabotage or direct attack to cause a major industrial "accident" with potentially wide-ranging health effects. One need only mention Bhopal, India to recall the potentially devastating effects industrial sabotage or accidents could engender.

🜪Summary

When it comes to BCN terrorism, clinicians' obligations extend beyond the immediate diagnostic and treatment needs of their patients into the realm of public health. Although public health is an unfamiliar arena for many physicians and health care workers, this is an historic anomaly that is increasingly untenable in today's world. The nation's public health system has been underfunded and understaffed for decades, the necessity of rebuilding that infrastructure and expanding the available pool of broadly trained public health professionals has resulted in a system that is better prepared to respond to a true national crisis than it was only a short while ago. Clinicians may be largely unaware of these critical changes and their potential impact on medical care and public safety during a bioterrorism event. In the end, clinicians remain central to any effective public health response. Communities, the media, and local health departments will turn to clinicians for guidance and expertise as part of the nation's preparedness efforts. This has already been the case during the smallpox vaccination compaign, anthrax attacks or scares, and the SARS epidemic. During any such event, clinicians must be mobilized rapidly in order to provide clinical care, and as importantly, reassurance to their patients and their communities. To do so, clinicians must be aware of the structure of the public health system and the roles played by local, state, and federal agencies. Specifically, they should be aware of key local and state resources that have been put into place to assist communities in the event of a bioterrorism attack, including the ICS, the newly created LRN, and how basic health and safety principles—such as PPE—can enable health care workers and first responders to safely continue their essential duties during an emergency. Clinicians should be familiar with the potential hazards in their communities and should be aware of the medical and public health responses needed to address the most likely hazards in the event of a terrorist attack on an industrial facility.

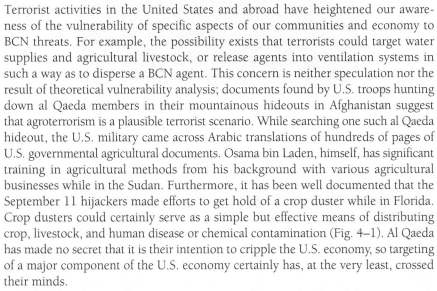

CHAPTER 4

Environmental Terrorism

Terrorist activities in the United States and abroad have heightened our awareness of the vulnerability of specific aspects of our communities and economy to BCN threats. For example, the possibility exists that terrorists could target water supplies and agricultural livestock, or release agents into ventilation systems in such a way as to disperse a BCN agent. This concern is neither speculation nor the result of theoretical vulnerability analysis; documents found by U.S. troops hunting down al Qaeda members in their mountainous hideouts in Afghanistan suggest that agroterrorism is a plausible terrorist scenario. While searching one such al Qaeda hideout, the U.S. military came across Arabic translations of hundreds of pages of U.S. governmental agricultural documents. Osama bin Laden, himself, has significant training in agricultural methods from his background with various agricultural businesses while in the Sudan. Furthermore, it has been well documented that the September 11 hijackers made efforts to get hold of a crop duster while in Florida. Crop dusters could certainly serve as a simple but effective means of distributing crop, livestock, and human disease or chemical contamination (Fig. 4–1). Al Qaeda has made no secret that it is their intention to cripple the U.S. economy, so targeting of a major component of the U.S. economy certainly has, at the very least, crossed their minds.

This chapter provides a concise overview of potential vulnerabilities in regards to water, air, and food resources and infrastructure with which clinicians ought to be familiar.

FIGURE 4-1   A crop duster would be a simple means of disseminating agents against crops, livestock, and humans.
Courtesy of the USDA.

Drinking Water Security

The nation's water supply system is comprised of an elaborate and multifaceted system involving public, private, and quasipublic entities that manage and maintain our water supply to serve industry, agriculture, municipalities, and the general population. The water system infrastructure is composed of four basic components: water sources (in the form of lakes and ponds), water treatment, storage, and distribution. Sources of water include rainwater, underground springs, or surface water reservoirs that are usually dammed by an earth, concrete, or masonry structure and that collect water through stream or surface run off. There are roughly 75,000 dams and reservoirs nationwide.

Water treatment is accomplished in a variety of ways, ranging from no treatment to basic treatment (e.g., chlorination, adjustment of pH, iron removal) to more complex systems where filtration and chemical treatment of surface water exist. Following treatment, water is stored in containment facilities, such as reservoirs, storage tanks, and water towers. From this point, water is distributed through progressively smaller conduits to consumers involving thousands of miles of pipe. Distribution systems include pumping stations, and large water mains running from the source or storage area to increasingly smaller pipes and ultimately ending up at hydrants or the kitchen tap. The complexities of such a system make it vulnerable to acts of terrorism at virtually any point.

The EPA is the primary federal agency responsible for protecting the nation's water supply. The 2001 Bioterrorism Act mandates that water utilities conduct vulnerability assessments and create emergency response plans. As part of this assessment, critical system components must be identified and procedures to prevent or respond to sabotage must be developed. Certain considerations make the water supplies a possible target; for instance, 75% of the population is provided water from 15% of national water systems. Consequently, a few key facilities vital to a majority of the population are more likely targets. But even an attack on a smaller site would have tremendous psychological and logistical consequences.

Water sources often encompass a wide land area, a fact that complicates security considerations (Fig. 4–2). As a practical matter, protecting a community's water

FIGURE 4–2 The water supply is vulnerable in a variety of ways, including contamination of reservoirs.
Courtesy of the USDA.

supply, like all of BCN terrorism, is a cooperative effort. The combined resources of local, state, and federal law enforcement, emergency response systems, the medical community, intelligence agencies, and to no small degree, the general public must all be engaged in securing our communities' water systems. Water companies have well-defined response plans to assess potential bioterrorist situations, identify any breaches in security and safety of the water supply, and act accordingly; nonetheless, they are asking the public to maintain vigilance around water supply sources. Suspicious activity—such as unauthorized individuals or vehicles in the vicinity of any drinking water supply source, dam, or gatehouses—must be reported immediately to local law enforcement, and, in most states, local or state public health officials. Individuals should not intercede when they observe suspicious activity; rather, they should notify local law enforcement or call 911.

Terrorists need not inflict illness to sow terror in our communities. The mere presence of a contaminant, even without its ever causing illness would likely cause widespread anxiety. Disruption of water service by tampering (whether genuine or faked), introducing contaminants, conventional attacks on dam or facilities, all would cause widespread concern or even panic. Another consideration is the possibility of cyber attacks on complex water treatment facilities' computer systems, which could cause disruption of water distribution for days. The impact on any of these scenarios

OF NOTE...

April 1993, Milwaukee, Wisconsin

Following an inordinate number of reports of debilitating gastrointestinal illness in the area, it was discovered that in April 1993, one of the two primary water treatment facilities in Milwaukee, Wisconsin, was malfunctioning and inadequately filtering the water supplied to residences and businesses in the area. In the weeks that followed, it is estimated that over 400,000 people developed diarrhea. This is thought to be an underestimate. Over one hundred people died. The largest waterborne epidemic in the history of the United States had just occurred. The water samples revealed that while levels of the most waterborne pathogens remained stable, there was a one hundred-fold increase in cryptosporidium in the water. Although the failure of the water treatment process was purely accidental, it does suggest the potential morbidity and mortality that could result from sabotaging of water treatment facilities. Although it may be more sensational, a bioterrorist attack with a waterborne pathogen would likely affect far fewer people than a strategic sabotage of a water treatment plant might.

It is significant, too, that no warning from the monitoring system at the facility indicated the problem. Instead, contamination was discovered after an observant local pharmacist noticed a boom in the purchase of antidiarrheal products. He contacted authorities and an investigation was begun. This episode underscores underappreciated but creative opportunities to develop surveillance systems for bioterrorism (see Chapter 5).

TABLE 4–1   Partial List of Potential Waterborne Agents

Vibrio cholerae
Campylobacter
Yersinia enterocolitica
Legionella pneumophila
Salmonella spp
Shigella spp
Escherichia coli (EHEC)
Rotavirus
Norwalk/Norwalk-like Hepatitis A
Isospora belli
Cryptosporidium parvum
Cyclospora cayetanensis
Microsporidia
Giardia lamblia
Entamoeba histolytica

on communities would be enormous, logistically, economically, and psychologically. Hardest hit would be urban areas and communities most dependent on public water systems.

The nature of the water treatment is such that, with the exception of *Cryptosporidium parvum,* it is unlikely that most biological agents would survive usual disinfection practices. However, treatment methods are not foolproof. Sabotage of treatment systems combined with contamination could occur. The other end of the spectrum, those communities with little to no treatment at all, are obviously more vulnerable to a one-strike bioterrorist attack. It is worth reemphasizing the reassuring fact that contaminating a water supply on a scale that could affect large numbers of people would require prohibitively large amounts of contaminants and comparable logistical and technical resources. A threat of this nature at such a scale is improbable, but not impossible.

Table 4–1 provides examples of biological agents that could be used to contaminate water supplies. Readers should remember that chemical and radiological agents could be used as well. With these, too, they need not necessarily inflict health injury as their mere presence would have a ripple effect within the community, disrupting usual activities dependent on the presumption of safe water.

❧Indoor Air and Ventilation Systems Security

Protecting indoor environments from airborne biological, chemical, or radiological attack has been the subject of intense study because of particular vulnerability to workplaces, schools, and other occupied buildings in which large numbers of people are situated. As with the water supply, indoor air quality is an enormous and complex subject deserving a discussion beyond the scope of this book. Health care providers should at least be aware of some general considerations relating to building ventilation and air quality systems as potential terrorist targets. The CDC promulgated detailed recommendations for those organizational units and individuals responsible for protecting the nation's heating, ventilation, and air-conditioning (HVAC) systems. Preventive strategies include putting into place security measures to protect vulnerable sites, such as air intakes or computer control systems. Anticipatory training measures and scheduled systems maintenance as well as an improved filtration system should all be considered in light of a system's potential vulnerabilities. Determining which building, which community, and which individual could be a terrorist target is an impossible task. However, buildings involved in large or critical industrial facilities, military facilities, laboratories, and hospitals may be particularly enticing targets for terrorism, and this fact alone mandates advance planning. As an example, hospital isolation rooms or emergency departments are very likely to be involved should a BCN event occur. Most hospitals and many institutions are retrofitting HVAC systems in light of the nation's security concerns. No building can be fully protected from a determined group or individual intent on releasing a BCN agent, but effective air filtration and air cleaning can help to limit the number and extent of injuries or fatalities and make subsequent decontamination efforts easier (Fig. 4–3).

HEPA filters and well-designed HVAC systems are effective means of increasing building cleanliness, limiting effects from accidental releases, and generally improving IAQ. Properly designed, installed, and maintained air-filtration and air-cleaning systems can reduce the effects of a BCN agent, such as botulinum toxin, released within a building, or outside, by removing the contaminants from the building's air supply. Consistent with the "dual use" philosophy by which bioterrorism preparedness supports general public health systems, these measures offer the added benefit of reducing cases of respiratory infection, exacerbations of asthma, and allergies among building occupants. Together, these accrued benefits may provide

FIGURE 4–3   Dust contamination after September 11.
Courtesy of the EPA.

greater protection in the event of contamination of the air, particularly if aerosols are involved. The danger from aerosols released into a building's ventilation system depends on the infectivity or toxicity of the agent, its ability to remain aerosolized, its dispersal properties, and concentration. Radiological hazards, such as alpha and beta particles, can be introduced as aerosols, carried on particulate matter, or present as gasses. Conventional explosives are most likely to be used to spread radioactive materials. In any event involving airborne introduction of these particles, these systems can collect the particulate carriers upon which the radioactive particles reside. Radioactive particles are readily removed, for example, by HEPA filtration. Filtration and air-cleaning devices are ineffective at stopping blast effects or radiation from detonation of a nuclear device. Any system where radiation has been introduced will subsequently require decontamination, an expensive and time-consuming task.

Assessing a building's HVAC system is important in and of itself, but also because many HVAC systems are poorly maintained. This assessment should be done as part of BCN preparedness and should consider which types of BCN agents are of most concern in that setting. This determination will be informed in part by local risks, such as nearby industries, railroads, and storage facilities for chemicals or nuclear materials.

In assessing a filtration system, the physicochemical properties of the different agents dictate how protective filtration systems are in protecting a building's occupants. Many filtering systems are based on adsorption, the electrostatic interaction between a molecule of gas or vapor and a surface. This is a common means of filtering unwanted particulates, vapors, or gases from the air. A variety of absorbent materials, including activated carbon, are available to absorb volatile organic compounds. Solid adsorbents—such as activated carbon, silica gel, activated alumina, zeolites, porous clay minerals, and molecular sieves—are useful because of their large internal surface area, stability, and low cost. Adsorbents lose reactivity over time and through saturation. Military research has provided guidance on the need for and frequency of scheduled maintenance programs for HVAC systems depending on the types of filters being used. For example, nerve and blistering agents are strongly adsorbed to carbon-based filtration systems, whereas choking and blood agents (phosgene and cyanide) do not adsorb well to carbon filters and must be trapped by filter impregnates that chemically bind an agent. A multicapacity filter developed by the military is now available commercially (ASZM-TEDA carbon filter). This material is a coal-based activated carbon that has been impregnated with copper, zinc, silver, and molybdenum compounds and protects against a wide range of chemical warfare agents.

Another means of filtration is through the use of impregnates. Since World War I, impregnates have been used to protect soldiers from chemical warfare agents, such as mustard gas and phosgene. Impregnates are usually metal compounds that interact with various chemical agents to prevent contamination of the air stream. Chemical impregnates supplement physical adsorption by an added chemical reaction and can aid activated carbon to remove high-volatility vapors and nonpolar contaminants. Isopropyl methylphosphonofluoridate (GB), which is a nerve gas (sarin); and bis-(2-chloroethyl) sulfide (HD), which is a vesicant—are effectively removed by physical adsorption. Reactive chemicals have been successfully impregnated into activated carbon to decompose chemically high-vapor pressure agents, such as the blood agents.

Agroterrorism

Agroterrorism includes the use of BCN agents to inflict damage on the agricultural industry and the food supply. Because of the immense complexities of both the agriculture and the food industries, a comprehensive discussion is beyond the scope of this text, but a brief overview is warranted.

Many experts believe that a significant agroterror act is more likely than a bioterror act primarily because attaining the agents to do so is much easier and dispersal requires minimal expertise and resources. The agroterrorism threat is two-fold: widespread illness or widespread economic damage to one or more of the following:

- Individual species—including direct attacks on livestock, poultry, and, crops
- Housing, storage, and other facilities
- Centers and means of food distribution

Many of the BCN threats against humans are equally and sometimes more damaging to animals. Diseases such as glanders, brucellosis, and of course, anthrax are examples of biological agents that could cause disease in both. There are also fungal, viral, and bacterial agents, many of which are thought to exist in weaponized forms, that target only crops (Fig. 4–4). Chemical weapons are equally harmful to animals, while radiation is damaging to all plant and animal species.

Economic damage from agroterrorism has devastating potential and does not necessarily require a large-scale attack. An isolated attack on livestock or a crop might, for example, result in domestic and international disruption of trade. Even the use of a plant pathogen that is harmless to humans would likely result in embargos or tremendous dropoff in sales both in the United States and abroad. Regardless of what scenario might come to be, the psychological effects of an attack on the food supply is immense. The food and agricultural industry is a $1.25 trillion a year industry, accounting for nearly 20% of all jobs in the United States. Agricultural and animal exports alone make up $50 billion per year. It is not implausible to envision a scenario whereby a significant sector of the U.S. economy could be crippled even without a single person becoming ill.

The U.S. Department of Agriculture (USDA) has implemented a number of changes to help protect against such events, including the newly created Office of Food Security and Emergency Preparedness. This agency focuses on training and implementation of inspectors, improved surveillance, increased testing and laboratory facilities, as well as strengthening biosecurity, physical security, cyber-security, and telecommunications in the nation's food supply and agricultural sectors.

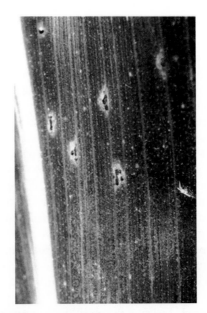

FIGURE 4–4   Corn infected with common rust.
Courtesy of Institute of Agriculture and Natural Resources Cooperative Extension, University of Nebraska-Lincoln.

FIGURE 4–5   Cattle grazing on land vulnerable to attack via aerosols, contaminated grass, feed, or water.
Courtesy of the Agricultural Research Service, USDA/Scott Bauer.

Although security improvements are necessary, even the most thorough of precautions are not enough, since many nations food and animal industries rely on imports from other countries. For example, nearly two million cattle, three-fourths of a million pigs, and some 30 million birds are brought into the United States from all over the world. In addition, there are tens of millions of tons of crops, grains, and agricultural produce, imported to the United States yearly, exceeding $46 billion per year. These figures highlight the additional challenges of protecting our food supply, as infection, exposure, and contamination of agricultural products consumed within the United States could easily be initiated long before these products even arrive. As an example of how agricultural disease affects an economy, in 2000, an outbreak of foot and mouth disease in Singapore resulted in losses approaching $20 billion. It takes only a single infected animal to cross into the U.S. to create agricultural, medical, psychological, and of course, economic havoc. Similarly, the prion disease, bovine spongiform encephalopathy (so-called mad cow disease), has devastated the British cattle industry and threatened both Canadian and American herds as well.

The USDA has compiled a list of the most likely threats to livestock and crops (Table 4–2). Included in the table are those agents deemed to be potentially of "High Consequence." Those that are also harmful to humans are designated as "overlapping" with the Department of Health and Human Services' (HHS') list of pathogens that are harmful to humans. The overlap designations indicate those

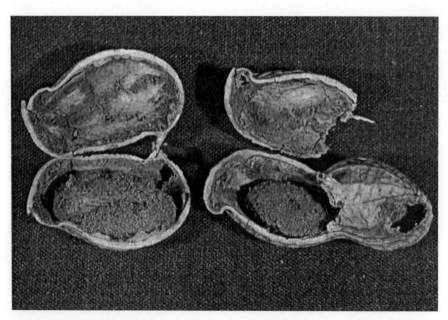

FIGURE 4–6 *Aspergillus flavus* **growing on one of its most common food, peanuts.**
Courtesy of the Department of Primary Industries Queensland, Australia.

TABLE 4-2   A List of the Potential Biological Agents That Could Affect Crops, Livestock, or Both

	Class	Designation
Livestock Pathogens		
Aflatoxins	Toxin	USDA-HHS overlap agents
African horse sickness virus	Virus	USDA high-consequence livestock pathogens and toxins
African swine fever	Virus	USDA high-consequence livestock pathogens and toxins
Akabane virus	Virus	USDA high-consequence livestock pathogens and toxins
Avian influenza virus (highly pathogenic)	Virus	USDA high-consequence livestock pathogens and toxins
Bacillus anthracis	Bacteria	USDA-HHS overlap agents
Blue tongue virus	Virus	USDA high-consequence livestock pathogens and toxins
Botulinum toxins	Toxin	USDA-HHS overlap agents
Bovine spongiform encephalopathy agent	Prion	USDA high-consequence livestock pathogens and toxins
Brucella abortus	Bacteria	USDA-HHS overlap agents
Brucella melitensis	Bacteria	USDA-HHS overlap agents
Brucella suis	Bacteria	USDA-HHS overlap agents
Burkholderia (Pseudomonas) pseudomallei	Bacteria	USDA-HHS overlap agents
Burkholderia (Pseudomonas) mallei	Bacteria	USDA-HHS overlap agents
Camel pox virus	Virus	USDA high-consequence livestock pathogens and toxins
Classical swine fever	Virus	USDA high-consequence livestock pathogens and toxins
Clostridium botulinum	Bacteria	USDA-HHS overlap agents
Clostridium perfringens epsilon toxin	Toxin	USDA-HHS overlap agents
Coccidioides immitis	Fungi	USDA-HHS overlap agents
Cowdria ruminantium (heart water)	Bacteria	USDA high-consequence livestock pathogens and toxins
Coxiella burnetii	Bacterial	USDA-HHS overlap agents
Eastern equine encephalitis virus	Virus	USDA-HHS overlap agents
Equine morbillivirus (hendra virus) Nipah virus	Virus	USDA-HHS overlap agents
Foot and mouth disease virus	Virus	USDA high-consequence livestock pathogens and toxins
Francisella tularensis	Bacteria	USDA-HHS overlap agents
Goat pox virus	Virus	USDA high-consequence livestock pathogens and toxins
Japanese encephalitis virus	Virus	USDA high-consequence livestock pathogens and toxins

(Continued)

TABLE 4–2   A List of the Potential Biological Agents That Could Affect Crops, Livestock, or Both (*Continued*)

	Class	Designation
Select Agent	Type	HHS, USDA classification
Lumpy skin disease virus	Virus	USDA high-consequence livestock pathogens and toxins
Malignant catarrhal fever virus	Virus	USDA high-consequence livestock pathogens and toxins
Menangle virus	Virus	USDA high-consequence livestock pathogens and toxins
Mycoplasma mycoides mycoides (contageous bovine pleuropneumonia agent)	Bacteria	USDA high-consequence livestock pathogens and toxins
Mycoplasma capricolum/M.F 38/M. *mycoides capri* (contageous caprine pleuropneumonia agent)	Bacteria	USDA high-consequence livestock pathogens and toxins
Newcastle disease virus (exotic)	Virus	USDA high-consequence livestock pathogens and toxins
Nipah virus	Virus	USDA high-consequence livestock pathogens and toxins
Peste Des Petits ruminants virus	Virus	USDA high-consequence livestock pathogens and toxins
Rift Valley fever virus	Virus	USDA-HHS overlap agents
Rinderpest virus	Virus	USDA high-consequence livestock pathogens and toxins
Sheep pox	Virus	USDA high-consequence livestock pathogens and toxins
Shigatoxin	Toxin	USDA-HHS overlap agents
Staphylococcal enterotoxins	Toxin	USDA-HHS overlap agents
Swine vesicular disease virus	Virus	USDA high-consequence livestock pathogens and toxins
T-2 toxin	Toxin	USDA-HHS overlap agents
Venezuelan equine encephalitis virus	Virus	USDA-HHS overlap agents
Plant Pathogens		
Liberobacter africanus	Bacteria	USDA
Liberobacter asiaticus	Bacteria	USDA
Peronosclerospora philippinensis	Bacteria	USDA
Phakopsora pachyrhizi	Bacteria	USDA
Plum pox potyvirus	Virus	USDA
Ralstonia solanacearum race 3, biovar 2	Bacteria	USDA
Schlerophthora rayssiae var *zeae*	Bacteria	USDA
Synchytrium endobioticum	Bacteria	USDA
Xanthomonas oryzae	Bacteria	USDA
Xylella fastidiosa (citrus variegated chlorosos strain)	Bacteria	USDA

Adapted from the USDA.

agents that are harmful to livestock and humans, and are therefore a multiple threat. For example, aflatoxin could be used in one of two ways (see Chapter 19) (Fig. 4–7). *Aspergillus flavus,* the fungus that naturally produces aflatoxin, could intentionally be used to contaminate crops or dairy supplies. Alternatively, aflatoxin harvested from *Aspergillus* or synthesized industrially could be aerosolized and used as a weapon directly against animal or human populations.

Historical Precedence

Although it may seem far fetched, threats of the nature discussed in this chapter are not without precedence. In 6th century BC, an alliance of Greek armies during the siege of Kirrha, used hellebore root to poison the city's water sources, debilitating the city and its army with diarrhea. In the modern era, the German army deliberately and secretly infected large numbers of livestock and horses using anthrax and Glanders in World War I. Hardest hit was the Russian army along the eastern front. As a result of the animal illnesses, supply, artillery, and troop movement were debilitated. In World War II, both the Allies and Axis powers were in possession of anthrax bombs, but no use beyond isolated testing was ever recorded. In the 1970s there was great controversy as to whether the use of Soviet-supplied T-2 mycotoxin was used by the Laos communist army against Central Intelligence Agency-backed Hmong rebels.

FIGURE 4–7 Wheat rust, one of the known plant pathogens to have been stock-piled by several countries for use as economic warfare.
Courtesy of the Agricultural Research Service, USDA/Les Szabo.

The very first act of bioterrorism in the United States in the modern era occurred in 1984 in the form of intentional contamination of restaurant food resulting in the infection of almost 800 people. A large religious sect based in Oregon, whose aim was to alter voter turnout for local elections, spread *Salmonella typhimurium* at popular area restaurants.

It is well documented that during the Cold War, both the Soviet Union and the United States researched and stockpiled tons of wheat rust spores; enough, it is believed, to contaminate all the wheat fields in the world. Although no evidence exists to suggest these were used militarily, the stockpiles reportedly still exist. Like much of the BCN agents remaining in the states of the former Soviet Union, the security and account of these agents are in doubt.

Summary

Pre-September 11 vulnerability to BCN attacks generally and in terms of air, water, and agricultural systems was real and disturbingly so. Since then measures have been implemented to address these environmental vulnerabilities. Although there is still much more that needs to be done, the good news is that the issues are presently and actively being addressed with new measures being implemented regularly. Although the primary attention relating to BCN terrorism has focused on the immediate risks to the general public, water, indoor air, and agroterrorism have received less attention. This oversight may be a grave mistake as some BCN experts consider these indirect threats more probable than direct attacks on communities. Such oversight may be a grave mistake as many would argue that these are more likely to be venues of attack than are human populations. An important distinction between human targets and the environmental targets focused upon in this chapter, is that economic and psychological after effects will likely overshadow the biological effects given the nature of the threat. Nevertheless, there are numerous environmental agents affecting humans as well as plant and animal species (such as T-2 mycotoxin and anthrax), so physicians should anticipate sharing with their veterinary and plant biology colleagues key roles in responding to the public's concern with these forms of bioterrorism. Physicians will be sources of education and reassurance for their patients and the community at large in the face of any form of environmental terrorism.

CHAPTER 5

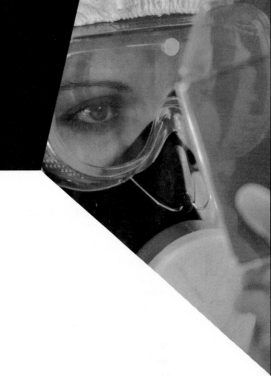

Surveillance Systems and Bioterrorism

Monitoring disease activity has been one of the primary responsibilities of government-supported public health units since they first came into being during the formative years of our nation. For much of our nation's history, surveillance has focused on the most important infectious disease threats, such as cholera, tuberculosis, or more recently HIV/AIDS. Today, developing accurate and reasonable specific surveillance systems capable of identifying bioterrorism accurately and swiftly is one of the most active areas of development and research relating to bioterrorism. Early detection of exposures could allow for containment and treatment that could greatly reduce morbidity and mortality of countless numbers. Unfortunately, implementation of such surveillance systems is no simple task. There are several general approaches to surveillance: clinical or case-based surveillance, environmental surveillance, or laboratory-based surveillance. This chapter summarizes general principles of surveillance systems, as well as various models for conducting surveillance as it relates to bioterrorism. The concept of sentinel health events is also introduced, and its potential as a useful tool for practitioners in diagnosing bioterrorism is considered. Last, the chapter ends with a discussion of the utility of syndromic surveillance and how this might be applied in order to distinguish naturally occurring sentinel health events (e.g., a case of plague) from a deliberate bioterrorist attack.

Developing and testing surveillance systems to detect at the earliest possible stage potential bioterrorist incidents ranks high as a national public health priority. In 2002 alone, the CDC distributed over $1 billion in funds to accelerate research and development of surveillance systems with two specific objectives. First, to create state-based surveillance systems capable of receiving and evaluating diagnoses, disease case reports, or illness syndromes that could signal a bioterrorist attack. Second, to establish communication networks capable of sharing quickly critical public health information to emergency departments, local and state public health units, and local and state law enforcement officials. Although progress is being made, the degree to which these goals have been met varies considerably from state to state.

☣An Overview of Bioterrorism Surveillance

The overall objectives of an effective surveillance system are defined readily enough. According to the CDC, ideal surveillance systems are capable of recognizing a BCN terrorist event quickly, with reasonable sensitivity and specificity, and capable of quickly disseminating these data to critical nodal points in the public health decision-making structure. Further, surveillance systems should monitor diseases or illness syndromes with manifest public health importance. The operational requirements of surveillance systems should be known as well, including direct costs, and personnel and information technology requirements. The accuracy, flexibility, and acceptability of the system should be known, preferably through actual field testing or computer-simulated modeling. A surveillance system that cannot be modified as existing threats evolve, or as new ones emerge, has limited long-term utility and the costs associated with implementing such a system will be difficult to justify. If a surveillance system has limited acceptability to those who are being asked to assume the real costs of the system, for example, onerous, unduly complex, or time-consuming data entry requirements, the system will be useless as a surveillance tool. Other components of a good system include timeliness of reporting to key stakeholders, being discriminating enough to detect genuine outbreaks, and being capable of monitoring a population or locations that are reasonably representative of the larger population at risk. In summary, surveillance systems are only good if they identify early and accurately a potential or evolving bioterrorism event; are user friendly, flexible, and secure; and provide rapid information to decision makers. The capabilities of an ideal surveillance system are summarized in Table 5–1.

At present, the specific structure, methods, and purposes of existing surveillance systems demonstrate a great deal of significant variability. This variety offers opportunity and obstacles to the implementation of effective surveillance for bioterrorism. On the one hand, diversity offers the possibility of developing multiple effective surveillance models, or mutually reinforcing surveillance systems that afford a wider scope of protection than a single approach might. On the other hand, given the urgency that exists relating to this issue, the abundance of options may delay the timely implementation of a workable and effective surveillance system. There is value in bringing some standardization to the type of data being collected, monitoring and analytic methods, as well as reporting mechanisms. Achieving some consensus on these issues will facilitate the creation of a

TABLE 5-1   Features of an Effective Surveillance System

A Surveillance System Should Have the Following:

The capability to quickly recognize bioterrorist events

Reasonable sensitivity and specificity with a good positive predictive value for detecting genuine outbreaks

The capability of disseminating these data quickly to key stakeholders in the public health decision-making structure

The capability of recognizing diseases or illness syndromes with manifest public health importance

Be fiscally, technologically, and logistically feasible

Be modifiable for use in new or unexpected threats or uses

The capability of monitoring a population or location reasonably representative of the larger population at risk

Adapted from the CDC.

bioterrorism surveillance system, or set of systems, that is consistent, reliable, and accurate.

Unfortunately, recent reviews of existing biological and environmental surveillance systems make the general point that few, if any, currently available systems meet either the CDC criteria outlined previously or the need for standardized approaches that can be disseminated within the medical and public health communities. There are a number of reasons for this assessment. First, consensus within the public health and medical communities as to what diseases, syndromes, or agents ought to be included in the surveillance systems does not yet exist. There are similarly divergent views on the best sources of data and optimal methods for collecting and analyzing data.

Although many of these barriers can be overcome with improved research, planning, and national public health leadership, there are other barriers to creating an effective surveillance system that are not so easily remedied. One inescapable problem is that bioterrorism events are still (one might add, thankfully) rare occurrences. Consequently, prevalence and incidence data do not exist, making direct comparisons between one surveillance system and another difficult. Without reliable reference standards, formal evaluations of test characteristics such as posttest probabilities, predictive values, sensitivity, and specificity becomes serious obstacles. A related problem is that a "gold standard" surveillance system has not been created to which newer or modified approaches can be compared. Existing military surveillance systems for BCN agents might meet many of these objectives, but these have not been fully described in the scientific literature. On a positive note, efforts to improve cooperation between military and public health organizations are underway.

As mentioned briefly, there is a paucity of literature describing the test characteristics of the majority of surveillance systems. To be more precise, we know little about their ability to detect early cases that might herald an epidemic (system sensitivity) or how many false alarms might result from the use of the system (system specificity). For example, numerous bioterrorism-related hoaxes have been perpetrated in recent years. Between April 1997 and June 1999—well before the actual U.S. Post Office anthrax attacks—there were over 200 documented anthrax powder hoaxes in the United States. These hoaxes triggered some thirteen thousand evaluations by the emergency response system with significant costs in time, resources, and money. Distinguishing a real bioterrorist event from a false alarm is clearly a priority, given the limitations of our existing resources. Because we know so little about these fundamental test characteristics, we cannot define other useful constructs such as predictive values, likelihood ratios, or accuracy.

Computer simulation provides an increasingly viable method by which to surmount some of these problems. Retrospective mining of documented bioterrorism events, for example, the U.S. Postal Service anthrax attacks, have proved useful in creating some diagnostic algorithms that may have generalizability to the design of surveillance systems. How these algorithms will work in a genuine attack remains to be seen.

Novel approaches to extend surveillance systems into the community to a greater extent than is possible at present with hospital-, laboratory-, or emergency room-based systems are being actively investigated. Among the possibilities are using over-the-counter medications and prescription data as early markers of unusual disease or illness activity. Additional markers of potential interest in developing novel surveillance systems have been examined in various contexts. These include school absenteeism, calls to nursing information lines, physician house calls, or sick leave prescriptions for influenza-like conditions diagnosed by community practitioners. Some studies have found, for example, that physician diagnosis of flulike illnesses is a more sensitive marker of an emerging influenza epidemic than viral isolates, whereas other researchers have found the reverse to be true.

Timeliness of reporting to key stakeholders is critical in any workable surveillance system. It would seem that information technology (IT)-based systems would have a natural advantage over manual systems in this regard, although this is not necessarily the case based on our current knowledge. Moreover, IT systems are vulnerable to cyberterror attacks and internal failures, and often cannot handle substantial demands. Put simply, existing technology infrastructure is insufficient to handle the sheer volume that an actual or presumed bioterrorist attack would almost certainly place on any system. The example of the CDC website crashing temporarily in the days following the post office anthrax attacks as a result of the volume of hits illustrates the vulnerability of IT-based information, communication, and surveillance systems. Efforts to build redundancy, fail safes, surge capacity, and security into the nation's health care and public health systems are underway, though nothing is fool-proof. The need for trained personnel at the local and state health department level in order to respond appropriately to the information being fed to them is obviously a critical component of current preparedness efforts nationwide.

Public Health Smoke Detectors: Surveillance Models

Notwithstanding the practical and theoretical problems noted previously, progress toward improved public health surveillance systems is being made. Public–private and philanthropically funded nonproprietary vendor-neutral collaborations are underway with the goal of creating a public health surveillance infrastructure with agreed-on norms for reporting and coding. A variety of surveillance systems are being tested in emergency departments because they are important points of first contact with the health care system. The province of Ontario, Canada, has linked its provincial telemedicine health information system and emergency department triage databases. Other public-domain software programs have been created to mine existing hospital or emergency room databases in real time for potential bioterrorism-related diagnoses.

The CDC has taken the lead in working with states and major cities to implement a national electronic disease surveillance system integrating data from hospital or regional laboratories, emergency room encounters, and hospital admission databases. Such programs can search hospital registration, emergency room encounters, or emergency room discharge data sources using predefined disease or illness categories or text words (such as chief complaint) in order to identify possible bioterrorism syndromes. Recent studies have substantiated the value of using admitting diagnoses, chief complaints, and ICD9 codes—or better yet, combinations of these variables—to create syndrome recognition algorithms capable of being used for bioterrorism and other public health surveillance. Chief complaints or emergency room admission or discharge diagnoses have all been found to be useful for surveillance systems, and because they can be obtained at the point of contact, they meet the criteria for timeliness of reporting.

Surveillance systems using this "drop-in" approach have already been used at a number of high-profile public events—such as the Utah Winter Olympic Games and the 2004 Republican National Convention. The CDC collaborates with several states on surveillance programs with selected hospitals using clinical and laboratory data systems to identify possible bioterrorism syndromes. The CDC is also collaborating with Poison Control Centers and medical toxicologists to create surveillance models whereby index cases, clusters of toxidromes, or illness patterns consistent with chemical agents can be identified quickly and the information passed along to key public health authorities. A lingering concern is the necessity of developing greater uniformity and standards in case and syndrome definitions, data analysis, and reporting. Standardization will facilitate the creation and expansion of surveillance systems and create a more robust public health information network.

One of the difficulties noted earlier is that normative incidence or prevalence data are needed in order to design surveillance programs that pick up excess rates of conditions that might have public health importance, such as bioterrorism syndromes. Typically, baseline rates are established by sampling these databases using variable "windows" of time (e.g., 2 weeks before an upcoming major event) and then comparing these expected rates to observed rates. These systems can set different thresholds for flagging an alert. This is a critical issue because an overly sensitive system runs the risk of "crying wolf" too often and draining public health resources in the labor-intensive follow-up that must follow any "positive." Bayes' theorem applies to bioterrorism surveillance: an overly sensitive system looking for rare or infrequent events will identify far more "false-positives" than "true-positives." Similarly, eliminating "noise" from the system using various statistical techniques may improve the system's specificity but may allow some genuine cases to slip below the level of detection. At the present time, even a good surveillance system is perhaps best viewed as a "smoke detector." They may pick up important public health diagnoses or syndromes, but only labor-intensive follow-up investigations will distinguish genuine fires from backyard barbecues.

As noted earlier, event-based surveillance systems have already been used to proactively monitor for a preestablished set of clinical syndromes in emergency rooms and hospitals in a geographically defined area during and immediately following major public events, such as international summits, sporting competitions, presidential inaugurations, or national party conventions. These systems provide participating hospitals a Web-accessible reporting form that includes fields for describing the patient's symptoms and signs, as well as whether they participated in the event being surveilled.

There are a number of military surveillance systems for which some information is known. For example, the U.S. military tracks clinical data daily from Department of Defense facilities worldwide as part of its Electronic Surveillance System for the Early Notification of Community-Based Epidemics (ESSENCE). Other military-based systems have been developed as well, including particulate monitoring systems aboard U.S. naval ships. Enhanced collaboration between the Department of Defense and the CDC relating to research and testing should prove very helpful in developing improved surveillance programs applicable to future public health surveillance systems.

The anthrax attacks and the international SARS epidemic underscore the importance of early identification and timely dissemination of such data in order to facilitate appropriate medical and public health response. Historically, case-based surveillance has proven to be a valuable tool. For example, an unusually large bump in a rare form of skin cancer—Kaposi's sarcoma—in the San Francisco area in the early 1980s alerted the CDC to the coming AIDS epidemic. Similar epidemiologically based diagnosis or syndrome surveillance efforts are currently in use in selected areas of the country, particularly larger metropolitan areas. A continual theme of this text is that underlying the clinical approach to diagnosing a bioterrorism event is identifying temporal or geographic clustering of cases that meet certain syndromic features (Tables 5–2 and 5–3). In anticipation of such attacks, a number of government agencies, most notably the CDC, are actively researching improved diagnostic and therapeutic modalities as well as lab accessibility and

TABLE 5-2   Clinically Based Syndromic Recognition

Respiratory Distress and Fever Syndromes

Category A:	Anthrax
	Plague
Category B:	Ricin
	Staph enterotoxin B

Rash and Fever Syndromes

Category A:	Smallpox
	Viral hemorrhagic fevers

Flulike Syndromes

Category A:	Tularemia
Category B:	Brucellosis
	Q Fever

Neurologic Syndromes

Category A:	Botulism
Category B:	Venezuela equine encephalitis
	Eastern equine encephalitis
	Western equine encephalitis

Blistering Syndromes

Category B:	T-2 mycotoxin
Chemical:	Arsenicals
	Mustards
	Phosgene oxime

clinical pathways to ensure diagnostic accuracy. State-based surveillance systems are under development for monitoring local emergency rooms, hospital admissions, and ambulatory care encounters for clustering of signs, symptoms, and other findings indicative of a biological attack.

Since the September 11 attacks, a number of states and cities have implemented hospital admissions syndromic surveillance systems. The aim of these systems is to provide a daily census of admittable diagnoses to acute care facilities and to track these admissions by clustering symptoms and signs into discreet syndromes. Collecting the data and analyzing it enables state and municipal health departments to be alert to an unexpected bump in particular disease clusters. Most of these systems have gone to web-based reporting, which facilitates data collection and real-time analysis. The systems are flexible, allowing changes in diseases or syndromes being surveilled as circumstances change. For example, when the anthrax bioterror attacks occurred in the late fall of 2001, anthrax was promptly added to these case-based syndromic surveillance programs. Case-based reporting through Internet-accessible systems has not proven to be an undue burden on hospitals and emergency departments that have agreed to participate in these surveillance systems.

Categories of disease or syndromes included in these case-based surveillance programs include three categories of respiratory illnesses, such as pneumonia,

TABLE 5–3   Clinical Syndromes Associated With Selected Bioterrorist Agents

Agent	Symptoms	Signs
Smallpox	High fever, myalgias, itching, abdominal pain, delirium, rash on face, extremities, hands, feet	Maculopapular rash central to extremities, synchronous hard, firm pustules
Nerve agents	Muscle cramps, runny nose, difficulty breathing, sweating, diarrhea	Hypersalivation, tearing, miosis, wheezing, diarrhea
Anthrax		
Inhalational	Fever, short of breath, cough chest pain	ARDS, hypotension, mediastinal widening
Cutaneous	Papule/macule on exposed skin	Pruritic macule/papule ulcerates into painless, depressed black eschar, local edema, +/− painful adenopathy
Plague	Fever, productive cough, chest pain, hemoptysis	CXR infiltrates
Botulism	SOB	Symmetric cranial neuropathies, respiratory distress, no sensory deficits
Tularemia	Acute flu-like illness, fever	Pleuropneumonitis on CXR
VHFs	Abrupt fever, headache, myalgias, nausea/vomiting, abdominal pain, diarrhea	Truncal maculopapular rash, hemorrhagic signs, e.g., petechia, bruising

Adapted from the CDC.

hemoptysis, and acute respiratory distress syndrome (ARDS) of unknown origin. These categories can be further subdivided, for example, adding a special subcategory in the respiratory cluster for occupations at particular risk, such as health care workers. Other syndromic categories cover gastrointestinal, neurologic, dermal presentations, as well as fever-rash, sepsis, and miscellaneous disease clusters, for example, multiple admissions for any given cluster of symptoms or signs.

Once the data are received from the hospital or emergency department, sophisticated biostatistical methods are employed to determine statistically significant differences between the average number of such diagnoses or syndromes over a defined period of time (e.g., 6 months) and the daily census reported to the surveillance database. If a potential cluster or sentinel diagnosis is identified, a follow-up investigation is necessary by the state or local health departments to determine whether a potential bioterrorism syndrome is present. This will often require direct contact between public health officials and clinicians taking care of the individuals whose diagnoses or syndromes triggered the investigation, or even of the patients themselves. Clinicians should be aware that all states and most, if not all, local public health agencies have broad statutory authority when it comes to protecting the public's health. Medical records, for example, can be reviewed or patients interviewed, although ethical and legal obligations to maintain confidentiality apply in these circumstances as well.

States are working to develop multiple surveillance tools in order to identify and manage public health emergencies or even nonemergency but important public health issues. One of the benefits of these surveillance mechanisms is that they not only are potentially applicable to bioterrorism syndromes, but also will enable states to better monitor other public health threats (e.g., SARS, influenza, West Nile virus) and may develop into a *de facto* public health database in which the state's health is continuously monitored. Generalizability to noninfectious diseases is both possible, and desirable. The New York City Department of Health has one of the more sophisticated and resource-intensive surveillance systems in place today. This system monitors emergency room chief complaints that are entered into a database when the person is triaged. The program sifts through the data for four diagnostic categories: febrile illnesses, respiratory illnesses, diarrhea, and vomiting. A recent report on the system noted that it identified an incipient influenza epidemic two weeks before city laboratories began reporting positive viral cultures.

Several case-based surveillance networks are in operation in Europe. Although these have been designed for surveillance purposes other than bioterrorism, they are potentially valuable in this regard. The Communicable Disease Network in France and EUROSENTINEL, a cross-national collaboration, are two examples of clinical surveillance systems that are being modified to fulfill new roles in bioterrorism surveillance. These surveillance networks differ in a number of respects from one another, including the diseases being monitored, the frequency and means by which monitoring occurs, the clinicians gathering data, and the speed with which data are analyzed and reported to key stakeholders.

Medical Systems Challenges

The following problems address an overarching concern with an inadequately integrated public health system. Each of the following points illustrates different aspects of this concern, and all point to the need for systemwide changes in the way medical and public health practitioners communicate and the way surveillance systems operate.

- Fragmented medical care delivery system with multiple points of entry—multiple concurrent cases of symptom clusters may be missed if patients present at distinct centers—a private office, an emergency room, an urgent care facility that has no connection to the others and lacks a surveillance system.

- Poor lateral communication—the lack of established channels between health care providers means that cases seen concurrently but in different locations may not be recognized as sentinel events by public health officials or health care providers.
- Lack of timely surveillance and reporting—although most hospitals are connected to some degree to the public health infrastructure, most community clinicians are not. Given the likelihood that index cases will be seen in an outpatient setting, recognition of a bioterrorism attack could be missed easily.
- Incomplete public health infrastructure—inadequate funding of the public health infrastructure means inadequate training for health care professionals including community clinicians, faulty surveillance systems, and inadequate planning for bioterrorism events.
- Disconnect between private and public health sectors—community physicians play pivotal roles in a bioterrorism events by seeing index cases in their offices, providing care in an acute community event, or helping to educate and reassure community members during and after these events. To function well in any of these capacities, community physicians must be integrated into the public health infrastructure.
- Lack of training and awareness by front-line professionals—at present, more resources and training are needed to enable those in the outpatient setting to diagnose, manage, and participate in bioterrorism events and preparedness efforts.

Laboratory-Based Surveillance

Laboratory-based surveillance programs have promise, but effective development requires a change in many standard routines. For example, traditionally, many laboratories view unidentified gram-positive bacilli growing on the standard agar gels as contaminants and throw them out. Such a finding in an individual in whom a bioterrorism history has indicated a potential high-risk occupation (e.g., mail-room worker) should instead prompt more detailed analysis looking for anthrax. Currently the CDC recommends that these "contaminants" be treated as a potential index case of anthrax in any previously healthy individual presenting with acute febrile syndromes. Similarly, laboratories are in unique positions—as are infection control personnel in hospitals—to recognize unusual bumps in the number of "contaminated specimens," or indeed unusually high numbers of received specimens from the same biological source (blood, cerebrospinal fluid, aspirates, or stool cultures) as potential clues to a bioterror attack. Similar monitoring can be done by private laboratories that receive large numbers of samples from community physicians offices, urgent care centers, or hospital facilities.

Environmental Surveillance

Surveillance done by sampling environmental exposures are familiar in one form or another to most clinicians. Examples range from the common (e.g., smoke or carbon monoxide detectors) to the sophisticated (e.g., continuous or intermittent sampling of public water systems or swimming areas or air pollution monitoring programs). For most bioterrorism agents, however, continuous environmental surveillance untethered to specific concerns for most biological or chemical agents is probably of limited practical value at the present time. Environmental sampling has greater utility in the case of radiation, where local radiation detection devices or air monitoring systems may be selectively or continuously engaged in surveillance. Radiation surveillance systems are already in place around the world. In fact, it was this system that alerted the world community to the emerging catastrophe at Chernobyl (see Chapter 27).

The same variability exists in surveillance systems that use laboratory-based and environmental monitoring as early detection strategies. In such systems, clinical samples or environmental samples are sent in for analysis. Approaches to monitoring environmental features differ substantially. For example, airports or politically visible public buildings might have continuous environmental monitoring of air or water systems. Alternatives to fixed-site environmental monitoring include point source sampling in an area of presumed contamination, or even from potentially contaminated food, animals, water, or even humans. At present, no adequate models of such methods are applicable for bioterrorism surveillance, though further development efforts continue.

Surveillance has also played a significant role in monitoring for health effects from harmful exposures and takes the forms of epidemiologic data and comes to light in the context of clustering of signs, symptoms, and diagnoses. Examples of such uses might include identification of higher than expected cases of childhood leukemia in neighborhoods pointing to environmental factors. Another common example is a sudden rise in infectious diseases diagnoses, such as Legionella or bacterial meningitis.

Bioterrorism as a Sentinel Health Event

The sentinel health event concept is an extension of the principles of case-based or clinical surveillance. Sentinel health events have their origin in the maternal and child health literature of the 1930s when the medical community increasingly came to believe that maternal or perinatal deaths were inherently preventable and that any such death or adverse pregnancy outcome was a marker for inadequate prenatal care. By viewing adverse pregnancy outcomes as preventable, physicians were able to improve the overall care of women in childbirth. The sentinel health event concept was generalized to the occupational health literature in the late 1970s and 1980s, but its conceptual base was the same: any adverse occupational event (e.g., worksite death or occupational disease) was by definition preventable and an indication that improved conditions at that worksite, or more broadly in an industry, were needed.

The concept of the sentinel event has become a staple in the public health literature as a result of its historical utility and is also a useful clinical tool. In this context, a sentinel health event is any diagnosis, illness, or adverse outcome that is preventable and that has identifiable risk factors that may have important health consequences for individuals who are not yet sick. The occurrence of a sentinel health event, for example—a low-birth-weight infant or silicosis—is a sign that something went wrong and should elicit questioning and/or diagnostic strategies to identify known risk factors for the adverse event. A trite example of this would be a person who is diagnosed with lung cancer. Nearly all clinicians in this circumstance would inquire as to smoking status, but the more astute ones would investigate other social, occupational, or environmental exposures associated with lung cancer, such as previous cancer treatments, home radon exposure, or occupational exposure to asbestos or radiation.

The sentinel health concept is applicable to bioterrorism as well and has utility as a clinical tool. An individual presenting to an outpatient clinic with a skin lesion resembling a black eschar can be a sentinel diagnosis that raises the possibility of cutaneous anthrax. In this case, the sentinel diagnosis or syndrome prompts further questioning as to whether the person is a postal worker, politician, or someone who works in the mail room of a large, high-profile, or government facility. Any positive finding increases the possibility of anthrax, and diagnostic testing would proceed from that assessment. Similarly, early on in the SARS epidemic, clinicians were advised that individuals presenting with flulike symptoms or respiratory illnesses must have a careful travel history taken. Individuals who had traveled to Southeast Asia, or were exposed to someone who had, or health care workers with exposure to such individuals were presumed to have SARS until confirmed otherwise. Similar public health-based clinical advisories were issued for Avian influenza in 2005. Travel to geographic regions, thus, is another example of a sentinel event, health, or datum, and serves to demonstrate the variety of forms that sentinel events might take. In the context of BCN terrorism, as with sentinel occupational diagnoses, clinicians need to recognize the link between occupational or environmental exposures and the disease or syndrome that may be present. This includes understanding both the nature of the work and the location where work is done.

The provider must be prepared to follow-up on all the necessary issues, should a positive diagnosis be confirmed. For example, if a clinician even suspects a deliberate biological attack or receives a positive laboratory result confirming the diagnosis of one of the many potential biological agents, it is imperative to notify the infection control officer or infectious disease specialist and public health official. In addition, there must be vigilant follow-up with the patient—contacting other patients who may have presented with similar symptoms to determine whether there is a disease cluster requiring possible isolation, further diagnostic testing, initiation of medicines for the patient, and institution of infection control practices—and possibly pre- and postexposure prophylaxis for contacts, medical staff, or clinicians themselves. Fortunately, if the index of suspicion is sufficiently high, the notification of

TABLE 5–4   Recognizing Sentinel Bioterrorist Events

Clinical Clues

Unexpected high rate of rare or nonendemic disease

Unusual age, occupational, or geographic distribution incongruent with known risk factors

Unusual seasonal distribution

High infection rates

High morbidity and mortality

Agent-specific pathognomonic clues (e.g., classic smallpox rash)

Public Health Clues

Epidemic curve showing an explosion of cases

Tight geographic or facility location

Predominance of respiratory symptoms

Simultaneous outbreaks of unusual diseases

Unusual resistance patterns or mode of transmission

Animal casualties

Absence of normal disease host

---

local and state public health officials will generate a response that will be immeasurably helpful to the clinician who has made the call.

There is reasonable consensus among bioterrorism experts as to the clinical and epidemiologic clues that should trigger an investigation into a potential bioterrorism attack (Table 5–4). A sentinel bioterrorism event may have occurred if any of the following are present. Depending on the disease, even a single rare or nonendemic diagnosis should be a red flag requiring consideration of a bioterrorism event, whereas a high rate of a rare or nonendemic disease should certainly be a tip off that something is amiss. For example, a family practitioner in California might occasionally see a case of tularemia presenting as an influenzalike pulmonary syndrome. On the other hand, it would be distinctly unusual if multiple cases were identified in the same community or within a state in a temporal cluster. Similarly, a case of plague in Michigan would be so unusual that this diagnosis would almost certainly prompt additional epidemiologic investigation and a call to the local health or state health department.

Along these same lines, if a disease occurred at an unusual age, outside of known occupational risk factors, or in a geographic distribution incongruent with known risk factors, this too should raise the index of suspicion for a bioterrorism event. Other factors that could indicate a bioterrorism event are an unusual seasonal distribution of a relatively common disease, unusually high rates of infectivity or transmissibility (such as occurred with the weaponized anthrax used in the U.S. Postal Service attacks), atypically high mortality or morbidity, and of course, pathognomonic presentations that clinch a diagnosis, such as the classic homogeneous centripetal pustular rash of smallpox.

In addition to clinical clues, there are epidemiologic clues that if appreciated could signal a bioterrorism event. These include multiple simultaneous outbreaks of unusual diseases in geographically dispersed areas, a rapidly rising epidemic curve or tight geographic clustering of known toxidromes, such as febrile acute respiratory distress, rash fevers, or neurologic syndromes. Additional epidemiologic clues include unusual resistance patterns or mode of transmission, large numbers of animal casualties, or the absence of a normal disease host.

A recurring theme of this text is that one of the challenges for clinicians and public health officials will be recognizing a bioterrorist event before such an attack has been confirmed unequivocally. This challenge is great because of the nonspecific findings of most of the bioterrorism agents. Clinicians are trained to think syndromically, though it is not always termed as such. Most clinicians, when evaluating a patient acutely, compile a differential diagnosis based on the patient's history and physical examination. The process of doing so is an unconscious use of syndromic surveillance. A general clinical pattern is recognized, and this recognition forms the basis of what etiologies are likely. For example, gastrointestinal distress with fever brings a set of possible etiologies: infectious, inflammatory, neoplastic, and so forth. Similarly, age, gender, diet, or travel history all factor into the narrowing of the diagnosis. With experience and wisdom, such a process becomes a clinical reflex.

The clinical approach to bioterrorism is analogous. Basic clinical patterns can help guide diagnosis, with the caveat that presentations do vary and aspects of the patient's history will factor into diagnostic considerations. There is no universal agreement on the syndromes, but following the CDC's guidance, clinicians should consider the following useful constructs with which to approach clinical syndrome recognition: respiratory distress and fever syndromes, rash and fever syndromes, flulike syndromes, neurologic syndromes, and blistering syndromes (Table 5–2).

Perhaps the greatest challenge for community physicians is the diagnosis of an index case after a small-scale or subtle attack with a biological, chemical, or (less likely) nuclear weapon. Diagnosis is complicated because of the nonspecific signs and symptoms seen early on with most of these agents. Furthermore, many of the biological agents exist naturally with outbreaks occurring with varying degrees of rarity. This can either simplify the recognition of an attack or complicate it. Smallpox, for example, should never be considered a natural outbreak; rather it is in itself proof of an attack and considered an international medical emergency. Tularemia, on the other hand, is an example of an agent that in addition to being a viable weapon sporadically occurs as a natural outbreak.

There are some general principles and considerations of which clinicians should be cognizant of that assist in making a bioterrorism diagnosis, or determining obstacles to such a diagnosis. The following examples are meant to serve as possible signals to investigate further, not for drawing conclusions. There may be perfectly legitimate alternative explanations, and barring positive identification of an attack, no universally agreed-on means of making a determination of bioterrorism events yet exists. Grunow and Finke developed an assessment tool

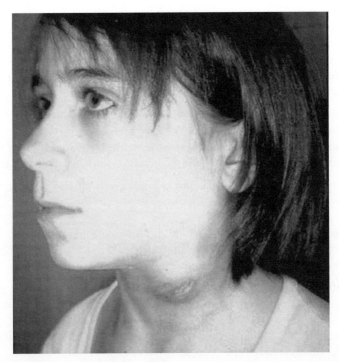

FIGURE 5-1 A patient with tularemia during the outbreak in Kosovo, 1999. Investigators were sent to determine if this was a natural outbreak or an intentional one. It was determined to be the former. Note cervical adenopathy, a clue to the microglanular form of tularemia.
Courtesy of the CDC.

for predicating the likelihood using a number of criteria (some discussed below) based on a Tularemia outbreak in Kosovo; however, the field assessment tool lacks specificity or sensitivity data. A detailed discussion of the tool, or more broadly, how to make such a determination in the field is outside the scope of this text, but understanding the central points and obstacles to diagnosis can help clinicians minimize the risk of missing an index case (Fig. 5–1).

Clinical Clues to a Potential Bioterrorism Event

The following are features that may alert clinicians to the possibility of a bioterrorism attack in various clinical settings, such as private offices or community health centers.

The appearance of nonindigenous bacteria or viruses, particularly if they are seen at a high rate, is suspicious. A finding of this type requires timely and accurate identification procedures.

Example: The identification of an *Ebola* strain in the United States or in Europe.

The appearance of unusual forms of bioterrorism agents. Again, this finding requires timely and accurate identification procedures.

Example: The identification of a botulinum toxin whose subtype differs from the known or even the most common "naturally occurring" strains.

The sudden appearance of respiratory symptoms (or cutaneous lesion) in an otherwise healthy person in a population in which no respiratory epidemic is present.

Example: A 40-year-old, otherwise healthy male and no other risk factors, with pneumonic symptoms refractory to usually effective antibiotics for community-acquired pneumonia.

Unusual age distribution among patients presenting with similar symptoms, particularly diagnoses or syndromes seen more commonly in older populations.

Example: Near-simultaneous admissions of pediatric, adult, and geriatric patients with ARDS (e.g., plague, tularemia).

Disease presenting in a traditionally low-risk occupational group or setting.

Example: Anthrax lesions seen in someone with no contact with livestock or their products.

Multiple cases of the same symptom pattern or diagnosis.

Example: More than one patient with cranial nerve palsies and descending progressive motor paralysis—the sine qua non of botulism.

Example: Multiple cases of ulceroglandular tularemia, which might suggest a contaminated food or water source.

Distinctive cutaneous features.

Example: An otherwise healthy, 35-year-old male with large blisters (painful or not) on his arms that appeared acutely and are associated with respiratory symptoms or lacrimation (mustard gas).

Atypical seasonal patterns.

Example: Cases of *Coxiella burnetii*, the causative agent of Q fever and a Category B agent, in the fall or winter rather than in the natural seasonal occurrences of spring or summer.

Public Health Clues

When public health surveillance systems are in place it should be possible to detect a bioterrorism attack. The following clues may help.

Pneumonia Cases

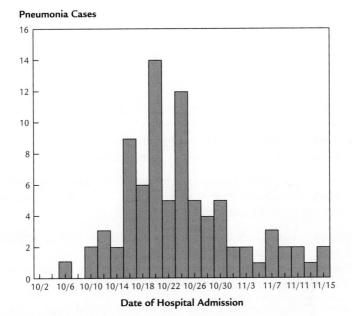

Date of Hospital Admission

FIGURE 5–2 **A typical epidemic curve: *Legionella* outbreak in Louisiana.** *Courtesy of the CDC.*

An epidemic curve showing a significant bump of simultaneous or near simultaneous cases. It is perhaps self-evident, but the recognition of such a pattern reemphasizes the importance and value of developing adequate surveillance systems and reporting syndromic findings (Fig. 5–2).

Example: A statewide surveillance of Emergency Room admissions in January shows a sudden increase in rash fever cases.

Animal casualties. The presence of high numbers of animal deaths, particularly livestock, or known disease vectors such as rats. This requires an integrated public health system in which environmental, veterinary, and medical agencies communicate and integrate information.

Example: City officials note multiple rat carcasses in areas where no pesticide efforts have taken place.

Tight clustering of an outbreak either geographically or occupationally.

Example: Employees working in a municipal building presenting with a similar constellation of nonrespiratory symptoms.

Absence of usual disease hosts, vectors, or natural reservoirs.

Example: A brucellosis outbreak in an area such as a large city where natural animal hosts or reservoirs are absent or well controlled and no evidence of arthropod infestation is present.

A natural outbreak is less likely, even if zoonotic sources are present, if distant outbreaks occur concurrently.

Example: Simultaneous bubonic plague diagnoses made in New York City and in Seattle.

Morbidity/mortality deviations. Significant variations from natural clinical courses seen in terms of virulence, susceptibility, and rate of antigenic shift.

Example: Increases in the expected number of cases, morbidity or mortality rates, or populations at risk in an influenza epidemic.

The sudden appearance of an unusual form of a naturally occurring disease.

Example: The identification of pneumonic forms of plague without any diagnosed cases of the bubonic form.

A disproportionately high concentration of naturally occurring agents present in soil, water, or air.

Example: Soil samples showing botulinum levels that are orders of magnitude higher than normal.

Unexpectedly high morbidity or mortality rates in populations within a relatively short period of time without any known explanation.

Example: Mortality rates in the U.S. anthrax attacks was higher than expected provided a clue to the use of weaponized anthrax.

Possible Indications of Weaponized or Modified Agent

Although a bioterrorism attack using a known form of a bioterrorism agent is a horrible enough prospect in and of itself, some agents can be weaponized, adding significantly to the risk that they pose. Weaponizing bioterrorism agents is usually associated with biological weapons and typically involves genetic modifications that enhance virulence, such as adding or changing genetic traits that code for antibiotic resistance or sensitivity. Weaponization can also be applied to chemical agents, either by creating more potent isomers (e.g., lowering the LD_{50}) or changing properties in ways that amplify the agent's environmental persistence (e.g., water solubility). Viewed in this context, radiation dispersal devices and radiation point sources represent weaponization of medical or industrial radiation sources.

Distinguishing weaponized bioterrorism agents from nonweaponized agents is most easily considered in the case of biological agents, in part because of a longer history of bioweaponry research and development. Features that might suggest that a weaponized biological agent has been used follow:

An epidemic that spreads dramatically faster than is typical of a naturally occurring outbreak. The rate of epidemic spread is a function of many factors, including virulence, concentration, and social factors such as mass transportation.

Example: Rapidly spreading tularemia would be highly unusual.

If the rate of infectivity, pace or severity of clinical manifestations, or mortality rate is uncharacteristic of the naturally occuring form of the disease.

Example: The U.S. Post Office attacks used weaponized anthrax that required far fewer spores to cause clinical disease than natural anthrax.

If atypical resistance patterns to appropriate and recommended antibiotic regimens are identified.

Example: A case of plague that doesn't respond to recommended antibiotics.

If person-to-person transmission is noted with an illness that does not naturally do so.

Example: The confirmation that botulism is spreading through close contacts.

Clustering of cases in a particular subset of a community or population; for example, religious, ethnic, or other unifying features of a group that seems to be exclusively presenting with, or conversely is protected from, an outbreak.

Example: An outbreak of anthrax in a community where no members of a supremacist group or their families become infected.

Impediments to Recognition

Although the preceding discussion focused on clues and indications suggestive of a bioterrorism attack, there is another side to the issue. As difficult as it is to identify an attack, understanding what makes it so difficult may serve to keep clinicians attuned to ways to overcome these inherent challenges.

- Delayed or insidious onset of symptoms. Agents such as anthrax spores can theoretically take weeks to manifest, whereas low-level radiation may take weeks before skin changes appear (see specific chapters for more detail).
- Ability to mimic endemic disease. Nonspecific symptoms, often respiratory in nature, can easily be mistaken for upper respiratory infections, or missed altogether. Some of the biological agents do cause natural outbreaks, so their occurrence (e.g., tularemia) must be differentiated from a bioterrorist attack.
- Delayed diagnosis because of disease rarity. Few practicing physicians have seen patients with any of the bioterrorism agents discussed in this book. The naturally occurring biological agents present so infrequently and in such small numbers that few have any clinical experience with them. Fewer

civilian, let alone military, physicians have experience with chemical or radiation weapons. The paucity of clinical experience makes early or quick identification of these agents all the more unlikely. Effective training courses and education mitigate this challenge, but only to a degree.

- Victims may be widely dispersed by time and place. Although an epidemic in a community facilitates the identification of a bioterrorism attack, the means and mode of exposure can alter geographic distribution. For example, an exposure to smallpox from a source in an airport may result in individual patients presenting at scattered national or worldwide clinics very early in the epidemic.
- Variable incubation period. Some agents, such as anthrax, have an incubation period of days to months, which would mean that there could be wide variability in the timing of presentation so the clustering effect that facilitates diagnosis may be missed.

TABLE 5–5   Syndromic Recognition: Example of an Emergency Department or Hospital Surveillance Data Form

Number of Admissions	Diagnosis or Syndrome	Bioterrorist Agents
1.	Pneumonia a. in Health Care Worker (?SARS)	Anthrax, tularemia
2.	Hemoptysis	Inhalation anthrax
3.	ARDS of unknown origin	Plague, tularemia

4.	Meningitis/encephalitis/
	  unexplained acute
	  encephalopathy
		West Nile virus, Eastern equine encephalitis, Venezuela equine encephalitis

5.	Nontraumatic paralysis/descending paralysis, Guillain–Barre syndrome	Botulism
6.	Sepsis/nontraumatic shock	
7.	Fever and rash	Smallpox, VHFs
8.	Fever of unknown origin	
9.	Gastrointestinal agents	Anthrax, Category B agents (e.g., Shigella, Salmonella, etc.)
10.	Skin infection, r/o cutaneous anthrax	Anthrax, vesicants
11.	Clusters of illness a. If yes, describe type	_____

Adapted with permission from Connecticut State Health Department, 2004.

TABLE 5–6 CDC Syndrome Case Definitions and Potential Category A Conditions

Syndrome	Definition	Category A Diagnosis
Botulismlike	Acute cranioneuropathies and descending motor paralysis	Botulism
Hemorrhagic illness	Acute febrile illness with multiorgan involvement. Abnormal CBC and clotting system (e.g., thrombocytopenia, neutropenia, + DIC screen)	VHFs
Lymphadenitis	Regional lymph node swelling or infection, painful buboes (especially in groin, neck, axilla)	Plague
Localized cutaneous lesions	Localized edema, cutaneous ulcer, eschar, vesicle (excludes diabetic skin manifestations and insect bites)	Cutaneous anthrax, or tularemia
Gastrointestinal	Acute GI distress, nausea, vomiting, diarrhea	Gastrointestinal anthrax
Respiratory	Acute upper/lower respiratory tract infection, cough, stridor, laryngitis, pharyngitis, pneumonia, short of breath, includes exacerbation of existing chronic lung conditions	Inhalational anthrax, tularemia, plague
Neurological	Acute CNS infection, mental status changes	None
Rash	Acute macular, papular, vesicular homogenous disseminated rash with face and extremity dominance	Smallpox
Specific infection	Acute infection by unknown cause, often with generalized signal symptoms; includes sepsis, miscellaneous febrile illnesses	N/A
Fever	Fever of unknown origin	N/A

Adapted from the CDC.

- Zebras. Bioterrorism agents fly in the face of classical medical training that preaches to not think of zebras when hoof beats are heard. Unfortunately, given the geopolitical landscape, consideration of bioterrorism exposures is now necessary for clinicians. Providers will be on the "front line" in recognizing and responding to a covert event and must maintain a high index of suspicion (Tables 5–5 and 5–6).

Summary

A variety of case-based, syndromic, and environmental surveillance systems have been developed to address the issue of BCN terrorism. Systems that integrate multiple sources of data—including hospital, emergency room, laboratory, pharmacy, physician office, or ambulatory sources—and that have been evaluated to determine if they meet the goals of surveillance are likely to emerge in the years to come. At present, considerable uncertainty exists about our present capacity to collect and analyze data from multiple sources in order to identify emergent public health threats. Advanced spatial and temporal analytic techniques show promise in this area. Any surveillance system must not only be effective—that is, capable of early detection and monitoring for bioterrorism events—but also must meet ethical and legal standards that balance the need to protect public health and the citizen's right to privacy. Both clinical and systemic issues delineated in this chapter highlight the challenges faced at every level of the health care system, yet too often clinicians and public health officials see themselves as entirely separate from one another. Although the complexities of health and health care do not warrant such hard distinctions to be drawn, bioterrorism threats demand, at least in this context, that interconnection occur. Interconnection needs to exist between clinicians and public health officials locally, statewide, nationally, and internationally among countries.

CHAPTER 6

Vulnerable Groups: A Summary of Relevant Concerns

In the setting of bioterrorism, the medical and public health community must be prepared in advance to address the unique needs of vulnerable groups, including children, pregnant and breast-feeding women, the elderly, and the immunocompromised. Vulnerabilities may derive from medical, social, or economic conditions, each of which impacts the planning and response to any BCN events. Virtually all bioterror education, training, and outreach has focused largely on hospitals and emergency response and done so with the general population in mind. Preparation must also include consideration of those individuals and groups whose ability to withstand, respond, or survive BCN terror is modified by age, developmental or chronologic age, co-morbidities, or social standing. This chapter briefly summarizes these considerations, although each is also included in each agent specific chapter. The information in this chapter, drawn from historical experiences, is largely based on our current understanding of and extrapolation from physiological principles, toxicologic and animal studies, and not a small amount of common sense.

Children

As outrageous as it may seem to most of us, terrorists have been more than willing to attack targets where injury and death of children are very likely to occur. It is also the case that pediatric experience with Category A bioweapons is limited so that guidance regarding unique susceptibilities in children is based on a general appreciation of children's unique physiologic, developmental, and social differences, extrapolation from animal studies, and general principles.

Children's organ systems physiologically differ from adults in ways that have specific relevance to BCN agents. First and foremost, children have smaller bodies and body weights. Consequently, doses that might not affect adults may induce illness in children (Fig. 6–1). In quite specific ways, their physical stature can be a risk factor in and of itself. For example, being closer to the ground puts them at risk for inhalation exposures, particularly from chemical weapons that are often more dense than air. Further, they have increased respiratory rates and more permeable skin and blood–brain barriers providing opportunities for enhanced absorption of chemical and biological agents. Children have more immature neurologic and immune systems and so may be more vulnerable to infectious agents or to complications of chemical agents. For example, children who develop smallpox demonstrate more neurologic symptoms than adults do, including delirium, seizures, and coma. Their increased surface area-to-volume ratio and smaller volume reserves make children at risk for dehydration caused by gastrointestinal toxins or hypothermia as a result of decontamination procedures. Children will often show clinical effects at far smaller doses or intensity exposures compared to adults. Childhood lead poisoning comes to mind as a relevant analogy. Neurodevelopmentl abnormalities occur at much lower blood lead levels in children

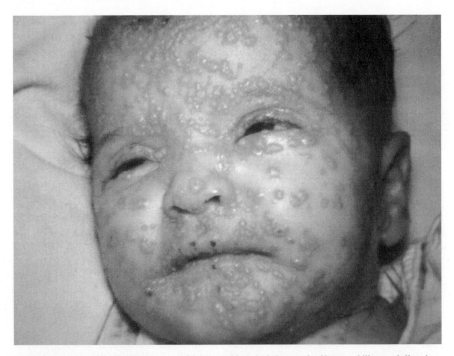

FIGURE 6-1 **The pediatric population suffers heightened effects of illness following BCN exposures.**
Courtesy of the CDC.

than in adults. Children's developmental state may have a disproportionate impact on their reaction to bioterrorism events, including the loss of parents or other caregivers, social isolation resulting from public health measures such as quarantine, and greater likelihood of PTSD following any significant life trauma. Caregiving adults may be called away to dangerous and unknown environments in order to respond to an emergency (health care workers, emergency response, and first responders), and they or members of their family may be forced into isolation or quarantine as a result of exposure to index cases or occupational risks. All of these possibilities may disproportionately and adversely affect children.

There are important systems issues that are often underappreciated as well. First responders are often affected emotionally when called to a scene with pediatric injuries. Treatments are often modified when a child is involved and overall experience is more limited. In the setting where a BCN attack may be involved, bulky and restrictive protective equipment, fear of the unknown, and worries about their own children and family may clearly impact the skill with which resuscitation, treatment, and triage are undertaken by individuals working under conditions of enhanced stress. Pediatric wards and specialized children's hospitals will be overwhelmed in the setting of a mass casualty bioterrorism event. National planning does not yet include detailed preparation for surge capacity in the case of pediatric casualties.

☙ Pregnant and Breast-Feeding Women

Significant hormonal and biochemical changes associated with pregnancy have important implications for clinicians. In order to accommodate the developing fetus, normal immune responses are blunted. Thus, it is easy to understand how pregnant women represent a uniquely vulnerable population. Diminished cell-mediated immunity, humoral immunity, chemotaxis, and natural killer cell activity all occur with pregnancy. Progesterone, which rises significantly throughout pregnancy, relaxes smoothe muscle not only in the uterus, but also in the respiratory tract. This has the effect of decreasing clearance of secretions and secondarily enhancing bacterial growth. Generally, live virus vaccines, such as the vaccinia vaccine for smallpox are contraindicated in pregnancy. This must be kept in mind when the screening programs for smallpox vaccination are implemented.

Specific BT agents represent added risk for pregnant and breast-feeding women. Smallpox appears to have greater propensity to present in the fulminant hemorrhagic form with toxemia and death in pregnant women or in infants. Mortality rates for pregnant women based on historical norms are about 50%, compared to 30% for nonpregnant women. In part, this increased mortality is due to vascular leak phenomenon and hematologic changes, such as coagulopathy and thrombocytopenia. Breast-feeding women who are vaccinated may inadvertently inoculate their infant not through the breast milk itself but by virtue of the physical intimacy involved in the act of breast-feeding (Fig. 6–2). Smallpox vaccination in pregnant women is associated with increased risk of fetal loss, low birth weight, premature delivery, and spontaneous abortions. A developing fetus exposed either

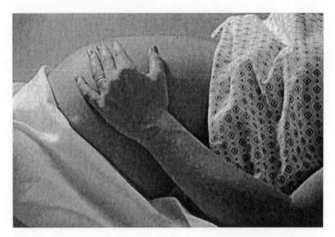

FIGURE 6-2   Pregnant women and their fetuses are subject to greater effects of BCN events.
Courtesy of the Department of Animal and Food Sciences, Texas Tech University.

to smallpox or to the smallpox vaccine is also at risk for intrauterine infections presenting as fetal variola or fetal vaccinia. The impact is highly variable, ranging from normal term delivery to disseminated congenital variola. Pregnant women should also avoid recently vaccinated individuals. Current screening for smallpox vaccination—in the setting of no-diagnosed cases—includes a mandatory exclusion for individuals who have a pregnant spouse at home, or who are planning to become pregnant within the next month. Recommendations for using contraception or abstention should be considered for fertile couples considering vaccination. Should pregnant women be exposed to smallpox, vaccination is recommended with the possible addition of vaccine immunoglobulin (VIG), although safety and efficacy data are limited.

Geriatric Patients

Older individuals have unique vulnerabilities. The reasons are multiple, including growing numbers, age-related decreased immunocompetency, co-morbidities, polypharmacy, enhanced risk of drug interactions, delayed response to treatment, and greater morbidity and mortality.

Systems issues are of equal importance in terms of preparedness for the geriatric population. For example, skilled nursing facilities and other geriatric care facilities represent potentially easy targets filled with vulnerable individuals. Loss of key caregivers (spouses, children, and staff) in the event of a mass casualty situation would be both disorienting and dangerous to patients who rely on these caregivers for medical, social, and psychological support.

FIGURE 6-3   The elderly are particularly vulnerable to the effects of BCN.
Courtesy of the White House.

Elderly patients in part due to decreased sensory acuity (vision, hearing) as well as social isolation, may have decreased access to normative communication, such as radio and television where warnings may be given related to BT events, advice on evacuation, mass vaccination clinics, and shelters. This is particularly true with shut-ins, regardless of age. Older individuals are less mobile, a fact that has implications for planning mass vaccination clinics, evacuations, or other public health mobilization (Fig. 6–3).

Immunocompromised Patients

Advances in diagnosis, treatment, and prevention, as well as the emergence of new diseases like AIDS, mean that the numbers of individuals who are immunocompromised has grown tremendously in recent decades. Hemodialysis, HIV, cancer treatments, bone marrow transplantation, and the widespread use of immunosuppressive and cytotoxic drugs all complicate preparedness and management in the setting of BT. Experience with immunocompromised patients exposed to BCN agents is limited, although they are assumed to be at greater risk. This risk assessment is based as much on reasoned generalization from experience with other more common diseases rather than through direct experience with bioterrorism agents. However, such generalizations may aid in providing useful guidance to clinicians and public health authorities. Diseases in which cell-mediated immunity is lost, such as HIV, place these individuals at risk for routine and opportunistic infections. Theoretically, these same individuals are both more likely to become infected and suffer greater rates of complication when exposed to many biological agents, such as smallpox. Diseases associated with defects in humoral immunity, such as primary immunoglobulin deficiencies, are also at greater risk from infectious agents. These individuals may also respond poorly to immune-based preventive strategies, such as vaccination, or may develop more severe complications from immunization, such as occurs with vaccinia vaccine. Immunocompromised patients may not respond as readily to treatment due to their immunologic deficits, or they may have a longer convalescence following the bioterrorism-related illness or treatment.

Medications that modulate or suppress the immune system are widely used today. Steroids, antimetabolites, cyclosporine, alkylating agents, and radiation are used routinely to treat a wide range of medical conditions. For example, steroids decrease cytokine-mediated reaction to endotoxins and inhibit the normative inflammatory response. This may result in atypical presentations for certain bioterror agents, for example, fewer and shorter prodromal symptoms, atypical skin lesions, or less localization of disease. Microbes may replicate faster and disseminate more readily in patients on steroids, and adverse vaccine-related events and diminished immunogenicity of vaccines have also been seen, for example, in smallpox vaccination. Radiation, chemotherapy, bone marrow transplantation, and the drugs used to prevent graft versus host disease are known to uncover latent infections with Q fever and tuberculosis specifically, and result in immunocompromised states more generally.

There are other risk factors for functional immunocompromised status in our society. Alcoholics, substance abusers, the homeless, and individuals with end-stage chronic diseases are often malnourished and demonstrate impaired host defenses. These individuals may not only have impaired host defenses, but their social isolation may make them tempting as targets, and further, when they do present for medical at-

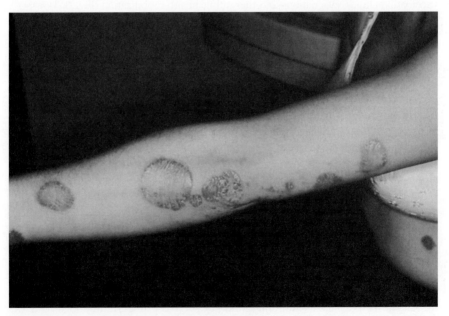

FIGURE 6–4 Dermatologic diseases that lessen the barrier quality of the skin increase the likelihood of potential BCN effects.
Courtesy of the CDC.

tention they may not be evaluated with the same level of alertness or vigilance. Similar concerns may be present for diabetics, who have defective neutrophil activity and are at enhanced risk for infections.

Patients with Skin Disorders

Because many biological and chemical agents come into contact with the skin and because prevention often incorporates vaccination strategies, individuals with significant dermatologic conditions have unique vulnerabilities that may not appear obvious at first glance. Children and adults with atopic dermatitis, eczema, and other exfoliative skin conditions (Fig. 6–4) may suffer enhanced or accelerated presentations of some biological and chemical agents. This is due to greater absorption through the skin due to loss of normal integumentary integrity, as well as to functional impairment in T-cell immunity. Drugs such as tacrolimus or other immunomodulants may increase risk from viral agents because of their affect on interleukins and cytokines. Adults who themselves are asymptomatic but who had childhood eczema or atopic dermatitis remain at enhanced risk for adverse events with smallpox vaccination, including generalized vaccinia and eczema vaccinatum.

Summary

The medical and public health community must address the unique medical, psychological, and social needs of vulnerable groups in our society in its bioterrorism planning and preparedness efforts. Although direct experience is lacking, basic physiological principles, generalization from analogous conditions, and a measure of common sense enable clinicians and public health care officials to anticipate some of these particular needs. Children's physiologic, anatomic, immunologic, and developmental immaturities impact directly on their response to real, threatened, or imagined terrorist attacks. Pregnant and breast-feeding women have significant hormonal changes that affect normal immunologic function and enhance the risk for infection, as well as complications from the infection or treatment and prevention strategies. The immunologic, physical, psychological, functional, and social changes associated with aging place elderly patients at greater risk compared to younger and healthier individuals.

Vulnerability need not be thought of strictly as a physical or physiologic issue. Populations such as the poor, the homeless, and the underinsured lack adequate access and participation in the health care system. They are at enhanced risk because they are less likely to receive timely attention, if they get any at all. They are less likely to receive adequate follow-up care. Last, advances in treatment mean that there are many more immunocompromised individuals with impaired ability to withstand exposure to BCN agents, or the therapies used to treat or prevent these diseases. Immunocompromised patients are at greater risk of acquiring infection, may show accelerated or atypical courses, and experience longer convalescence and more complications than immunocompetent patients.

CHAPTER 7

Psychological Effects of the BCN Threat

Our understanding of the psychological effects of BCN terrorism events is limited, but extrapolation from these few episodes can help us better prepare for such events in the future. What has been found from studying the effects of terrorist acts is that reactions follow those seen in other traumatic events, such as natural disasters. Recommendations and likely clinical effects are largely extrapolated from these more "usual" disaster scenarios.

Based on studies prior to September 11, primary care providers are said to manage roughly 70% of all mental health problems in the United States and that upward of 75% of all patient visits to physicians' offices have significant or primarily psychological issues. These statistics are particularly relevant in the context of BCN terrorism. Following the September 11 terrorist attacks, a survey of primary care physicians found that nearly 80% identified terrorism-related psychosocial complaints in their patients, particularly in those areas geographically close to where the events transpired. The psychological fallout from traumatic events typically exceeds the medical consequences, in some instances by an order of magnitude. Following the 1995 Tokyo subway sarin attack, for example, 80% of those seeking medical care had no exposure to the gas. This phenomenon is seen commonly with any perceived public health or nonmedical emergency as well. During the 2003 SARS epidemic in Toronto, nearly 200 individuals sought medical evaluation for every diagnosed case of SARS. Clinicians should anticipate that anxiety and fear will result in a large number of individuals seeking care from the medical, hospital, and public health community following major disasters, public health

FIGURE 7-1 One-year memorial service at Ground Zero.
Courtesy of FEMA/Andrea Booher.

emergencies, and of course, terrorist attacks. Although the majority of survivors experience only mild reactions, and recover fully, as many as a third may meet the *Diagnostic and Statistical Manual of Mental Disorders*, Fourth Edition (DSM-IV) criteria for anxiety, depression, or PTSD (see Table 7–3).

Increased anxiety in the context of bioterrorism may be explained by risk perception theory. Risk perception theory suggests that risks that are voluntary, controllable, distributed fairly, imposed from a known or trusted source, have the potential to benefit others, or are familiar or even natural, are handled with far greater aplomb. The willingness of American soldiers to climb the cliffs at Normandy in the face of Nazi guns, or a patient participating in a clinical trial of a new cancer drug, might be understood in light of this theory of risk perception. In contrast, BCN terrorist attacks are involuntary, imposed in the time, place, and manner chosen by an unknown, unfamiliar, or threatening source, have no goal other than harm, are not equitably distributed, and may involve the stuff of Hollywood movies such as unseen microbes, chemicals, or nuclear material. The very nature of terrorism amplifies the perception of risk and thereby influences greatly both individual or a community response to the event.

The BCN agents discussed in this book are all plausible and potentially dangerous threats. Creating a climate of anxiety and fear may be even more effective as a weapon of terror. Fear and anxiety regarding the use of such weapons or of other attacks since September 11 have become a part of our individual and collective psyche. This fear is neither unwarranted nor inappropriate, but it is a double-edged sword in the fact that while making us more appropriately vigilant and cautious, it creates anxiety, higher stress levels, and ethnic and religious distrust as well.

Fear is not the only consideration. The events of September 11 were traumatic for those who survived and for many who watched the events from a safe distance—the death and destruction now immortalized on video and in our memories. For many, the post office anthrax attacks had similar effects, particularly coming on the heels of September 11. PTSD affected many in the wake of these events and would surely affect many more should another event occur. Given all this, it is incumbent on clinicians to be prepared for how to deal appropriately with a spectrum of psychological issues that could result from a similar trauma in the future.

PTSD is frequently seen in those immediately affected by any traumatic event and in those who identify with the event or its victims. Millions of Americans felt (and continue to feel) that they too were victims of the September 11 attacks, whether a few miles or a few thousand miles away. Numerous studies after September 11 demonstrate a rise both in PTSD features and overall psychological stress in Americans. Not surprisingly psychic distress appears to increase with increasing proximity to the attack sites. In the six months following the WTC and Pentagon attacks, the nation experienced a significantly elevated perception of risk and immanency regarding future attacks, as well as the likelihood of themselves or loved ones being victims of an attack. Such findings were independent of both

FIGURE 7–2   Mourners at a candle vigil at Ground Zero.
Courtesy of FEMA/Michael Reiger.

physical proximity and personal connection to the attacks that occurred in New York City, Washington, DC, and elsewhere. Though no attacks have occurred since the anthrax attacks of 2001, the psychological impact persists: bioterrorism has sensitized us as a nation. Upgrades in the homeland defense advisory system, flights cancelled or rerouted, suspicious packages, powders, and letters causing evacuations all make the headlines on a regular basis and contribute to an already elevated level of psychic distress. Not uncommonly, any major accident (e.g., power outages, plane crashes, industrial accidents) or new infectious diseases (e.g., SARS) once presumed to be accidental or coincidental in origin is now assumed at first to be a terrorist event. The uncertainty with which we now live has direct psychosocial effects and is a powerful reminder of the lingering effects of terrorist actions.

Although the emphasis of this book is on diagnosing and treating the patient exposed to one of these weapons, it should not be interpreted to mean that these are the only patients affected, or that only physical signs and symptoms result. Because of post-September 11 tensions, if there should be another such attack, significant numbers of patients will seek care from their doctors without having been exposed to BCN weapons. Some will seek counsel about protective and preventive measures to keep themselves and their families safe, whereas others will be concerned about possible exposures. In both cases, educating your patients regarding the likelihood of events or the actual risks, helpful steps, and precautions will likely be of great benefit. Hopefully, information contained in this book can be a resource for clinicians. Additional resources to address psychological consequences of BCN terrorism are provided in the bibliography at the end of *The Bioterrorism Sourcebook*.

Signs and Symptoms

In naturally occurring or manmade disasters, the majority of individuals handle the circumstances without any major psychological distress, but inevitably some will not. Loss of valued possessions, community, or home; loss of loved ones; lack of easy communication with family, friends, and colleagues; fatigue; adverse weather conditions; hunger or sleep deprivation; and prolonged emotional strain are all predictive of more severe reactions to traumatic events. Although the spectrum of psychological responses varies tremendously, there are a few clinical diagnoses for which practitioners should be on the alert. The diagnostic criteria for common disorders are listed at the end of the chapter. Formal diagnostic criteria may or may not be met, but even if not, patients may still be suffering and warrant some degree of management or intervention.

Anxiety is the most common psychological response to traumatic events. For most individuals, this will resolve without any formal intervention. Indeed, patients and communities should be reassured repeatedly that temporary hypervigilance, anxiety, and somatic symptoms are expected following a major trauma and in time resolve without sequelae in the vast majority of cases. That being said, generalized anxiety disorder, acute stress disorder, depression, and PTSD may all be anticipated following a disaster. In addition to the formal DSM-IV diagnoses, patients will commonly experience less dramatic and subtle psychic distress. Common nonspecific complaints suggestive of a psychological component include the following: sleep difficulties, persistent fearfulness, anxiety, and somatic features of back pain, myalgias, gastrointestinal discomfort, and other nonspecific symptoms. Somatization is very common following trauma. It occurs more frequently in vulnerable groups, such as children with chronic medical or psychological conditions, but it can occur in any group or individual. Physical manifestations affect multiple organ systems, although GI, nervous system, musculoskeletal, and constitutional symptoms predominate. A partial list of common physical symptoms found in survivors of disasters are noted in Table 7–1.

Vulnerable Groups

Certain individuals may be a greater risk for long-term psychological trauma following a BCN event. Certainly those individuals directly affected by the event, who may have witnessed graphic events, or who were made ill but not incapacitated by the event will experience greater distress. The ferocity of media coverage of any such event will inevitably result in playing and replaying of graphic images that can also evoke similar

TABLE 7-1 Physical Symptoms Commonly Seen with Psychological Stress

Gastrointestinal	Dizziness
Nausea	Anorexia
Diarrhea	Insomnia
Constipation	Anhedonia
Cardiac	Loss of sexual interest
Chest pain	**Musculoskeletal**
Palpitations	Myalgias
Elevated blood pressure	Back pain
Neurologic	**Neurohormonal**
Discoordination	Sweating or
Headaches	Diaphoresis

symptoms. Individuals with prior psychological traumas, existing or latent psychological diagnoses, chronic medical conditions, and recent or significant past life stresses are more vulnerable. Young children, the elderly, and those medical and emergency response personnel dealing with the aftermath are also at enhanced risk (Table 7–2).

Management

Although immediate medical needs must take priority, clinicians should be aware of the potential psychological sequelae following a BCN event. Identifying and appropriately triaging to mental health professionals any individual with severe stress responses is an essential skill that clinicians will be expected to demonstrate.

TABLE 7-2 Developmental Stage and Psychological Response to Trauma

Age Group	Signs and Symptoms
Pre-K to 2nd graders	Separation anxiety
	Avoidance
	Regression
	Fear of dark
3rd to 6th graders	Reenactment through traumatic play
	Social withdrawal
	Atypical aggressiveness
	New and atypical hyperactivity
Adolescents/young adults	Increased risky behaviors
	Decline in responsible behavior
	Social withdrawal and apathy
	Unusual rebelliousness

Adapted from the University of Washington Northwest Center for Public Health Practice.

Evaluation of somatic symptoms and reassurance that these are physical manifestations of the emotional distress should be done with maximal attention to the patient's justifiable worry. Finally, communication regarding both immediate and potential long-term risks to the individual, their family, and their community will aid both affected and unaffected persons to adjust more readily to what the future has in store. Of course, no one can predict the future with certainty, and empty promises can be detrimental. A fair statement that "we don't know" is better than false reassurance in most instances. Active listening and knowledge-driven reassurance will go a long way to allaying fears that may be imagined, as well as to framing fears that are appropriate and real (Table 7–3).

Talking openly and honestly and offering reassurance are essential to enabling individuals and communities to cope with significant traumatic events. When children are affected, it is often useful to encourage them to express their anxiety not only by talking in a safe environment with a trusted adult, but also by using alternative methods, such as drawing, playing games, or music. Children should be asked what they think happened and why, and will respond well to an honest statement about your own worries and responses to what occurred. Repeated exposure to the media, either directly or indirectly, should be actively discouraged for those suffering posttraumatic stress responses, regardless of their intensity.

Patients thought to be suffering from PTSD should be referred to a mental health professional for some form of cognitive–behavioral therapy or exposure therapy, which has been shown to be of benefit. Timeliness is an issue. Psychological debriefing soon after an event is highly therapeutic. Also of benefit are the use of selective serotonin reuptake inhibitors (SSRIs) in managing symptoms and the addition of mood stabilizers when SSRIs seem to be of no value (assessed after an eight-week trial). From this discussion it should be evident that primary care providers should maintain close ties with mental health professionals in their communities and consult them in particularly severe cases.

Regardless of whether patients are working with mental health professionals, there are a number of ways that primary care providers can assist patients in coping with the psychological aftermath of terrorism or other disasters. Studies of successful coping mechanisms for traumatic events include a number of strategies that can be implemented by the patient and need to be shared with them as early as is appropriate. These strategies include the following:

- Keeping a reasonable perspective. Encourage individuals to maintain small but achievable short-term goals without ignoring the overall presence of this event in the context of their lifetime.
- Encouraging patients to adopt a positive outlook and maintain their self-esteem.
- Encouraging utilization of support systems rather than the "go it alone" approach; individuals who talk to others similarly affected or who allow others to provide support recover more quickly than those who feel they

have all the resources they need to handle the circumstances on their own.

- Discourage patients turning to stimulants (caffeine) or other more harmful substances, such as alcohol or drugs.
- Utilizing mental health resources including professional counselors and judicious pharmaceuticals whose impact is monitored carefully.

Patients who have undergone any life trauma should be seen on a more regular basis by their primary care physicians, whether they need professional mental health services or not.

Despite great progress, mental health and mental illness is still stigmatized in this country. Most will not seek out mental health assistance, so health care providers must be vigilant about making an assessment of psychological consequences following a BCN event. Clinicians should remind their patients that it is common for psychological distress to take more than one form. Reassure your patients that these normative responses do not indicate psychopathology and will in all likelihood resolve with minimal or no intervention.

☣Mental Health Strategies for Clinicians

Patients are not the only ones who suffer after a terrorist attack. Caregivers are likely to be overworked, exhausted, frightened, and anxious. When caregivers are suffering mentally, physically, or spiritually, it diminishes their ability to help others. It is important to be conscious of one's own limitations particularly in a crisis setting. Fearfulness and anxiety may impair clinical judgment and impede proper patient care if not properly recognized and addressed. For this reason, physicians must be equally aware of their stress responses and those of their colleagues and staff now and during a BCN event or similar disaster. It has been reported, for example, that 61% of health care providers have heightened anxiety regarding the likelihood of future bioterrorism attack. Some useful coping strategies for health care workers are to ensure a working environment where personal safety is assured. Optimism, charity, collegiality, and ready complements directed to overworked staff all help sustain morale. Establishing a safe haven where you and your staff can go to relax, talk, and communicate with loved ones is essential. Being aware of and providing respite from physical and emotional fatigue, stress reactions, and also getting adequate sleep and nutrition will be necessary to continue to provide care to injured individuals or traumatized communities. This goes for clinicians and public health officials alike. During the SARS epidemic in Toronto, many public health officials worked nonstop for days under enormous pressures from the media, the public, the government, and the health care community itself. Only later on was the detrimental impact of the public health crisis on these public health officials fully recognized. When working in a team setting, frequent and honest communication among the group will be necessary for the community of workers to continue to function well in circumstances that may be

TABLE 7–3 Psychological Effects Associated with Traumatic Events

Acute Stress Disorder Symptoms: Present for Less Than One Month and Lasts Less Than Four Weeks

Association with upsetting threat or action

Sense of or intrusive reexperiencing event—nightmares, flashbacks

Avoidance behaviors (hypervigilance or substance abuse)

State of extreme psychological arousal (rage, severe irritability, agitation)

Social withdrawal

Anhedonia

Dissociation

Heightened startle response

Depression Symptoms

Major depression disorder—Four of the symptoms in addition to depression or anhedonia, for 2 weeks or more

Dysthymic disorder—Two of the first six symptoms, in addition to depression, for two years or more

Sleep disturbance—increased or decreased guilt/diminished self-worth

Decreased energy/poor concentration

Hopelessness

Appetite disturbance decreased or increased

Anhedonia

Psychomotor changes—retardation or agitation

Suicidal ideations

Memory problems

Naming apraxias

Repeated visions of the event

Vivid or distressing dreams

Decreased attention span

Problem-solving difficulties

Generalized Anxiety Disorder

Persistent sense of fear that is disproportionate to a known source or no source identified

Physiologic symptoms—shakiness, palpitations, tingling, dizziness, or syncope

Gastrointestinal symptoms—nausea, vomiting, diarrhea, discomfort, urinary frequency, paralyzing worry, helplessness, obsessive or compulsive behaviors

difficult and stressful, that may require temporary disruption of normal social ties, or that may be a genuine threat to the workers themselves (Fig. 7–3).

For most members of the community, a traumatic event such as a terrorist attack will not result in long-term psychological problems; rather, they will show resilience with any initial symptoms resolving in time. The pace of resolution will differ from person to person. Patients need varying degrees of assistance in working through these difficulties. Primary care providers will play a vital role in assessing what level of intervention, if any, is needed for their patients and for their communities.

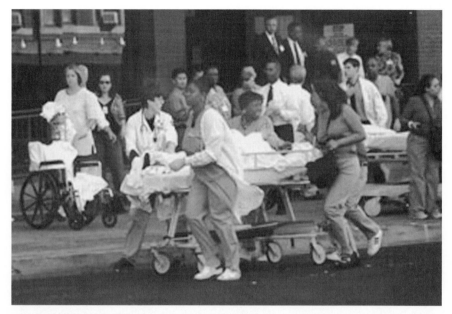

FIGURE 7-3 Medical staff at St. Vincent's Hospital in Manhattan tending to victims of the WTC attacks.
Courtesy of SEIU.

FIGURE 7-4 Disaster relief planning in New York City after the attacks on the WTC.
Courtesy of FEMA/Michael Rieger.

The transition back to a normative social life can be promoted in a number of ways. Many of these require the assistance of outside agencies skilled in disaster management. Providing a work environment that meets basic needs for security, food, shelter, warmth, and human contact are paramount. Focusing on problem solving rather than on hand wringing is critical to providing situational leadership. Reconnecting with close contacts, family, and friends to share feelings and retell the story are healing strategies as well. Identify key resources to help with the rebuilding process, including the Red Cross, FEMA, or the Salvation Army, organizations with significant experience in disaster management. Utilizing established agencies for help with important practical considerations, such as loans to rebuild homes or provide temporary financial support, can help individuals and families and communities begin to move forward again (Fig. 7–4).

☣Summary

Experience with natural and manmade disasters, public health emergencies, or terrorism enables us to predict that the psychological fallout from such events will typically dwarf the immediate medical consequences. Acute and long-term psychological sequelae are dependent greatly on the traumatized individual's physical, emotional, developmental, and social experience. The same may be said of communities themselves. Most will experience temporary effects and soon resume or indeed maintain normative social roles. A minority will experience a range of somatic complaints as part of an acute stress response. Anxiety, depression, and PTSD do occur and may require referral to mental health professionals. Children, the elderly, and individuals with preexisting psychological trauma, history of childhood abuse, or psychological diagnoses are more vulnerable to the immediate and chronic effects of BCN terror. Patients, families, and communities need reassurance that psychological reactions are expected following such events, do not indicate serious psychopathology, and will fade over time. Primary care clinicians are far more likely to be approached by their patients who are struggling with these issues and must be prepared to reassure, guide, treat, or refer, as circumstances dictate.

Disasters affect the resilience and physical capacities of communities as well as of individuals. This fact has implications for those responsible for disaster planning and management. By disrupting physical, economic, and social resources, terrorism in any guise tests communities sense of security and togetherness. Support from governmental and nongovernmental agencies will play a vital role in helping to restore normalcy to the traumatized communities, and in turn, promote a sense of normality and security for individuals.

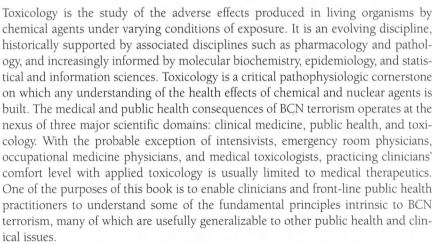

CHAPTER 8

Toxicology and Bioterrorism

Toxicology is the study of the adverse effects produced in living organisms by chemical agents under varying conditions of exposure. It is an evolving discipline, historically supported by associated disciplines such as pharmacology and pathology, and increasingly informed by molecular biochemistry, epidemiology, and statistical and information sciences. Toxicology is a critical pathophysiologic cornerstone on which any understanding of the health effects of chemical and nuclear agents is built. The medical and public health consequences of BCN terrorism operates at the nexus of three major scientific domains: clinical medicine, public health, and toxicology. With the probable exception of intensivists, emergency room physicians, occupational medicine physicians, and medical toxicologists, practicing clinicians' comfort level with applied toxicology is usually limited to medical therapeutics. One of the purposes of this book is to enable clinicians and front-line public health practitioners to understand some of the fundamental principles intrinsic to BCN terrorism, many of which are usefully generalizable to other public health and clinical issues.

Basic knowledge of the principles of toxicology is important for a deeper understanding of the pathophysiology from BCN agents. Understanding toxicologic principles is important when considering the health effects of BCN terrorism for several reasons. First, knowing these principles will facilitate an understanding of how cells and cellular structures, tissues, organs, and ultimately, individuals respond to these agents. This short overview of toxicology summarizes physicochemical properties common to many of the chemical agents covered in this book.

111

☣ Toxicology: The Study of Poisons

The study of poisons, as toxicology is rightly viewed, stretches back into antiquity. The scientist and alchemist Paracelsus conducted extensive experimentation on metals from which basic concepts, such as dose and dose response, were placed on more scientific rigorous ground. Indeed, it was Paracelsus (Fig. 8–1) who observed that all chemicals may be injurious given the right circumstances, an aphorism widely quoted but infrequently sourced. Careful clinical observations by the Italian physician Ramazzini (1710–1767), the English surgeon Sir Percival Pott (1780–1843), and many others established the connection between trades and a wide variety of disease processes. Propelled by Newtonian scientific principles, the Industrial Revolution of

FIGURE 8–1   A woodblock print of Paracelsus, the father of modern toxicology.
Courtesy of the National Library of Medicine.

the 17th and 18th centuries vastly expanded the number, utility, and adverse health and environmental consequences of a wide range of hazardous materials and led directly to the creation of entire industries, including the chemical and pharmaceutical industries from which many of the chemical agents discussed in *The Bioterrorism Sourcebook* emerged.

In time, the health hazards brought about as a result of this remarkable economic transformation were recognized and led to the creation of the discipline of industrial toxicology in the 1930s. Indeed, some of the earliest pioneers in this area were physicians, such as Harvard's Alice Hamilton, whose work integrated field investigation, clinical assessment, and laboratory-based toxicologic evaluations. The social, economic, and public health consequences of the Industrial Revolution led not only to the recognition of new diseases, but also to substantial changes in the prevalence of many known diseases. Advances in mathematics and physics led in time to the discovery of radiation, and a host of medical and public health consequences have followed over time. The commercial, industrial, and medical applications of atomic physics generated a host of previously unknown medical diseases. For example, in the early 1900s, case reports of a progressively destructive mandibular cancer ("phossy jaw") were described in women painting radioactive phosphorus on the watch hands of time pieces. Twentieth-century developments in atomic and quantum physics led to the creation of thermonuclear bombs and the nuclear energy industry (see Section IV). Radiation burns, radiation sickness, and hematopoietic and thyroid cancers are a few of the medical conditions that resulted from the new technologies arising from these scientific discoveries. Diseases such as asbestosis or silicosis that were known in ancient times occurred in increasing numbers as technology transformed many processes such as mining, foundry work, or shipbuilding from ones involving relatively few skilled tradesmen to huge industries where exposure levels and the number of exposed individuals were increased dramatically.

More often than not, the adverse human and environmental consequences of chemicals and processes occur only after a delay that can be of considerable length. This biological fact has considerable implications in terms of prevention and regulation of industry. Further, the human body's innate ability to deal with a wide number of toxicants interjects substantial variability in the clinical manifestations of toxicity, a fact that complicates the study of toxicologically mediated disease. Even at present, detailed knowledge of adverse effects is confined largely to toxicants with which our experience has been longest. Mechanisms of injury, distribution, metabolism, excretion, and chronic sequelae are far better understood with metals such as lead, arsenic, and mercury, for example. One need only point out that of the four million chemicals now known—of which approximately 50,000 are in regular use—only a tiny fraction of them have been evaluated according to best modern toxicologic practices.

This chapter confines itself to basic principles of toxicology with the aim of informing subsequent chapters on chemical and nuclear agents, their acute and chronic health effects, and medical and public health control strategies. Each of these aspects, in part, is based on toxicologic properties and our body's response to toxicants.

Principles of Toxicology

Understanding of the toxic effects of chemicals is predicated on an understanding of the physical state of compounds, how they enter the body and are transported to sites where they exert their effects, how they are broken down or metabolized, and finally how they are ultimately eliminated from the body. The composite description of these actions—absorption, distribution, metabolism, and excretion—is termed *toxicokinetics*.

The physical form in which a substance exists under ambient environmental conditions plays a significant role in determining what, if any, toxic effect it might have. Changes in these environmental variables may alter the physical state and therefore the potential toxicity of any given substance. The three most basic physical states are solids, liquids, and gases, but these in turn may be modified under the right environmental conditions into fumes, mists, or vapors.

Solids present minimal risk when undisturbed, but they may create airborne dust or fibers if subject to disruption or mechanical attrition by crushing, grinding, or detonation. Due to the progressively narrow branching structure of the lung, dust particles or fibers ranging from three to seven micrometers find their way to the distal airways and alveoli of the lung where they have their characteristic health effects. Asbestos fibers, for example, are typically 5 microns in size and when inhaled settle into terminal bronchioles and alveoli and initiate an inflammatory process that may culminate in the pneumoconiosis that bears the name of this mineral.

Liquids are free-flowing incompressible fluids that may be pure, or come in the form of mixtures (if a solid is dissolved in it) or solutions (if another liquid is dissolved in it). To some degree, all liquids are in equilibrium with the atmosphere. This equilibrium is affected by the intrinsic volatility of the liquid and by ambient conditions of temperature and pressure. For example, gasoline is a volatile liquid that evaporates more readily as temperature or pressure increases. Partially evaporated liquids at the interface between the liquid itself and the air are vapors. Consequently, vapors are potentially hazardous through dermal and respiratory routes of exposure. Vapors may also be generated from solids through a process known as sublimation. Vapors are diffusible in air, with essentially gaseous properties, and may reform (or condense) under suitable conditions.

Gases are compressible fluids that expand freely and uniformly to occupy the space available under ambient environmental conditions. The principal health risk from gases arises via inhalation. Gases and vapors may be denser than air, a fact that has direct health consequences. Gases and vapors that are denser than air will preferentially seek the lower ground and will collect in pits and other confined spaces. "Ideal" chemical warfare agents, including VX and mustard gas, are designed with this property in mind and cause greater injury to those exposed as a result.

Another aspect of this property is that the encroaching gas or vapor may displace oxygen and thereby cause asphyxiation. Asphyxiation is an important clinical syndrome from both an occupational and BCN perspective. There are two types of asphyxiation: hypoxic asphyxiation and histotoxic asphyxiation. The former occurs when a gas or vapor simply displaces oxygen from the environment and the oxygen-poor atmosphere lowers the partial pressure of oxygen in the blood. Hypoxic asphyxiation can occur with both inorganic gases (e.g., sulfur dioxide) and volatile organic solvents. Simple asphyxiants displace oxygen from respirable air such that the higher the concentration of a simple asphyxiant, the lower the concentration of oxygen. At some level, the resulting oxygen deprivation begins to have adverse clinical consequences (Table 8–1). With simple asphyxiants, the concentration of oxygen in air is the primary determinant of its physiologic effect. Many of the clinical effects of an oxygen-poor atmosphere cluster around the central nervous system. The adverse health effects of simple asphyxiants also depend on the presence or absence of other variables. For example, the effects of asphyxiants can be accelerated or exacerbated by personal factors such as increased work pace (i.e., increased minute ventilation), underlying medical conditions (e.g., chronic lung disease), and the absence or improper use of PPE. Environmental factors such as inadequate dilution ventilation, high ambient temperatures, and elevated altitude also impact susceptibility. Ultimately, all these factors operate either to decrease the amount of oxygen available to tissues or to augment the tissues' oxygen requirements, which cannot be matched by the available oxygen in the inspired atmosphere.

In contrast to simple asphyxiants, histotoxic asphyxiation is due to disruption of cellular metabolism at the molecular level. A chemical or toxic asphyxiant (carbon monoxide, cyanide, acrylonitrile, or hydrogen sulfide) exerts its effects by interfering with cellular metabolism, thereby causing cells to become starved for oxygen. Toxic asphyxiants, such as hydrogen cyanide and hydrogen sulfide function through a variety of mechanisms, discussed separately later. An example of histotoxic asphyxiants are cyanide or hydrogen sulfide, two chemical agents discussed in Chapter 25. Chemical asphyxiants produce symptoms by interfering with cellular oxygen transport or utilization. The mechanisms by which each toxic asphyxiant interrupts cellular metabolism vary and are addressed individually. However, as with simple asphyxiants, other variables influence the severity of clinical effects of

TABLE 8–1 Physiologic Effects of Oxygen Deficiency

Percentage Oxygen Concentration	Symptoms
16–21	None[a]
12–16	Poor coordination, increased heart and respiratory rate
10–14	Dyspnea on exertion (DOE), fatigue, irritability, emotional lability
6–10	Lethargy and somnolence, nausea/vomiting, possible unconsciousness
<6	Seizure, apnea, asystole

[a] Individuals with underlying cardiac or respiratory conditions are more vulnerable and may be symptomatic with even minor decreases in inspired oxygen concentrations.

Used with permission from JL Weeks, BS Levy, GR Wagner, KR Rest. Preventing Occupational Disease or Injury: Washington, DC, APHA Press, 2004, 2nd edition.

any given toxic asphyxiant. Concentration of the gas, length of exposure, and adequacy of ventilation are important environmental conditions that influence toxicity. Individual factors, such as baseline health status (e.g., cardiac or pulmonary disease) and improper use or nonuse of PPE (see later), can modify toxicity as well. Finally, concurrent exposure to other asphyxiants or gases may augment the potential toxicity of any given asphyxiant. Such is the case for fire fighters, who are exposed to high levels of carbon monoxide and hydrogen cyanide in fires in which pyrolysis of plastics occurs.

Fumes are formed when volatilized solid materials, such as molten metals, recondense in air into minute solid particles capable of being inspired. Their solid nature distinguishes them from vapors, which exist as gases. Occupational examples that are familiar to many clinicians are fumes generated from soldering, smelting, or welding. Mists are suspended liquid droplets condensing from the vapor state or formed by liquid dispersion due to splashing or atomization. Occupational examples are oil mists from cutting or grinding metals, acid mists from electroplating, and organic solvent mists from paint-spraying operations. From the perspective of BCN terrorism, chemicals that can be aerosolized as a gas, fume, vapor, or mist posed the greatest risk. Aerosols are solid particles or liquid droplets that may remain suspended in air for a considerable length of time, unlike mists or fumes that usually settle out more quickly. Particulate aerosols may become "carriers" for materials that become adsorbed to them. Gases or vapors condensing on particulate materials may be deposited on the respiratory epithelium at correspondingly higher concentrations than would arise from simple diffusion from the air into the lung, thus magnifying their effect. Thus, the irritancy of sulfur dioxide (SO_2) is markedly increased when it is inhaled following condensation or carriage on smog particles. Aerosols may also facilitate the carriage of microorganisms that may lodge in upper or lower respiratory tracts. This is seen in such diseases as humidifier fever (from thermophilic bacteria or fungi that are carried in liquid aerosols) or *Legionella* transported in building humidification systems.

Toxicokinetics

Toxicokinetics is the study of the dynamic (kinetic) relationship between the concentration of a chemical in body fluids and tissues and its biological effects. The factors that affect the toxokinetics of a particular substance are the rate of absorption, distribution, metabolism, and excretion. Dermal, respiratory, and GI absorption are the most relevant for our purposes. Both dermal and respiratory routes of exposure occur with aerosolized agents, whereas GI exposures would be the preferred pathway for agroterrorism or waterborne terrorism. Parenteral routes of exposure are familiar to clinicians as a common therapeutic route of delivering medications. For terrorists, however, parenteral routes are of limited value as they require proximity and can only be directed at a single individual. One of the more celebrated cases of parenteral use was the 1972 assassination of a Bulgarian defector by Soviet spies through the injection of ricin from a specially engineered umbrella.

Once absorbed, the toxin may have local or systemic effects. Injury may be caused directly by the toxin or following metabolic transformation by the body's normal detoxification mechanisms. A full discussion of metabolic detoxification mechanisms is beyond the scope of this book. However, the organ systems in which these processes occur help explain the common site for toxicity for a wide range of chemicals. These include the liver, hepatobiliary tree, and hematopoietic systems. Excretion occurs through the urine, biliary tract, feces, sweat, or expired air.

A chemical injury requires contact with cellular components. Thus, the route of exposure and absorption of a toxicant has a profound impact on the nature of injury that may result. Major routes of absorption are inhalation, ingestion, and skin contact (dermal or percutaneous absorption). From a BCN perspective, inhalation and percutaneous absorption are of primary importance in selected instances, whereas ingestion is of secondary importance. Simple contact may provoke a local toxic effect due to direct cell damage, such as the tracheobronchitis that exposure to chlorine gas may cause.

Absorption is dependent on intrinsic properties of the chemical (concentration, water, and lipid solubility) and host factors (diet, nutrition, and health status). For example, the water solubility of the pulmonary agent chlorine results in the chemical being absorbed very readily in the water-rich milieu of the upper respiratory tract epithelium. Consequently, the clinical effects of chlorine gas occur rapidly and are characteristically an upper respiratory tract syndrome. Similarly, the skin functions as a protective barrier by virtue of its water impermeability and lipophilicity. Highly polar compounds or toxicants that are water soluble are not avidly taken up through the skin. An example of diet and health status affecting absorption of toxicants is lead: lead is absorbed through the same GI mechanism as calcium and so will be avidly absorbed in pregnant women more than in nonpregnant women or in men.

Inhalation is an important route of absorption, and toxicity is contingent on the physical form of the toxicant (particle, fume, and gas), as well as on the size. The location for deposition is largely responsible for the physical effects seen. This principle is well worked out for the classic pneumonconioses such as asbestos, but it is equally applicable to selected bioterror agents. Smaller particles (less than 7 microns) reach distal airways and alveoli, whereas larger particles settle out in the upper respiratory tree. Soluble particulates may be absorbed into the bloodstream or taken up by the reticuloendothelial system for transport to distant sites. A relevant example is anthrax. Inhalational anthrax is characterized by a mediastinitis. The size of the anthrax spore enables it to penetrate to the distal alveoli where it is taken up by macrophages and transported to hilar lymph nodes.

Many chemicals penetrate healthy intact skin, and gases and vapors may also penetrate, to some degree, through the skin. Penetration of chemicals through the skin are partly dependent on the chemical's polarity and lipid solubility. Chemicals that are lipid soluble and have low polarity penetrate the skin barrier more readily. Importantly, dermal absorption is enhanced by anything that diminishes or breaks down its normal protective mechanisms. Individuals with compromised skin protection—for example, eczema, atopic dermatitis, psoriasis—may absorb more toxins through the nonintact skin. This has implications for chemical agents in particular because individuals with dermal conditions may suffer enhanced effects. Ingestion is an important occupational route of exposure, whereas for BCN agents this is less significant. There are a few notable exceptions. For example, inhaled anthrax spores may settle out in the upper respiratory tree where they are cleared by the lung's normal mucociliary escalator and subsequently swallowed. GI anthrax may then be the primary clinical manifestation of the anthrax exposure. The same process has been well described for *Mycobacterium tuberculosis*. Routes of exposure may also occur through inoculation, the deliberate or inadvertent introduction of a toxicant or biological agent directly into the tissues, or through mucus membrane exposures such as the conjunctivae. This is an uncommon route of exposure.

Once a toxicant comes into contact and is absorbed, it may have local or distant effects. The distribution of a toxicant will influence its ability to impact systemically, and may also have important long-term consequences such as carcinogenicity, teratogenicity, and mutagenicity. Distribution usually takes place via blood or lymphatic transport to the tissues and organs of the body. Similar to the variables that control absorption, the extent and efficiency of distribution is determined by a variety of factors, including the toxicant's properties, whether the toxicant is compartmentalized, and by the processes involved in uptake and elimination. Lipophilic substances, such as mustards, will readily cross membrane barriers such as the skin and may then be stored preferentially in certain body compartments where their effects may linger or even cause chronic changes. In toxicokinetics, a "compartment" refers to those organs and tissues that take up and release or clear the toxin or its metabolites. Rapid equilibration into a "central" compartment such as the bloodstream may be accompanied by slower equilibration into "peripheral" compartments (which might include organs such as the liver or brain, or tissues such as bone or fatty tissue). This effect is sometimes described as a "multicompartment" model.

Whether the chemical is protein bound and to what degree will determine how much "free" or unbound toxin is available to cause damage. Bioaccumulation is an important process as well. Lipid-soluble materials that preferentially accumulate in fatty tissues and are slowly metabolized exhibit persistence. Materials such as dioxins, polychlorinated biphenyls (PCBs), and organochlorine pesticides, as well as other chlorinated compounds, fall into this category (Fig. 8–2). Metals such as mercury, cadmium, strontium, and lead demonstrate prolonged storage and slow excretion in bone, with lead and strontium sequestering for many years and even decades. Associated toxic damage to metabolic and excretory processes that arise through impairment of liver, renal, or respiratory function may also contribute to persistence.

Metabolism, Biotransformation, and Detoxification

The dynamics of metabolism and biotransformation and repair are also critical factors. The principal site of metabolism of xenobiotics is the liver, with secondary activity occurring in the kidney, lung, gut, and skin. Metabolic reactions generally are

FIGURE 8-2 Pesticides are among the most commercially abundant poisons worldwide and could be used as a chemical agent, exhibiting in the case of organophosphates or carbamate pesticides, a syndrome similar to nerve agents. *Courtesy of the U.S. Department of the Interior, U.S. Geological Survey.*

carried out by enzymes located on hepatic microsomes and are designed to detoxify parent chemicals by rendering them more polar, and hence more water soluble, to assist in their excretion. This process normally follows a two-step pathway. The most common metabolic transformation involves a first step of hydroxylation (addition of an –OH), oxidation (e.g., carboxylation or oxidation to a carboxylic acid –COOH) or reduction of the compound; the result is generally a more polar compound. This is then followed by conjugation with endogenous chemicals such as glucuronic acid. The resulting metabolites are much more water soluble and thus more easily excreted through the urine. This two-step process is a fundamental principle of biotransformation and one of the key mechanisms by which the body protects itself from environmental toxicants. When metabolic or excretory pathways are saturated, toxicity is enhanced as tissue concentrations rise. A familiar example of this is acute alcohol intoxication. If enough alcohol is ingested, hepatic alcohol dehydrogenase becomes saturated and can no longer metabolize the alcohol, resulting in greater toxicity. Unfortunately, biotransformation of toxicants does not invariably result in less toxic metabolites. When metabolic pathways are saturated, persistence is prolonged. Additionally, metabolites may be more reactive than the parent compound. When this occurs, biotransformation enhances toxicity of a chemical.

Mixed exposures can result in greater toxic effect than simple addition, as seen when a metabolic pathway is altered by a second toxin. For example, carbon tetrachloride (CCl_4) interferes with the conjugation (and thus clearance) of the carcinogen benzene (C_6H_6), resulting in greater overall benzene toxicity. Alternatively, enzyme induction caused by a second chemical may increase the rate of production of a more toxic metabolite of the first chemical. Toxicity can also be enhanced when normal protective mechanisms are inhibited. Tobacco smoke, for example, interferes with normal mucociliary clearance, resulting potentially in greater risk from inhaled particles or microbes.

The biological half-life of any chemical agent is an important determinant of how long the individual may be exposed to the toxic effects. The half-life of a chemical is dependent on absorption, metabolism, and elimination. Elimination occurs through several routes, again dependent in part on first-pass metabolism and polarity.

Water-soluble or highly volatile toxicants are largely eliminated unchanged through the kidney, biliary tract (from the liver), or lungs. More volatile or lipophilic substances must undergo metabolism first before they are excreted, largely in urine or bile.

Tissue Injury

Specific clinical effects of the different BCN agents are dealt with in each chapter. Effects of xenobiotics may be local or systemic, immediate or delayed. Local effects involve injury at the site of contact, and range in severity from simple primary irritation to corrosive injury to the skin or respiratory tract by acids or alkalis. Examples of acute local effects are skin, conjunctival, ocular, or respiratory tract irritation; asphyxia from inert gasses caused by simple oxygen displacement; or central nervous system narcosis due to inhalation of organic compounds. Systemic effects follow absorption and distribution via the general circulation and can result in both acute and chronic effects.

Irritant gases pose an interesting circumstance: highly soluble gases and vapors such as ammonia or chlorine provoke intense and immediate irritation of the upper respiratory tract prompting victims to leave the vicinity of exposure quickly and generally escape serious injury. Gases of intermediate solubility, such as phosgene or methyl isocyanate, are a more insidious threat causing deeper lung injury. The cloud of methyl isocyanate released at night in Bhopal, India, caused 2,000 immediate deaths, but delayed pulmonary injury was seen in many more individuals after hours and even days. As with all disease processes, variability of response and presentation occurs with toxic exposures and, undoubtedly, the same is true with BCN agents.

The dose–response relationship is the demonstration of a gradient of risk or effect that is associated with the "dose" or degree of exposure. The concentration of material absorbed over a given period (the dose) varies according to the circumstances of exposure. Thresholds, or the point on an exposure continuum where a toxic effect is first seen, are of particular importance in determining risks associated with exposure to chemical carcinogens.

FIGURE 8-3   Toxicologist doing field assessments after chemical exposures.
Courtesy of the NLM.

Chronic Toxicity

In addition to their acute effects, toxins may have delayed or chronic sequelae as well. Relatively little is known about the chronic health effects of many BCN agents, but where such information exists either from human epidemiologic or lab animal data, it is included in agent-specific chapters to follow. Some comments are applicable in a general discussion of toxicology, however. For example, inflammatory processes may result in local fibrosis or scarring, or even hypersensitivity or easily triggered reactive inflammatory response. Irritant gases may leave the affected individual prone to asthma or reactive airways dysfunction syndrome (RADS). Alterations in genetic material may also occur, result in carcinogenic, teratogenic, or mutagenic effects. A carcinogen is any agent that is capable of causing an increased incidence of malignant neoplasms in exposed versus nonexposed persons. The generally accepted theory of carcinogenesis identifies three stages: initiation, promotion, and progression. Carcinogenic agents or toxins have their effects on different stages of carcinogenesis. Some will cause direct and heritable changes in genetic material, such as ionizing radiation or mustard gas. Mustard gas has also been shown to work as a promoter, accelerating the clonal expansion of previously mutated genetic material. The blistering agent, nitrogen mustard, is a direct-acting carcinogen. It requires no metabolic transformation before causing carcinogenic transformation in the cells with which it comes into contact. Alternatively, some toxins require metabolic transformation by the body before they can have their carcinogenic effects. Typically, this occurs through the stripping of electrons from a toxin. The highly reactive chemical that results seeks out electron-rich macromolecules, such as DNA or RNA. The resultant disruption of cellular protein synthesis and reproduction may slow cell growth, or worse, cause cell death or apoptosis. In addition, such cells are also prone to mutagenic or carcinogenic transformation.

Experimental Testing and Regulatory Issues

There are a variety of methods by which toxicity of chemicals are tested. These include experimental animal testing and *in vitro* and *in vivo* testing, such as the well-known Ames test. A wide range of toxicities are looked for, covering different organ systems (skin, eye, and organs), as well as carcinogenic, developmental, chronic, mutagenic, and reproductive effects. Experimental toxicologic protocols are very well standardized to determine specific physiologic effects. A short-term measurement, such as an lethal dose 50 (LD_{50}), will give an indication of lethality. Longer-term tests are used to predict chronic or genotoxic effects. Newer methods, including sister chromatid exchanges, DNA adducts, and others, may provide not only means for testing chemicals for potential genetic effects, but also means of surveilling those exposed to BCNs.

There is a common nomenclature relating to toxicology that has parallels in pharmacology—indeed the two disciplines share many fundamental principles. Some of these terms are quite familiar to clinicians, such as dose–response, first-pass metabolism, or volume of distribution. Others are perhaps less familiar, although from time to time clinicians may come across them and for that reason they are worth describing briefly. Toxicologic tests are designed to answer different questions, such as acute lethality, chronic effects, carcinogenesis, mutagenesis, and teratogenesis. Terms often used to give a rough measure of relative acute toxicity are the LD_{50}, the LC_{50}, the LCt_{50}, and the ICt_{50}. The LD_{50} is probably the most familiar of the terms. The term refers to that dose of a chemical, drug, or toxin that is lethal to 50% of the lab animals tested. The LC_{50} is analogous to the LD_{50} but, as opposed to ingestion or parenteral administration, measures concentration in air or water that is lethal to half of the study animals. MCt_{50} is calculated from the product of concentration (C) and duration of exposure (t) and measures the lethality in air or water of a given chemical. The last measure of lethality is the ICt_{50}, which measures the dose and duration of exposure needed to incapacitate 50% of tested animals. These measures are useful for comparative purposes, but they are too crude for use in a clinical setting for a host of reasons. For example, there are often significant interspecies and intraspecies differences in these laboratory-based measures. Host factors, such as age, treatment of the animals, diet, stress, the medium in which the chemical dose is administered, and other factors may influence the outcome of lethality studies and undoubtedly have the same implications for human toxicity. This is to say that the impact of a given chemical on human health cannot always be accurately judged from lab animal testing. Further, as in humans, there are individuals who are uniquely resistant or uniquely vulnerable at the same concentration. Although relatively invulnerable and hypersusceptible populations complicate the applicability of lab animal data to the human toxicology, they probably represent intrinsic genetic polymorphisms that have a powerful, and one might add reassuring, evolutionary advantage.

Biopersistence is an important term that is mentioned frequently in the context of BCN terrorism. It refers to the tendency of a compound to remain available environmentally and therefore to be a potential hazard. Biopersistence is inversely related to volatility; that is, more volatile substances will evaporate and be diluted in the surrounding atmosphere. Chemicals that are lipophilic tend to have longer periods of biopersistence. A useful example of this is the nerve agent VX, an oily and dense compound that offers advantages from a chemical weapons perspective. Biopersistence has clinical relevance because the longer an agent remains viable in the environment, the greater the potential for contact by humans, the greater the difficulty of remediation, and the greater the difficulty of decontamination of victims or the environment. Each of these has significant implications in responding to a BCN event for victims, first responders, health care workers, and those responsible for environmental cleanup. The most obvious example of long-term biopersistence is environmental or tissue contamination by nuclear material, such as might occur with a radiation dispersion device or following a nuclear reactor accident. It is estimated, for example, that the 100-mile radius surrounding the Chernobyl nuclear facility is effectively a "no entry" zone (at least not without full protective gear) for a millennium.

CHAPTER 9

Legal and Ethical Issues

The "war" against terrorism is fraught with inescapable legal and ethical dimensions that evoke heated debate among constitutional lawyers, ethicists, civil libertarians, and the ordinary public. Strong differences of opinion regarding constitutionality of antiterrorism efforts exist as evidenced by the multiple legal challenges being brought against selected elements of the Patriot Act. These conflicts have and will continue to be played out in the area of bioterrorism, particularly the tension between protecting civil liberties and the public's health. This conflict has a long and fascinating history that has attracted the interest of historians of medicine and public health, as well as legal scholars, novelists, and health policy makers. In Colonial America, for example, medical, religious, and political leaders fought over variolation as a means of stemming smallpox epidemics. Echoes of this debate are evident in the more recent national smallpox vaccination campaign.

Many of these same issues remain today. During the TOPOFF 1 trial (described in Chapter 1), many participants called for imposition of a quarantine when the "plague" breached established perimeters of defense. Post-hoc analyses of TOPOFF 1 found that decision-makers made this recommendation without fully realizing the enormous logistical, ethical, and political dimensions of that decision. Further, it was not at all clear just who had the authority in these circumstances to implement or maintain such a quarantine. How contemporary societies would respond to public health measures (used with far greater frequency historically) is uncertain. For one thing, our experience with quarantine—arguably the most stringent public health control measure—is rather thin. Second, in most, if not all, Western society's individual liberties are often defended against even the

OF NOTE...

Oran, Algeria, 1940s

One of the most striking consequences of the closing of the gates was, in fact, this sudden deprivation befalling people who were completely unprepared for it. Mothers and children, lovers, and husbands and wives who had a few days previously taken it for granted that their parting would be a short one, exchanged a few trivial remarks, sure as they were of seeing each other again after a few days, or at most, a few weeks, duped by our blind human faith in the near future and little if at all diverted from their normal interests by this leave-taking—all these people found themselves, without the least warning, hopelessly cut off, prevented from seeing each other again, or even communicating with each other. . . . While our townsfolk were trying to come to terms with their sudden isolation, the plague was posting sentries at the gates and turning away ships bound for Oran. No vehicle had entered the town since the gates were closed. . . . Only a few ships, detained in quarantine, were anchored in the bay. But the gaunt, idle cranes on the wharves, tip-wagons lying on their sides, neglected heaps of sacks and barrels—all testified that commerce, too, had died of the plague . . .

—Albert Camus, *The Plague*

most minor intrusions of the state. For our purposes, it is not feasible or necessary to provide a comprehensive discussion of the legal and ethical aspects of bioterrorism. However, clinicians should be familiar with a few areas relating to public health law which this chapter presents. In addition, using the example of the recent smallpox vaccination program, we hope to illustrate some of the more salient liability issues that are likely to emerge should a bioterrorist attack occur at some point in the future.

The Public Health Contract

Clinicians may be unaware of the fact that well-established legal precedents give enormous power to state and federal authorities in regards to public health. This legal authority allows governments to institute a wide range of public health control measures, temporarily laying aside existing civil statutes, as part of what is sometimes referred to as the social contract between individual citizens and their government. Put simply, citizens are not entitled to exercise certain individual liberties if doing so threatens the health and welfare of the larger society.

Historically the legal boundaries and jurisdiction for these decisions have been areas of contention; pitting local, state, and federal authorities against one another. Over time, such jurisdictional conflicts have resolved gradually into what we have today. In most states, local public health issues are handled at the local level, and authority to implement public health control measures lies with local municipal health authorities. When these issues cross municipal or county boundaries, state governments assume responsibility. Last, when a public health issue crosses state lines or international borders, the federal government becomes the ultimate authority responsible for public health control measures. As noted earlier, President Clinton signed an Executive Order giving primacy to the CDC to manage all national public health emergencies, with assistance as requested by a number of federal agencies, including FEMA, the FBI, the Department of Defense (DOD), the Federal Aviation Administration (FAA), or any other federal agency whose resources and expertise are requested by the CDC. It is the CDC's responsibility, for example, to impose national or international travel restrictions in the event of an infectious disease threat, such as occurred with SARS. Similarly, the CDC can mandate the isolation of sick passengers arriving in any U.S. port of call and can restrict the movement of any person who is deemed a potential health threat. This authority extends to U.S. nationals as well as to foreign visitors.

In relation to bioterrorism, these jurisdictional issues have not been fully tested, and it is unclear whether the agreed-on delegation of authority will withstand a genuine bioterrorist threat. Experience to date suggests that confusion and chaos regarding this authority are probable and should be anticipated. Efforts to conduct field tests (like TOPOFF) or "table-top" exercises can help "train up" those involved in bioterrorist response, but they are no substitution for the real thing. Even in these exercises, lines of authority were unclear to many participants and draconian measures—including the decision by Colorado officials to quarantine the entire population (2 million) of Denver—were impossible to implement.

✸Model State Public Health Legislation

Bioterrorism raises legal and ethical issues that differ fundamentally from past public health issues. Why is this so? Unlike traditional public health threats, bioterrorism is simultaneously a public health issue and a federal crime. Some have argued that bioterrorism is an outright act of war. Viewed either as a criminal act or as an act of war, bioterrorism clearly is a federal responsibility. In contrast, jurisdictional traditions and precedents in public health law have been contingent historically on the nature and locus of the particular threat. Bioterrorism has altered, perhaps irrevocably, the locus of public health authority because it is simultaneously a national security issue—and so the purview of the federal government—and a public health emergency. Addressing the historical tension between state and federal authorities and the fundamental shift in responsibility as a result of the threat of bioterrorism, the noted bioethicist George Annas observed pithily that "the creation of these federal agencies, however, did not alter the state's responsibility for public health; the anthrax attacks did."[1]

Bioterrorism, or, more accurately, legal and regulatory responses to the threat of bioterrorism break new and highly controversial ground in public health law. Well before the September 11 attacks, however, the limitations of existing public health laws to address the modern realities of public health were recognized, and calls for substantive reform were under way. In 1988 and again in 2002, the Institute of Medicine issued two reports calling existing state public health laws antiquated and inconsistent. Among the measures recommended by the Institute of Medicine were the urgent need to revise existing public health statutes so that lines of authority and responsibilities regarding public health were clearly defined, and second to give these organizations the procedural, regulatory, and administrative tools needed to address modern public health threats and emergencies. Notwithstanding the consensus regarding the limitations of existing state and federal public health regulations and laws, what steps are justified to address these deficiencies remains hotly debated. The danger of not clarifying and modernizing existing laws was stated clearly and unambiguously by the legal scholar David Fidler who wrote in 2001 that "the ineffectiveness of existing legal frameworks in a real bioterrorism crisis would exacerbate pressure on governments to take drastic actions that might sweep away the rule of law in the midst of panic and uncertainty."[2]

Steps were taken to address the nation's system of public health law in the aftermath of September 11, including a national multiorganizational public–private collaboration whose charge it was to offer model language with which local, state, and tribal governments could update outdated statutes and regulations. The collaboration included representatives from five states, nine national organizations and government agencies involved in public health (including local, state, federal, and tribal groups), as well as consultants representing legal, public health, and public health law communities. The purpose of this collaboration was to modernize the dated legal framework in order to guide state, local, and tribal governments in their efforts to modernize existing public health statutes and regulations, thereby protecting both public health and national security interests. The original draft model legislation was first promulgated in 2001, and it quickly gained strong supporters and determined critics. Subsequent comments and revisions resulted in changes that have been incorporated into a document released in September 2003, referred to as the Model State Emergency Health Powers Act (MSEHPA).

According to the legislative proposal, MSEHPA "grants public health powers to state and local public health authorities to ensure a strong, effective, and timely planning, prevention, and response mechanisms to public health emergencies (including bioterrorism) while also respecting individual rights." The most important features of

---

[1] GJ Annas, "Bioterrorism Public Health and Civil Liberties." *Legal Issues in Medicine* 346(17): 1337–1342, 2002.

[2] DP Fidler, The Malevolent Use of Microbes and the Rule of Law: Legal Challenges Presented by Bioterrorism." *Clin Inf Dis* 27:186–189, 2001.

the model law address changes in the public health authority granted to states, designated state public health agencies and officials, and the responsibilities of health care providers, including hospitals, physicians, and other clinical practitioners.

Under MSEHPA, state governors may declare a public health emergency using their executive powers. A public health emergency is defined as the following:

> . . . an occurrence or imminent threat of an illness or health condition that is believed to be caused by any of the following: (i) bioterrorism; (ii) the appearance of a novel or previously controlled or eradicated infectious agent or biological toxin; or (iii) a natural disaster, a chemical attack or accidental release, or a nuclear attack or accident; and poses a high probability of any of the following harms: (i) a large number of deaths in the affected population; (ii) a large number of serious or long-term disabilities in the affected population; or (iii) widespread exposure to an infectious or toxic agent that poses a significant risk of substantial future harm to a large number of people in the affected population.[3]

During a declared public health emergency, state governors' broad powers include calling out the militia, changing state agency functions and authority, and temporarily suspending existing statutes that could, if followed, interfere with actions needed to control a public health emergency.

Earlier versions of the model public health legislation contained particularly contentious language, such as provisions for compulsory testing and compulsory participation by physicians and hospitals. It also included criminal misdemeanor penalties for noncompliance. Comments on the proposal since its initial promulgation resulted in a less controversial version, although this, too, has its share of critics among constitutional scholars, bioethicists, and health care organizations. Critics of the proposal insist that it gives public health officials historically unprecedented powers, far beyond what they possess currently. In the event of a declared public health emergency, for example, the act includes provisions for mandatory treatment by an individual who "has or may have been exposed to a contagious disease" and requires health care providers to inform individuals they are treating about how to prevent further spread of the disease and the need for treatment. Further, health care providers may be required to assist in "vaccination, treatment, examination, testing, decontamination, quarantine, or isolation of any individual as a condition of licensure . . . or the ability to continue to function as a health care provider in this state." In the 2001 version, health care workers, hospitals, and individuals potentially exposed were liable for criminal misdemeanor penalties if they did not abide directives issued by public health authorities to control the disease. In the 2003 proposal, health care workers and hospitals may still suffer consequences for noncompliance, including revocation of state licensure.

Legal organizations and scholars have pointed out that having coercive language in the legislation presupposes noncompliance and flies in the face of a recent

---

[3] Center for Law and the Public's Health at Georgetown and Johns Hopkins Universities. *The Model State Emergency Health Powers Act.* Washington, DC; 2001. (http:www.publichealth. law.net)

OF NOTE...

Model State Public Health Privacy Act (MSPHPA)

Alabama, Arizona, Connecticut, Delaware, Florida, Georgia, Hawaii, Idaho, Illinois, Indiana, Iowa, Louisiana, Maine, Maryland, Minnesota, Missouri, Montana, Nevada, New Hampshire, New Mexico, North Carolina, Oklahoma, Oregon, Pennsylvania, Rhode Island, South Carolina, Tennessee, Utah, Vermont, Wisconsin, and Wyoming and the District of Columbia

tradition of voluntary participation in public health control efforts on the part of the public, medical providers, and health care organizations. Most of the evidence in regard to voluntary compliance is to the contrary. Health care organizations, physicians, and other health care providers have a good record of participation so long as the rationale is compelling, the intervention is safe, and protection against liability and adverse consequences are unambiguous. These critics argue that coercive language could backfire and complicate rather than promote the efficient implementation of emergency public health control measures. Since the release of the model legislation, forty-four states have either introduced legislative proposals or adopted provisions of the act in their entirety or in part. At the time of this writing, thirty-three states have passed legislation based on the model legislation.

Other provisions of the Model State Public Health Privacy Act of interest to health care practitioners are those sections addressing privacy issues, use of "identifiable health information," medicolegal liability, and compensation. Briefly, the act enables public health agencies and officials to acquire, use, disclose, and store identifiable health information when it is in the interest of protecting the health of the public; however, it also proscribes inappropriate and unnecessary uses of this information, including criminal penalties for the illegitimate use of this information. In regards to liability, public health authorities as well as nongovernmental persons or organizations acting under the provisions of the act are exempted from liability for injuries caused to individuals or organizations as a result of actions taken to address the public health emergency. The only exceptions are acts of gross negligence or misconduct. Compensation for injuries caused to individuals or organizations by public health authorities, or designated agents of the public health authority, is addressed in the act, but these sections lack specifics and largely defer to existing state laws in regards to determining whether compensation is due and at what level.

Quarantine Versus Isolation: What's the Difference and Does It Matter?

Quarantine and isolation are used loosely and interchangeably, but this is inappropriate. Clinicians should be clear on the distinction between the two. Quarantine refers to *compulsory physical separation* of nonsick individuals who have been exposed to a contagious disease. The term itself comes from the Latin *quarante* denoting the forty days used traditionally by medieval Italian port cities to keep trading ships anchored in the harbor to prevent the spread of bubonic plague. Isolation, in contrast, refers to the separation or confinement of individuals who are known to be infected—or presumed to be infected—in order to protect those not yet infected. Quarantine, a highly restrictive public health measure, is rarely imposed; whereas isolation is an infection control practice familiar to clinicians and implemented commonly in hospitals. In the setting of an imposed quarantine, however, sick individuals may be isolated from the rest of the quarantined population. Further, public health measures short of quarantine may include such things as bidirectional travel restrictions (no one comes in and no one leaves) as well as restrictions on public gatherings such as school and work.

The imposition of quarantine brings with it many unintended but well-described consequences, ranging from accelerating disease transmission within a quarantined community to acts of individual and group resistance, or even outright violence. Equally destructive to the goal of protecting the public health is that the imposition of quarantine has too often been driven by social and ethnic prejudice rather than by scientific reasoning, an historical reality demonstrated repeatedly by historians of medicine and public health. Even if implemented on scientifically defensible grounds and in an equitable and nonbiased fashion, quarantines are likely to evoke public panic, followed by resentment, resistance, and efforts to bypass the restrictions. Many infectious disease epidemics—from the Black Death of 13th-century Europe to the 1918 Influenza Pandemic—elicited quarantine measures that had enormous social, economic, political, and public health consequences for the communities held hostage to the ravages of the disease. There are more than a few examples where efforts to implement quarantine measures have evoked public resistance, including threats, injuries, and an infamous episode in 1912 where a local Missourian public health official charged with maintaining quarantine measures during a smallpox epidemic was murdered.

Given the practical and metaphoric high stakes involved whenever quarantine is being considered as a public health measure, legal and ethical experts have articulated some of the issues that should inform such a decision. Risk assessment forms the basis for determining whether quarantine is justified or not. Bluntly stated, a biological agent may be dangerous without being communicable, and without person-to-person transmission the primary aim of quarantine becomes moot. For example, some Category A agents—anthrax, tularemia, botulinum toxin—are not transmitted from an infected to a noninfected person. Similarly, victims of chemical and nuclear agents may pose a risk to those who come into direct contact with them if proper decontamination has not occurred, but quarantine is an ineffective public health measure in these situations. Using this reasoning, quarantine could be justified with smallpox because the disease is highly contagious. Conversely, bioterrorist attacks involving anthrax, tularemia, or ricin would not meet this criteria because they are not spread from person-to-person. A second level of consideration relating to the implementation of quarantine is whether the resources exist to impose and maintain it. Answering this question requires knowing not only the properties of the bioterrorist agent (e.g., disease incubation), but also the population at risk for infection, which must be defined unambiguously. If a disease has a long incubation period, the mobility of most modern societies makes it difficult to know just how wide a quarantine perimeter must be established. Whether one has a sufficient number of appropriately trained and protected policing personnel—without which maintaining an effective quarantine would be impossible—is another resource question that must inform any decision regarding quarantine. Last, decision makers should ask themselves whether the benefit of quarantine outweighs the likely adverse social, economic, psychological, and potentially even medical consequences such a declaration would entail. In other words, less draconian public health measures might achieve the same objectives as quarantine with far fewer unintended consequences. Possibilities for these less restrictive measures include voluntary limits on public gatherings, travel, or mass transportation; the imposition of curfews, distribution of masks, and rapid and sharing of reliable, consistent information through public service announcements; and improved communication between organizations responsible for maintaining order and public safety, such as local and state law enforcement, health care, and public health officials. As described in Chapter 3, the latter effort is well underway. In all probability, a range of options can and should be considered—short of quarantine—to handle most bioterrorism events. The MSPHPA includes a number of provisions relating to infection control measures, including quarantine, isolation, and compulsory testing. The act also specifies the protections offered to individuals and health care providers affected by the implementation of these public health control measures (Table 9–1).

Table 9–1 Limited Public Health Measures to Control Epidemics

Voluntary limits on travel, mass gatherings, mass transportation

Voluntary or mandatory curfews

Distribution of PPE

Rapid information sharing infectious disease control measures

Rapid bidirectional communication between agencies responsible for law and order and public safety

Bioterrorism Preparedness and Medicolegal Liability

Liability issues relating to treatment and prevention of bioterrorism is an area that has generated a substantial amount of concern in the health care community. Recently this issue arose in the context of adverse reactions to prophylactic vaccination against smallpox vaccine, although the anthrax vaccine has raised similar concerns. These vaccines, as well as other devices or treatments, are often subsumed under the name of "countermeasures." Countermeasures are considered vaccines or drugs used to treat complications of vaccinia vaccination, such as vaccinia immune globulin (VIG). Liability is an issue that matters significantly to health care organizations, individuals, physicians and physician groups, as well as to the manufacturers of these countermeasures. It is not clear to what extent this term covers "countermeasures" used for non-smallpox bioterrorism-related preparedness. The following section considers briefly the liability concerns raised during the smallpox vaccination program in order to highlight several of the more pertinent medicolegal liability issues presented by the nation's preparedness efforts.

At present, guidance relating to liability and bioterrorism preparedness is found within the Homeland Security Act (HSA) that Congress passed into law in November 2002. The MSPHPA described earlier also contains language addressing the medicolegal liability under the provisions of the act as well as compensation for injuries caused by the use of the act's provisions. Because the HSA is an act of Congress, the legislative language used in its liability and compensation features is summarized next.

The HSA was developed in response to the events of September 11 in order to focus local, state, and federal government efforts on prevention, management, and recovery from further terrorist attacks. Section 304 of the bill deals specifically with bioterrorism and the preparedness and emergency response to such events. Included are clauses regarding smallpox vaccination, as well as liability of those involved in its administration. These points were included to allay concerns of health care providers about potential adverse side effects of the vaccinia vaccine. Although Section 304 of the HSA aimed to alleviate many of these concerns, as becomes clear in the subsequent discussion, not all potential claims and lawsuits are protected under the statute and will still be governed by applicable state laws. It should also be said that the liability and compensation provisions addressed within Section 304 are limited to smallpox countermeasures. It does not address explicitly medical interventions regarding other bioterrorism preparedness efforts and planning efforts, or actual bioterrorist events. For example, adverse reactions to antibiotics used to treat a victim of a BCN attack, other vaccines, or poor outcomes from supportive measures are not addressed by the HSA, although as noted earlier provisions of the Model Public Health Act do address these issues.

Section 304 of the HSA provides specific protections to "covered" individuals and organizations directly participating in the smallpox vaccination program. Claims must be brought against the United States rather than against covered individuals or entities. "Covered" is inclusive of individuals and organizations—such as hospitals, clinics, and health care entities—under whose auspices specific "countermeasures" are administered. Also covered under these provisions are licensed health care professionals and other individuals licensed to administer smallpox countermeasures, employees or agents of these covered entities, and individuals who are vaccinated but not employed by a nonparticipating entity. Finally, vaccine makers and the distributors of countermeasures are also protected from direct liability. Protection under Section 304 should also be afforded to hospitals that volunteer to have their staff members vaccinated but may elect to send them offsite for the vaccine (Fig. 9–1).

Coverage both for physicians who have hospital privileges but are not hospital employees and for the hospital should also be made explicit under the act's provisions, although this remains a somewhat murky area. It is worth noting that coverage under the HSA is provided while acting within the scope of employment, potentially limiting protection for vaccinated persons who inadvertently spread the infection outside the hospital setting. The HSA was originally interpreted not to extend compensation or liability protection to co-workers, patients, or household contacts of vaccinees who are injured as a consequence of exposure to a vaccinee. Most legal authorities believe that secondary contacts who contract vaccinia may apply for compensation under the Federal Tort Claims Act (FTCA).

Section 304 does not establish a compensation program that pays for any injury. Such claims must be brought under the FTCA. An FTCA claim can be brought by individuals who received a countermeasure as part of the Stage I vaccination program or by those who themselves did not receive a countermeasure but were injured as result of living with or exposure to someone who did receive the countermeasure. For example, the provision affords some protection to a family member who was inadvertently inoculated with vaccinia by a recent vaccinee. To prevail under the FTCA, a claimant must prove "gross negligence, recklessness, illegal conduct, or willful misconduct" by those individuals or institutions responsible for administering the vaccine, and submit the claim within 2 years. If a claimant wins compensation from the government because the person who administered the vaccine acted with gross negligence, recklessness, illegal conduct, or willful misconduct, the government can recover the costs of the suit from the health care provider. If no action occurs within 6 months of a claim, or if the claim is denied, the claimant may then take the claim to a federal court.

If our nation's bioterrorism preparedness efforts are to succeed, state and federal legislatures will have to expand liability protections and clarify existing areas of uncertainty regarding health, disability, and life insurance even as they claim greater policing authority to prepare for and respond to such events. Clearly in the event of a bioterrorist event or as part of expanded preparedness efforts over time, it is in the nation's best interest to clarify all liability and compensation issues equitably and unambiguously.

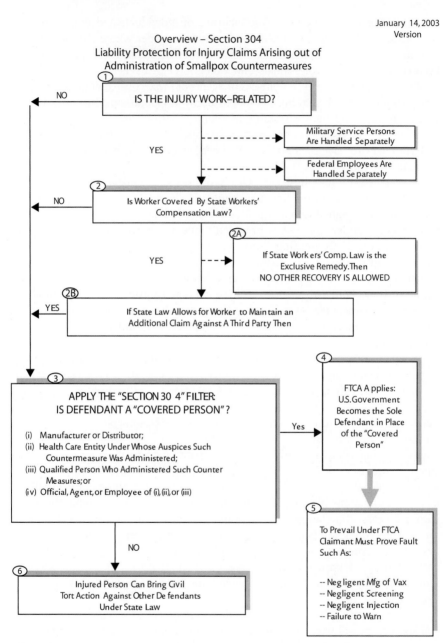

January 14, 2003
Version

FIGURE 9–1   Overview—Section 304 Liability Protection for Injury Claims Arising Out of Administration of Smallpox Countermeasures.

Courtesy of the CDC.

Workers' Compensation

Federal employees may only file claims pursuant to the Federal Employees' Compensation Act. For all other employees, HSA Section 304 does not bar pursuing claims in more than one jurisdiction, for example, workers' compensation. A minority of states have passed legislation making workers' compensation an exclusive remedy, thereby prohibiting an employee from also bringing a lawsuit against a covered entity. However, it is important to note that the HSA itself does not preclude filing a claim under both the FTCA and state workers' compensation programs.

The Stage I preevent smallpox vaccination campaign was a voluntary program. As a result, concerns were raised that vaccine-related injuries might not be covered under state workers' compensation statutes. During one of the vaccination clinics held in the state of Connecticut, every single physician who had signed up to be vaccinated declined to undergo vaccination because a memorandum received the previous night from their hospital's human resources department stated that the hospital's workers' compensation carrier had advised the hospital that they interpreted the voluntary nature of the Stage I vaccination program to mean that injured vaccinees were not eligible for workers' compensation coverage. Fortunately, the majority of legal experts believe that employees who develop complications would be compensated. Moreover, it is difficult to imagine a successful contestation of any claims in which well-described, vaccine-associated complication occurred. All the same, workers' compensation is a state-based system, so that employers and employees who elect to participate in vaccination programs should obtain specific guidance from their state workers' compensation agency or human resources or legal departments.

Another workers' compensation issue relates not to liability but to the extent of coverage. Whereas medical coverage for work-related illness and injury has no limits, workers' compensation does not provide full coverage for lost wages, precludes recovery for pain and suffering or impact on others, and prohibits negligence tort suits against employers. Most workers' compensation statutes also establish limits on the duration of payments and the maximum payments allowable. For many health care professionals, the maximum payment is far below their usual income. For example, workers' compensation payments would not provide adequate income protection for most physicians in the event of serious life- or career-threatening vaccine complications. Individuals considering vaccination should specifically ask their employer or personal disability insurer whether vaccine-related disability is covered.

Health Insurance Coverage

That the smallpox vaccination program targeted health care workers and was voluntary, created ambiguity with regard to health insurance coverage. In response to an inquiry from one participating health care organization, a major health insurer indicated that it would *not* cover health care costs associated with vaccination in vaccinees because such costs were assumed to be work-related. Individual states reacted to this ambiguity by passing statutes to protect vaccinees. For example, the governor of Connecticut at the time forwarded a bill to the legislature providing explicit workers' compensation protection for employees with vaccine-related illness and injury and extended workers' compensation statutes to individuals not covered by workers' compensation (e.g., contract or per diem nurses). The same bill also identified the failure to cover adverse events related to smallpox vaccination as an "unfair" insurance practice. Close household contacts or family members who are inadvertently inoculated and experience complications, and those who have health insurance coverage are also included in the coverage afforded by the bill. In addition, the bill would preserve the right of individuals to make a claim under the FTCA for medical bills, lost wages, pain and suffering, and impact on others. This bill provides health benefits to the public and to medical response team members injured as a result of receiving the smallpox vaccine. The program is analogous to compensation programs currently available to police officers and firefighters injured in the line of duty.

Relevant details of the HSA proposal are that it would be administered by the HHS and that compensation would be retroactive to cover individuals who have been recently vaccinated. There are four key elements of the legislation. The HSA provides approximately $262,000 for permanent and total disability caused by the administration of the vaccine. This payment is made regardless of other benefits available to the individual through personal disability insurance or workers' compensation. A $262,000 death benefit for deaths caused by administration of the vaccine is also paid regardless of other death benefits available to the individual. The bill also provides for temporary or partial disability benefits to compensate individuals for lost wages up to a maximum of $50,000. This sum is in addition to any workers' compensation or disability insurance benefits that might be available to the individual. The final provision offers full coverage for reasonable out-of-pocket medical expenses for major injuries, secondary to any health insurance benefit that might be available to the individual. The HHS program compensates third parties who contract smallpox from public health and medical response team workers who have been vaccinated. In the end, this legislation went a long way toward answering some of the most pressing concerns relating to compensation, liability, and insurance coverage for smallpox vaccination, but not all bioterrorism countermeasures.

Risk Management Strategies

Health care organizations already have extensive experience with a number of mandatory OSHA standards addressing various occupational health issues, including infectious diseases threats such as BBPs or tuberculosis. Incorporating a similar systems-based approach to employee bioterrorism preparedness is well within the capacity of most organizations and is unlikely to represent an unacceptable financial or administrative burden. Based on cumulative experience with the recent smallpox vaccination campaign, some guidance can be offered to physician and health care workers who may be recruited to work on these issues.

Health care risk managers should be proactive and keep abreast of the rapidly changing medical–legal issues related to bioterrorism preparedness, including preevent smallpox vaccination. Sound strategies to include in any organizational approach are identifying someone in the legal or human resources department who accesses regularly the American Health Association and CDC Web sites for current recommendations relating to legal and liability issues. Organizing a multidisciplinary team to decide whether your organization will participate in any voluntary program is an appropriate first step, followed by a systems-based approach to implementing bioterrorism preparedness programs. Although these steps may not prevent future medicolegal challenges or even liability, certainly the absence of these advanced planning efforts will make such challenges and liability more problematic to defend against.

Organizations should offer comprehensive education to all staff involved in providing—as well as receiving—vaccination and bioterrorism preparedness training. All such programs and materials should include a lucid and unambiguous discussion of medical–legal and insurance issues, highlighting any areas where the individual employee may inadvertently be assuming risks without being fully aware of the potential downstream consequences. In regards to vaccination, or other countermeasures that may emerge in the coming years, organizations should develop policies and procedures for candidate selection as well as informed consent policies and forms. When in doubt, it is better to delay vaccination or voluntary participation until any uncertainties are addressed. Implementing appropriate infection control procedures as part of the preparedness efforts is necessary.

Organizations that implement selective furlough policies for employees who have volunteered for these activities and programs, particularly those who work in high-risk areas, such as organ transplant units or pediatric intensive care units, would provide a measure of job protection. Few individuals experienced major side effects during the recent smallpox vaccination program, and fortunately many of the early concerns that the program would cause loss of employee productivity and entail significant adverse events in vaccinees were not realized. However, organizations should be prepared to address each of these issues in advance of any bioterrorism event so that employees and their families will have unambiguous guidance as to what protections exist in regards to salary, benefits, and liability.

Individual clinicians, too, share responsibility in regards to risk management whether they work in their own private offices or they are part of larger organizations. Maintenance of professional competency is an issue that is gaining increasing attention in the United States, driven in part by increasing public demands for physician accountability and the related issues of litigation and malpractice. No one would argue that there is a professional duty to maintain competence although the best means by which to achieve, regulate, or monitor this goal remains an area of heated controversy. For our purposes, it is reasonable to remind clinicians that even in medical or public health emergencies, they are bound by the same professional obligations and remain potentially liable for any failure to meet standards of care. Keeping abreast of the medical, public health, and systems-based approaches to bioterrorism preparedness and planning is a difficult task, but clinicians will be expected to make reasonable efforts to do so. In the clinical setting, clinicians are still responsible for taking an appropriately detailed history, examining their patients, ordering necessary ancillary laboratory or radiologic tests, and most importantly, documenting their clinical encounter as fully as possible. Even in the setting of mass casualties, where difficult life or death choices may have to be made and where liability cannot be precluded a priori, clinicians are well advised to approach their duties with the same diligence that they would in a nonemergency situation.

Summary

Medicolegal and ethical issues relating to bioterrorism are complex, evolving, and contentious. Legislation has been introduced or passed in nearly all states to address the perceived limitations of existing public health statutes. Although there is a general consensus that these efforts are long overdue, the specifics of the various laws being considered or implemented vary from state to state. Further, concern has been raised regarding the scope of these legislative efforts, in particular some of the more coercive aspects of the laws that require participation by health care organizations and health care workers in the event of a declared public health emergency. The bottom line for clinicians is that the authority of public health agencies in most states has been vastly expanded in order to address the issue of bioterrorism. Clinicians and hospitals may, in essence, be drafted into participating in public health control measures and if they defer may suffer sanctions relating to licensure. The good news is that clinicians and health care organizations acting under the direction of public health authorities are immune from civil liability and are (probably) eligible for compensation should they, too, become injured as a result of this involvement.

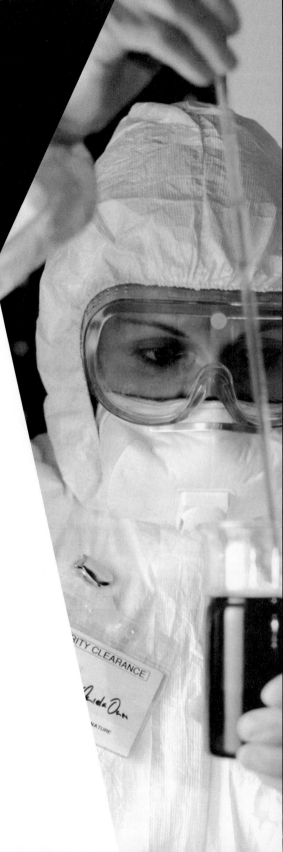

SECTION II

Infectious Agents

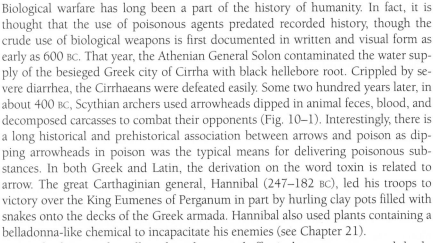

CHAPTER 10

A Brief History of Biological Weapons

Biological warfare has long been a part of the history of humanity. In fact, it is thought that the use of poisonous agents predated recorded history, though the crude use of biological weapons is first documented in written and visual form as early as 600 BC. That year, the Athenian General Solon contaminated the water supply of the besieged Greek city of Cirrha with black hellebore root. Crippled by severe diarrhea, the Cirrhaeans were defeated easily. Some two hundred years later, in about 400 BC, Scythian archers used arrowheads dipped in animal feces, blood, and decomposed carcasses to combat their opponents (Fig. 10–1). Interestingly, there is a long historical and prehistorical association between arrows and poison as dipping arrowheads in poison was the typical means for delivering poisonous substances. In both Greek and Latin, the derivation on the word toxin is related to arrow. The great Carthaginian general, Hannibal (247–182 BC), led his troops to victory over the King Eumenes of Perganum in part by hurling clay pots filled with snakes onto the decks of the Greek armada. Hannibal also used plants containing a belladonna-like chemical to incapacitate his enemies (see Chapter 21).

Both plague and smallpox have been used effectively to sow terror and death. During the Middle Ages, attacking armies sometimes catapulted the corpses of plague victims into besieged cities. In 1346, for example, the city of Kaffa became pestilent following a Tartar attack using this early form of biological warfare. During the French and Indian Wars in colonial America, Lord Jeffrey Amherst ordered that smallpox-contaminated blankets be sent to Native American tribes allied with the French. This type of practice became so common and so effective a military tactic that during the Revolutionary War, General Washington ordered variolation of the

FIGURE 10-1 **Solon of Athens, credited with using black hellebore root to debilitate and thus defeat the Cirrhaeans.**
Courtesy of Architect of the Capitol.

entire Continental army. Variolation required immunization with live vaccines taken from lesions of smallpox victims. Although effective, variolation also caused disease at a rate of about one for every 2,000 immunized individuals (Fig. 10–2).

World War I is remembered more for the utilization of chemical weapons than for that of biological ones; however, anthrax was used extensively to disrupt economic and political life by targeting enemy livestock. The brutal consequences of chemical and biological warfare during the Great War galled the international community into agreeing to the first multilateral agreement to ban the use of biological and chemical weapons in 1925, the so-called Geneva Protocol.

During World War II (WWII), improved scientific understanding of microbes and microbiology spurred the creation of even more efficacious biological weapons. The Japanese military engaged in an aggressive biological weapons program from the 1930s to the end of WWII. Unit 731 (Fig. 10–3), as the notorious program was known, employed over 3,000 scientists and technicians at its peak. Japanese airplanes dropped plague-infected rice and fleas across mainland China on at least 11 different occasions as part of its effort to conquer that nation. During the war, Unit 731 conducted inhumane experiments on Chinese, Korean, Mongolian, Soviet, American,

FIGURE 10-2   A political cartoon critical of smallpox vaccination of the Continental army.

Courtesy of the National Library of Medicine.

FIGURE 10-3   All that remains of the Unit 731 Crematorium.

Courtesy of Advocacy and Intelligence Index for POW-MIAs, Inc.

British, and Australian prisoners of war. The prisoners are known to have been experimented on with biological weapons such as anthrax, botulism, brucellosis, cholera, dysentery, gas gangrene, meningococcal infection, and plague. All told, over 1,000 prisoners were killed during these experiments, as documented during the war tribunals of the mid- and late 1940s. Japan was by no means alone in their efforts to develop bioweapons; nearly every major industrial power was attempting to develop such weapons. Fearing Germany's use of aerial anthrax, England tested its own version on sheep herds on several islands off the coast of Scotland. Later, they stated the intended use of such a bomb was as a retaliatory measure only.

In the United States, the biological weapons development began in force after WWII. The development, done in secret, was led by George W. Merck, of Merck Pharmaceuticals and the well-known *Merck Manual*. War crime charges against the Japanese scientists who had developed and tested the biological agents on prisoners of war (POWs) were ultimately dropped in exchange for technical help with development of the U.S. program. The focus of the research was the weaponizing of those agents that later would come to be classified by the CDC as Category A agents. As the Cold War developed, another major focus was the development of molds and bacteria intended for Soviet wheat crops. The purpose, of course, was to destroy their agricultural base and cause food shortages and economic strife. Recognizing that the Soviets were developing similar weapons, the U.S. Army ran experiments to assess American vulnerability to bioweapons attack by simulating these attacks on major cities such as New York, Saint Louis, and San Francisco, using "harmless" pathogens. After one such test in which Serratia marcesens was dispersed in and around San Francisco, eleven people became ill and one died. The government claimed it was merely coincidence. The U.S. military also wanted to assess the feasibility of using bioweapons against the Soviets and so ran simulations in Alaska because it best simulated the climatic and landscape conditions of the Soviet Union. Similar testing was done in Okinawa to determine feasibility for use in Southeast Asia.

In the mid-1950s, the U.S. bioweapons program shifted toward research and development on viral pathogens. From a strategic perspective, viruses held several advantages over bacterial and fungal pathogens. Viruses are impervious to traditional antibiotic therapy—a major obstacle in ensuring a successful attack with bacterial agents. Viral vaccines often require weeks or months to develop, limiting their usefulness in protecting soldiers and populations. There are a far greater number of viruses to choose from, and in fact, the U.S. government conducted research on some fifty viral agents compared with only sixteen bacterial agents. In the modern era, experts cite forty-three viral agents, nineteen bacterial, four rickettsial, and fourteen biotoxins in the arsenal of viable biological weapons. Finally, this paradigm shift was driven by the desire to promote public and political support for bioweapons research. Bioweapons producing targeted incapacitation, rather than indiscriminate death, became a more defensible goal. Today, advances in therapeutics and molecular and genetic engineering have obviated some of the theoretical advantages of viral bioweapons.

Another military concern influencing the direction of bioweapons research was the unpredictability of infectious agents. The risk of infecting our own troops was genuine. Agents that could not be transmitted person-to-person were more controllable and more strategically valuable. The exception to this rule was smallpox, where research continued throughout this period. Officially, the U.S. government denied smallpox was part of the biological arsenal, claiming it made an unsuitable weapon. In reality it was intended for "special actions" and the Central Intelligence Agency (CIA) maintained its own smallpox stock to be used at its discretion in its special operations. During the 1960s, the U.S. government continued research into weaponized smallpox while participating in efforts to eradicate smallpox from the globe. Smallpox's "useful" qualities were noted by other governments and organized nongovernment groups as well. As early as the mid-1960s, in fact, several federal agencies recognized smallpox's potential in a bioterrorist attack against the United States.

Breakthroughs in molecular biology and recombinant DNA technology altered the landscape of bioweapons research in the latter decades of the 20th century. Molecular biology was poised for the great leap forward into gene manipulation which would inaugurate the development of more medicines and greater understanding of the science of life. However, these technologies can be applied as easily to bioweapons research and development, allowing for the development of "super bugs" with enhanced virulence and decreased susceptibility to vaccines or antibiotics. The availability and increasing sophistication of current technology complicates enormously our ability to mount an effective medical and public health response.

Setting the Stage

The current threat of bioterrorism is rooted in the economic and political landscape of the Cold War. The peak of bioweapons development in the United States occurred in the 1960s with as many as thirty-five hundred people working on research and development. In 1969, President Nixon joined Great Britain and the Soviet Union in proposing a ban on continued bioweapons development as well as the destruction of all stockpiled weapons. The stated reason was that the viability of such weapons was minimal, but a major motivation was the realization that such weapons could be readily developed or obtained by potential enemies of the superpowers. The ease of development and relative low costs associated with bioweaponry meant that many nations could develop bioweapons programs, and many in fact did. This was in stark contrast to nuclear weapons that require immense financial and technological resources available to few countries. For these reasons, biological weapons have been referred to as the "poor man's" atomic bomb.

Nixon mandated that a new lab be designated in which small quantities of these agents could be maintained for developing adequate protective countermeasures, diagnostic procedures, and therapeutics in preparation for bioweapons attacks. The new lab was named. USAMRIID (Fig. 10–4) is housed at Fort Detrick, Maryland. From this time forward, the U.S. biological weapons program was confined to research

FIGURE 10–4 The U.S. institution primarily focused on biodefense.
Courtesy of USAMRIID.

on strictly defined measures of defense, such as immunizations. The Department of Defense (DoD) was ordered to draw up a plan for the disposal of existing stocks of biological agents and weapons. Under President Nixon, the United States made a unilateral decision to forego the first use of chemical weapons and renounced biowarfare unconditionally.

Evolving international positions on biological warfare—particularly the United States, Great Britain, and the USSR—gave impetus to an influential United Nations (UN) document: The Convention on the Prohibition of the Development, Production, and Stockpiling of Bacteriological and Toxin Weapons and Their Destruction, usually referred to as the 1972 Biological and Toxin Weapons Convention (BTWC). The BTWC was written with the purpose of stopping the development of biological agents as weapons and to ensure the destruction of stockpiled weapons internationally. It continues to be the premiere document steering international law on this subject and has, to date, been signed by 144 countries, including all the permanent members of the UN Security Council. The BTWC specifies that no nation is to:

> . . . produce, stockpile, or otherwise acquire or retain microbial or other biological agents or toxins, whatever their origin or method of production, of types and in quantities that have no justification for prophylactic, protective, or other peaceful purposes, and weapons, equipment, or means of delivery designed to use such agents or toxins for hostile purposes or in armed conflict.

Not surprisingly, there have been great difficulties in ensuring the BTWC is enforced. Every signatory nation is bound to submit a list of all bioweapons facilities, to list all meetings held at the facilities, and to provide an exchange of information on biological warfare agents as well as on any disease outbreaks. However, the BTWC treaty contains no provisions for oversight or enforcement of these guidelines by the UN Security Council or by neutral nations rendering it of questionable value in truly limiting bioweapons research and development.

Much of the problem lies in the unclear distinction between military and legitimate public or corporate development. Thirty years after the signing of the BTWC, such ambiguities are still being debated. Nations engaged in the production of biological (and chemical) weapons have many opportunities to sell or trade them internationally. The first documented occurrence of bioweapons being given to a third party occurred in the 1980s when the Soviet Union gave mycotoxins to the communist governments of Vietnam and Laos for use against CIA-supported resistance movements. Mycotoxins, derived from fungi, are known to be mutagenic, teratogenic, and carcinogenic. These agents are believed to have been dispersed from crop dusters over Southeast Asian villages and cities. International relief workers witnessed the characteristic "Yellow Rains" and the subsequent elevations in distinctive morbidity and mortality patterns in the affected population. These reports prompted U.S. accusations that the Soviets were in violation of the BTWC. Soil samples positive for the mycotoxin, recovered documents, confessions by local authorities, and other evidence were dismissed at the time by the Soviets who pointed out that mycotoxins were indigenous to these regions. Doubt remained in the West until a top Soviet biologist working in their bioweapons program acknowledged the transgression in the late 1980s.

Despite being a signatory to the BTWC, the Soviet Union actually expanded their research into both biological and chemical weapons throughout the 1970s and 1980s. At its zenith, the Soviet program, known as Biopreparat (Fig. 10–5), employed some sixty thousand individuals in fifty facilities. By way of comparison, the Soviet program was nearly twenty times the size of the U.S. program at its respective peak. Experts have expressed the view that the Soviet government used the 1972 BTWC as an opportunity to gain an advantage over its Cold War foes.

The U.S. and worldwide intelligence communities were deeply suspicious that the Soviets were actively developing bioweapons, but it took an accident to confirm what many had long believed. In 1979, an accidental release of weaponized anthrax from a bioweapons research facility in Sverdlovsk caused fatal inhalational anthrax in 70 people. The Soviet government denied any violations of the BTWC, claiming that the infections came from contaminated meat and accused the West of anti-Soviet propaganda. In 1992, however, the Russian government admitted its culpability. Biopreparat continues to operate, although it is thought to have been scaled down considerably.

In 1989, the United States and United Kingdom teamed up in an effort to force the closure of Biopreparat. Inspection teams were sent periodically into Russia from 1989 to 1994. In 1994, President Yeltsin decreed that no further offensive work

FIGURE 10–5   One of many sites of Biopreparat, the former Soviet bioweapons facility.

would be done. Although Biopreparat was maintained after the fall of the Soviet Union, many of the bioweapons facilities were left in disrepair, and many of the researchers who had once been well rewarded by the Soviet government were suddenly without a job. There were—and continue to be—grave concerns that many of these scientists could agree to work for "rogue" countries, extremist or paramilitary groups to help them develop biological weapons programs or provide them with stockpiled biological agents. In particular, it is feared that smallpox may already have been smuggled out of Russia or other former Soviet states into the hands of anti-Western governments or terrorist groups. Many feel that this is the greatest threat to keeping bioterrorism out of the hands of those who might use them. This threat seems ever greater as Russian laboratories experience ever worsening financial difficulties, emigration of substantial numbers of scientists, and increasingly lax security. Which countries and groups have actually hired these scientists is unknown, but it is believed that governments in Libya, Iran, Syria, Iraq, and North Korea have been actively recruiting former Soviet bioweapons scientists.

In order to help ensure peaceful use of technology and resources, Western funding has been sent to Russia to provide bioweapons scientists with financial alternatives to accepting jobs from potential enemy states or terrorist organizations. Analyses by numerous nongovernmental agencies estimated that the number of countries developing weaponized biological agents is fourteen, most of these are in Asia, North Africa, or the Middle East. Excluding China, the majority are "developing" nations and few are open societies, including China.

According to the National Defense University, there have been more than one hundred documented cases of biological agents as weapons. Of these, nineteen

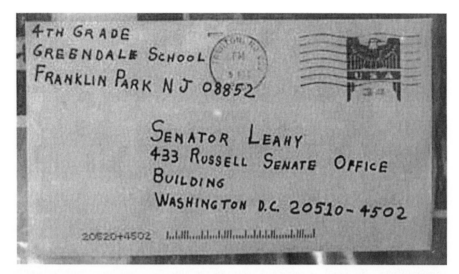

FIGURE 10-6 One of the letters containing anthrax that was found in the Capital building.
Courtesy of the FBI.

FIGURE 10-7 Drums filled with contaminated mail from the Capitol building.
Courtesy of the FBI.

were used by nongovernmental entities for "biocrimes." In the months following September 11, the U.S. Postal Service anthrax attacks awakened the nation and the world to the relative ease with which a motivated individual or individuals could spread both infective agents and fear. All told there were twenty-two cases of anthrax and five deaths (Figs. 10–6 and 10–7). Although the American public found it hard to conceive of how such things could happen, the DoD was all too clear how such a thing could happen after the "success" of a scenario it created: three "nonexperts" were asked by the DoD to determine how quickly and cheaply bioweapons could be developed. In less than thirty days, at a cost of less than $1 million, the three created a lethal arsenal of biological weapons capable of wiping out entire cities."*

An attack on the scale of a few individuals rather than whole cities is a macabre bargain. Recently, U.S. Army scientists determined that the post-September 11 anthrax attacks used an anthrax powder that was made using "simple methods, inexpensive equipment and limited expertise." The costs involved in those attacks are thought to be approximately $3,000. Clearly, bioweaponry can be developed without significant financial or scientific resources, making it a legitimate public health threat.

---

*New York Times, September 4, 2001.

CHAPTER 11

Introduction to Biological Agents

More than two years before September 11 and the anthrax attacks that followed, the CDC had been designated by Congress to prepare for the likelihood of a bioterrorism attack. Among numerous other tasks related to this directive, the CDC examined a myriad of bacteria and viruses that are pathogenic to humans and determined which of these might be used as biological weapons. A panel was assembled with experts in public health, infectious disease, military bioweaponry, members of the intelligence communities, and law enforcement to determine the agents that were viable threats as biological weapons (Table 11–1). Four criteria were considered in selecting these agents. As stated by the CDC, the four points were:

- Public health impact using two measures: morbidity and mortality
- Delivery potential to large populations based on two primary considerations: ability to mass produce and distribute a virulent agent and potential for person-to-person transmission of the agent, and the stability of the agent was considered
- Public perception as related to fear and potential civil disruption
- Special public health preparedness needs based on stockpile requirements, enhanced surveillance, or diagnostic needs

Based on these considerations, the working group developed a list of seventeen pathogens or types of pathogens that were felt to pose legitimate threats. These pathogens were subdivided subsequently into Category A, B, or C agents according to the degree and nature of the threat posed (Table 11–2).

TABLE 11–1   CDC Panel Criteria and Weighting[a] Used to Evaluate Potential Biological Threat Agents

Disease	Public Health Impact		Dissemination Potential		Public Perception	Special Preparation	Category
	Disease	Death	P-D[b]	P-P[c]			
Smallpox	+	+ +	+	+ + +	+ + +	+ + +	A
Anthrax	+ +	+ + +	+ + +	0	+ + +	+ + +	A
Plague	+ +	+ + +	+ +	+ +	+ +	+ + +	A
Botulism	+ +	+ + +	+ +	0	+ +	+ + +	A
Tularemia	+ +	+ +	+ +	0	+	+ + +	A
VHF	+ +	+ + +	+	+	+ + +	+ +	A
VE	+ +	+	+	0	+ +	+ +	B
Q fever	+	+	+ +	0	+	+ +	B
Brucellosis	+	+	+ +	0	+	+ +	B
Glanders	+ +	+ + +	+ +	0	0	+ +	B
Melioidosis	+	+	+ +	0	0	+ +	B
Psittacosis	+	+	+ +	0	0	+	B
Ricin toxin	+ +	+ +	+ +	0	0	+ +	B
Typhus	+	+	+ +	0	0	+	B
Cholera	+	+	+ +	+/−	+ + +	+	B
Shigellosis	+	+	+ +	+	+	+	B

[a] Agents were ranked from highest threat (+ + +) to lowest (0).

[b] Potential for production (P) and dissemination (D) in quantities that would affect a large population, based on availability, BSL requirements, most effective route of infection, and environmental stability.

[c] Person-to-person transmissibility.

Courtesy of the CDC.

The Category A agents are sometimes referred to as "the Big Six." These are smallpox, VHF, anthrax, plague, tularemia, and botulinum toxin. These particular agents share characteristics that make them the greatest threat to public health. Specifically, they can cause mass casualties that warrant a large-scale public health preparational response, including implementing surveillance systems, improving and expanding diagnostic methods, and maintaining adequate stores of antibiotics and other pharmaceutical interventions (Table 11–3).

However, there are a number of less virulent or dangerous microorganisms that may be more than, or certainly as adaptable as, bioweapons such as the Big Six. Table 11–4 provides the list of these Categories B and C agents, and they are discussed in more detail in Chapter 19.

TABLE 11–2   Criteria for Categorizing Biological Threats

Category A Diseases/Agents: High-priority agents. These include organisms that pose a risk to national security because they:

Are easily disseminated or transmitted from person to person
Result in high mortality rates and have the potential for major public health impact
Might cause public panic and social disruption
Require special action for public health preparedness

Category B Diseases/Agents: Second highest priority. These include organisms that:

Are moderately easy to disseminate
Result in moderate morbidity rates and low mortality rates
Require specific enhancements of CDC's diagnostic capacity and enhanced disease surveillance

Category C Diseases/Agents: Third highest priority agents. These include emerging pathogens that could be engineered for mass dissemination in the future because of their:

Availability
Ease of production and dissemination
Potential for high morbidity and mortality rates and major health impact

Courtesy of the CDC.

As mentioned, the designation of these groupings is based in part on their effectiveness as a weapon of mass infection, debility, and mortality. An infective weapon, therefore, would be most dangerous if it were easily accessible, highly virulent, highly infective, easily disseminated, easily transmissible, and hardy enough to withstand environmental changes such as sunlight, cold, or arid conditions. While no single agent possesses all of these features, a number of them possess some of the features that justify their designation as Category A agents.

The Bioterrorism Sourcebook provides individual chapters on the Big Six category A agents because of their preeminent importance and greater likelihood of used. These

TABLE 11–3   Category A Agents

Category A
Anthrax (*Bacillus anthracis*)
Plague (*Yersinia pestis*)
Smallpox (variola major)
Tularemia (*Francisella tularensis*)
Viral hemorrhagic fevers
Filoviruses—Ebola, Marburg
Arenaviruses—Lassa, Machupo
Botulism (*Clostridium botulinum* toxin)

Courtesy of the CDC.

TABLE 11–4   Category B and C Agents

Category B

Brucellosis (*Brucella* species)
Epsilon toxin of *Clostridium perfringens*
Food safety threats
  Salmonella species
  Escherichia coli O157:H7
  Shigella
Glanders (*Burkholderia mallei*)
Melioidosis (*Burkholderia pseudomallei*)
Psittacosis (*Chlamydia psittaci*)
Q fever (*Coxiella burnetii*)
Ricin toxin from *Ricinus communis* (castor bean)
Staphylococcal enterotoxin B
Typhus fever (*Rickettsia prowazekii*)
Encephalitis (VE)
Alphaviruses
  Venezuelan equine encephalitis
  Eastern equine encephalitis
  Western equine encephalitis
Water safety threats
  Vibrio cholerae
  Cryptosporidium parvum

Category C

Nipah virus
Hantavirus

Courtesy of the CDC.

agents are the most intensely studied and covered in medical literature; consequently, there is more of consensus on treatment recommendations, response plans, and preventive and safety efforts for these six agents. Based on the criteria such as availability and facility of use, such attention and detail is warranted.

On the other hand, the sixteen or so Category B agents are less likely to be used and are less virulent agents. Consequently, they are less rigorously studied and planned for in the context of an attack than are the Category A agents. Category C agents function more as a general class—infectious agents that fall within the umbrella term *emerging pathogens*. Members of this group, because of their ease of access and use, all have the potential to be used, particularly if they were to be genetically altered to increase their virulence. Although many such agents could be considered as falling within Category C, Hanta virus and Nipah virus are the only ones discussed specifically in *The Bioterrorism Sourcebook*.

The degree of attention given by most experts to Category A agents far outweighs that given to Categories B and C. Many resources on bioterrorism do not even bother to acknowledge these "lesser" categories. In this text, we choose to give considerable attention to the Category A agents, with each receiving its own detailed chapter. Because the other two categories lack equivalent analysis as weapons of terror, they receive less discussion in this text as well and are presented in a more concise format. In addition, although the agents selected by the CDC panel are appropriately chosen, many biological weapons experts believe that other pathogens are just as viable as those in the three categories provided under the CDC classification.

Although the chapters in this text basically follow the delineation of Categories A, B, and C and their respective agents, there is one exception: ricin. Designated as a Category B agent by the CDC, ricin is given its own chapter in the biotoxins section, included among the Category A agents for the purposes of this text; hence it receives a more detailed discussion than the Category B agents are allotted. Although all categories, even those made by the CDC, are to some degree subjective, the decision to "promote" ricin was based on the way it is portrayed in the literature on biological weapons. Ricin's toxicity, availability, and stability makes for an accepted view that it is a viable threat, certainly as viable a threat as some of the Category A agents. Thus, it receives equal status to the Category A agents in this text, thereby turning "the Big Six" into "the Big Seven."

Although the CDC categorization of agents is the standard for biological weapon discourse, other sources and experts have some differences about potential agents that warrant attention. Reviewing the available literature on potential biological weapons, four agents beyond the CDC's list were deemed enough of a threat to warrant inclusion in this text. For more discussion of these agents as well as of the Category B and C agents, see Chapter 20.

CHAPTER 12

Smallpox
(Variola major)

Clinical Vignette

You return to your office from lunch to find your office staff around the TV. A nurse states that CNN is reporting a confirmed case of smallpox in the Chechnyen Republic. You realize immediately that your office will be flooded with calls from anxious patients worried about smallpox, concerned about rashes they've just noticed, and demanding immediate vaccination. You go to the CDC website and confirm indeed that the first case of Variola major in over thirty years has been diagnosed and that a worldwide public health emergency looms. From the CDC website you download their fever rash algorithm and plan to meet with your staff to review the protocol. Within two hours you receive calls from your local health department and the hospital where you have admitting privileges asking you to report immediately for a planning meeting for a townwide mass vaccination clinic.

Background

Variola major is a double-stranded DNA virus of the Orthopox family. There are two recognized forms of smallpox: Variola major and Variola minor. Variola major is the more virulent strain with mortality rates 30% or higher in vaccine-naïve populations. Variola major has historically been more prevalent. Variola minor is considered rather mild with mortality rates less than 1%. Like its close relative, chickenpox, smallpox historically followed seasonal patterns of outbreaks, peaking in the late winter and early spring. Such a pattern is likely secondary to the sensitivity of the aerosol droplets to higher temperatures and humidity. It is not known what sensitivities (or lack thereof) weaponized forms might have. All Category A, B, and C biological weapons occur sporadically in nature with the exception of smallpox. Therefore, barring an extraordinary occupational history (i.e., laboratorian working in a Biosafety Level-4 (BSL-4) facility, diagnosing smallpox equates with diagnosing bioterrorism and constitutes an international medical emergency of the first order.

In what is one of the finest public health achievements of the 20th century, the virus that had killed more people than any other pathogen in the history of humanity was declared eradicated in 1980 thanks to a global vaccination program spearheaded by the WHO (Fig. 12–1). At that time, it was held that nations need not continue vaccination programs. It was not entirely accurate to claim the virus was eradicated as the WHO approved two sites to maintain smallpox: the CDC in Atlanta and the Institute for Viral Preparations in Moscow. Archives unearthed following the collapse of the U.S.S.R. demonstrated that the Soviets engaged in more than simple storage: they were actively developing large quantities and possibly vaccine-resistant strains of the virus that could be fitted to intercontinental ballistic missiles and bombs. The WHO called for both nations to destroy all stored viruses in 1999 and again in 2002; both declined. Reportedly, Russia continues to maintain smallpox research and active development of strains with greater resistance to standard vaccines amid reports of woefully lax security at these labs.

☣ Smallpox and Bioterrorism

A number of factors make smallpox one of the most feared biological warfare agents. There are *numerous* considerations that make an outbreak likely to spread more quickly and with medical and public health consequences far more concerning now than would have been the case even thirty-five years ago.

- Smallpox spreads easily from person to person and is transmissible through aerosols, droplets, or fomites.
- Discontinuation of routine civilian smallpox vaccination in 1972 has left 85% of the U.S. population with no prior immunologic exposure to smallpox or to the vaccinia vaccine used to protect against the disease. Should smallpox be introduced into an immunologically naïve population and clinically naïve medical community, the risk for disease and its transmission would be enormous.
- The loss of clinical familiarity with smallpox is likely to cause delay in recognition *and* diagnosis, implementation of isolation and vaccination efforts, thus allowing more time for the disease to become epidemic.
- The world's population is far more mobile than was the case in the early 1970s, and living conditions worldwide are more crowded.

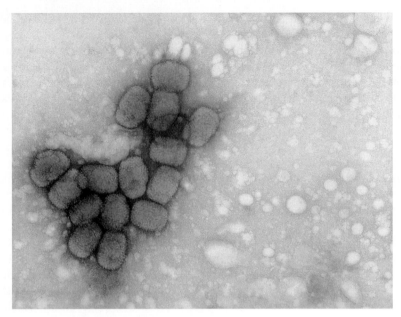

FIGURE 12–1   Electronmicrograph of Variola major.
Courtesy of the CDC.

OF NOTE...

Al-Rhazes: Description of Smallpox

The eruption of the smallpox is preceeded by a continued fever, pain in the back, itching in the nose, and terrors in the sleep. These are the more peculiar symptoms of its approach, especially a pain in the back with fever; then also a pricking which the patient feels all over his body; a fullness of the face, which at times comes and goes; an inflamed color, and vehement redness in both cheeks; a redness of both the eyes, heaviness of the whole body; great uneasiness, the symptoms of which are stretching and yawning; a pain in the throat and chest, with slight difficulty in breathing and cough; a dryness of the breath, thick spittle and hoarseness of the voice; pain and heaviness of the head; inquietude, nausea and anxiety (with this difference that the inquietude, nausea and anxiety are more frequent in the measles than in the smallpox; while on the other hand, the pain in the back is more peculiar to the smallpox than to the measles); heat of the whole body; an inflamed colon; and shining redness, especially an intense redness of the gums.

Al-Rhazes, Persian physician (865–923 AD)

- Finally, new diseases (e.g., HIV/AIDS) and advances in medical care (e.g., cancer treatment and immunosuppressive drugs) mean there are far more immunocompromised individuals at risk for rapidly progressive variola.

In essence, the virulence of smallpox and communicability, along with expanded opportunities for rapid worldwide dissemination and changes in world demographics, culture, and medical standards, offer many advantages to terrorists should they obtain access to the virus.

Countering this more pessimistic view are several considerations. First, many veterans of the smallpox eradication effort, infectious disease specialists, and public health officials believe that an aggressive ring vaccination strategy, or ring vaccination followed by mass vaccination, would contain an epidemic provided multiple concurrent releases did not occur. Second, mortality from smallpox historically may have reflected poor nutritional status, coexisting infectious conditions, inadequate health care infrastructure, slower communications, and the absence of antivirals. Each of these conditions are less prevalent now than in the past and could work in favor of epidemic control.

Pathogenesis

Smallpox virus is usually inhaled and enters the mucosal lining of the oropharynx. It is then taken up by macrophages and transported to regional lymph glands where the virus multiplies, activates cytotoxic T cells and B cells, and provokes a host antibody response. On about the fourth day a subclinical viremia seeds the spleen, bone marrow, and distant lymph nodes. Day 8 signals the start of the clinical phase, heralded by a viremia-associated prodrome of fever and malaise. During this period, the virus infects the dermis and oropharynx causing the classic skin manifestations of smallpox (Fig. 12–2). As the immune response occurs, the virus is taken up by white blood cells and it is transported through the small vessels of the dermis and pharyngeal mucosa where it infects the perivascular tissues (day 14). The classic pitted scarring that is left when the scabs form are secondary to destruction of the sebaceous glands, which shrink and are replaced with granulation tissue and soon thereafter with scar tissue. The lesions in the oropharynx ulcerate the quickest because the tissue there, unlike the dermis, lacks a stratum corneum. As a result of ulcerating, oropharyngeal lesions, the saliva contains an enormous amount of virus.

Smallpox stimulates cytotoxic T-cell responses, neutralizing B-cell antibodies, and the production of interferons. These lost responses restrict viral replication and induce prolonged immunity in the patients who recover. Smallpox is more virulent in infants, the elderly, and immunologically impaired hosts, particularly those with T-cell deficits (see Vulnerable Groups).

Means of Transmission

Patients are most infective from the onset of the rash (days 8–14) until the first scab forms. At this point of infectivity drops precipitously. Infected individuals spread

FIGURE 12-2 Members of a Niger community line up for smallpox vaccination as part of the campaign for global eradication.
Courtesy of the CDC.

smallpox through aerosolized droplets from the oropharynx. Although saliva may be positive for virus up to six days prior to the onset of skin lesions, patients do not transmit the virus until visible lesions form and salivary viral content begins to peak. Infection also occurs through contact with contaminated objects, such as clothes, bedding, and surfaces, or through bodily fluids such as urine, sweat, or sputum. These facts dictate the infection control strategies outlined below. Infectivity among those in contact with a primary case is estimated to be between 40% and 80%.

According to some epidemic models, fifty people initially infected from a bioterrorist attack with smallpox would infect secondarily somewhere between 2,500 to 5,000 individuals. Each successive generation would increase the number of infected individuals 50- to 100-fold. In large part, this exponential expansion is due to the delay between exposure and onset of transmissibility, which occurs when skin lesions erupt.

Sources

Unlike other Category A agents, no naturally occurring smallpox exists. The United States and Russia are known to have smallpox stores. Given the security issues in the former Soviet Union, the possibility exists that the virus has been sold on the black market. It is also possible that heretofore unidentified sources of the virus exist elsewhere in the world, secretly contained in countries such as North Korea, France, and Iran. Other than deliberate release, for which everyone who is not vaccinated would be at risk, laboratorians and research scientists probably represent the only reasonable occupational risk groups. The last known case of smallpox in the West occurred in a lab worker in England, resulting in her death and the suicide of the research scientist whose lapse in technique was responsible for the exposure.

Signs and Symptoms

It is important to bear in mind that there has been no clinical or public health experience with Variola major since the mid-1970s. What is known about the clinical presentation and complications of smallpox draws necessarily on historical experience and that largely from developing nations, which may or may not be generalizable to the United States or other Western nations.

Symptoms generally begin at the end of an incubation period of between one and two weeks. At the end of this period, the patient experiences high fever sometimes with mental status changes, malaise, exhaustion, headaches, backache, and occasionally abdominal pain. A maculopapular rash will appear within three days of prodromal onset, and fever will be noted (though not as high as the prodromal fever). The rash occurs simultaneously on the oropharyngeal mucosa and the face and then spreads outward to the forearms and the legs, relatively sparing the trunk.

The synchronous development of a centrifugally spreading viral exanthum is one of the classic diagnostic clues to smallpox. This is in contrast to another orthopox virus, chickenpox or vaccinia, which begins on the trunk and face and characteristically has lesions in different stages of development. Once the rash appears, it vesiculates and then assumes firm papular pustules over a period of one week. Approximately one week after the rash begins, the lesions umbilicate and then encrustate, leaving smallpox's signature pitted scar. Scabs are not highly infectious, although virus can be recovered from them. Patients are potentially infectious until the last scab falls off with infectivity peaking from rash onset until scab formation. The progression of a symmetric centrifugal homogeneous rash through the four stages of vesiculation, pustulation, umbilication, and encrustation is characteristic of smallpox and nothing else (Figs. 12–3 and 12–4, Table 12–1). Death from smallpox occurs usually around the second week. Encephalitis occurs infrequently, and secondary infection is surprisingly rare. Milder infections are seen in previously immunized patients.

Two significant clinical variants of smallpox occur in approximately 10% of all cases and may result in delayed diagnosis. In both instances, the dermal manifestations differ while the incubation period and prodromal symptoms are unchanged from the classic presentation summarized above. In the hemorrhagic form of Variola major—for which pregnancy appears to be a risk factor—the skin becomes dull and darkly erythematous, but vesiculation and pustulation do not occur. Petechiae and frank mucosal and dermal bleeding are followed by death in virtually 100% of cases, often within 6 days of the onset of clinical symptoms. In the second variation of

Variola major, known as the "flat" type, vesicles never form into the characteristically firm deep dermal papules. Instead, vesicles become confluent, flattened, and soft to the touch. Skin hemorrhages may also be seen. Presuming the patient survives, the lesions of flat smallpox leave no scar after they eventually peel away.

Variola minor presents with similar symptoms to the major form, but the manifestations are less intense, with a milder prodrome and a scattered, albeit typical rash. A similar clinical presentation may be seen in previously vaccinated persons in whom an amnestic response limits both morbidity and mortality.

Microscopy and Laboratory Diagnostic Tests

Collection of viral specimens is extremely important but should be done by a previously vaccinated individual with appropriate PPE including mask, gloves, and gown. Adequate collection requires pustule or vesicle fluids. This may necessitate manually opening a lesion with a sterile instrument and then transferring the fluid directly or soaking it up into sterile material. Because a BSL-4 laboratory is needed to analyze any potential smallpox specimen, samples must be transported in a water-tight, vacuum-sealed container to either a state health department

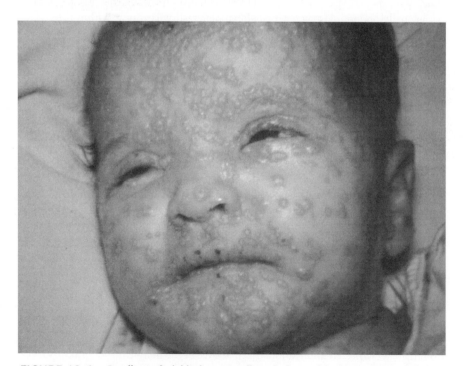

FIGURE 12-3 **Smallpox: facial lesions. Smallpox lesions with characteristic facial concentration. Notice the homogeneity of the pustules.**
Courtesy of the CDC.

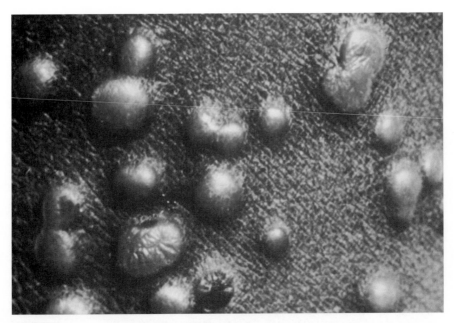

FIGURE 12–4   Smallpox: thigh lesions on the sixth day of rash.
Courtesy of the CDC.

laboratory where it will be sent to the CDC, or to the CDC directly. Establishing a definitively positive result for an orthovirus requires electron microscopy. Determination can also be made through viral culturing in egg chorioallantoic membranes or cell culture. Polymerase chain reaction (PCR) techniques and restriction fragment-length polymorphism are also available at selected LRNs, the CDC, and USAMRIID. Rapid turnaround, usually less than 1 hour, is available

TABLE 12–1   Making the Diagnosis: Smallpox

Pattern of Presentation	Multiple, concurrent acute cases of a febrile prodrome followed by generalized vesicular or pustular rash.
Features	Fever ± headache, abdominal pain, nausea/vomiting, prostration. Must have well-demarcated vesicular or pustular poxlike rash in same stage of development. Rash begins in mouth and face and moves outward to arms and legs.
Findings	Clinical presentation and epidemiology determine diagnosis. Must differentiate from chickenpox. Rapid PCR testing for definitive diagnosis.

both through the CDC and USAMRIID, although efforts to disseminate the technology to LRN Level C laboratories is underway.

Radiographic Findings

There are no key radiographic findings associated with smallpox.

Differential Diagnosis

The differential diagnosis for smallpox includes varicella, disseminated herpes zoster, impetigo, erythema multiform, scabies, and hand, foot, and mouth disease. Hemorrhagic smallpox may be confused with acute leukemia, meningococcemia, drug reactions, or cutaneous syphilis.

Diagnosis

Clinical findings of a synchronous centripetally spreading poxlike rash associated with fever should prompt consideration of smallpox. The CDC has developed a useful "rash fever" algorithm to assist clinicians. The algorithm is available online at the CDC website and is designed to offer a likelihood of a diagnosis.[1] The algorithm is intended for use in the absence of an identified outbreak.

Varicella, or chickenpox's characteristic pox rash, and its relatively high prevalence make it the disease most likely to be mistaken for smallpox (Fig. 12–5). Several distinguishing features are useful (Table 12–2). Varicella tends to have lesions at varying stages of maturation, with new lesions presenting over the course of several days. The lesions tend to be more superficial and are concentrated on the trunk with relative sparing of the extremities, especially the soles of the feet and palms of the hands (Fig. 12–6). Furthermore, children and adults (for the most part) are usually not prostrate when infected with varicella.

In contrast to varicella, the smallpox rash tends to be uniform in its stage of maturation and develops within one to two days. The lesions are firmer and deeper because they are intradermal. Smallpox lesions are concentrated on the face and extremities. As always, the distinction is made easier by taking a thorough history. Identification of an infected contact, for example, makes varicella more likely. Current standards for pediatric immunization include preschool varicella vaccination. If followed universally, this would have the effect of making varicella an increasingly less likely cause of poxlike rashes and fever, although in practice universal vaccination does not occur and varicella will remain in the differential diagnosis. Diagnosis of either the flat or hemorrhagic forms of smallpox is extremely difficult due to the absence of the classic pox lesions.

Chronic Effects

Recovery from smallpox confers lifelong immunity to the disease. Many individuals are left with deeply pitted pock marks. The mechanism for smallpox's characteristic

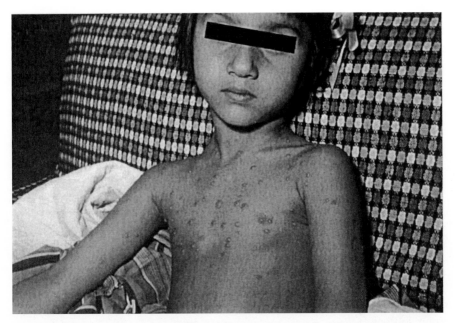

FIGURE 12-5   **Young person with varicella (day 6). Note the heterogeneity of lesions and the predilection for the trunk. This contrasts with smallpox (see text).** *Courtesy of the CDC.*

scar is unknown, but may have to do with destruction of sebaceous glands. Studies indicate that children younger than a year old who survive smallpox experience an unexplained excess of mortality between the ages of fifty and eighty. Ectropion, nasal stenosis, hearing loss, and blindness are also known complications of smallpox. Obstructive oligospermia and possible teratogenic effects are chronic sequelae that have also been described, but the latter is generally not considered likely.

Vulnerable Groups

Unvaccinated individuals are at risk for the disease. Mortality rates for pregnant women (50%) and fetuses are much higher. Early fetal loss and prematurity occur as well. Vaccination is not contraindicated for pregnant women postexposure. Spontaneous abortions, low birth weight, and prematurity have been described during mass vaccinations as well, but data are conflicting. Immunosuppressed or immunocompromised individuals, possibly including diabetics, may experience a rapid and more aggressive form of variola. The body's response to smallpox is through cell-mediated immunity; therefore, individuals with T-cell defects are probably at increased risk for morbidity and complications from both smallpox and the

---

[1] Rash fever algorithm:http://www.bt.cdc.gov/agent/smallpox/diagnosis/riskalgorithm/index.asp

TABLE 12–2 Making the Diagnosis: Chickenpox Vs. Smallpox

Feature	Chickenpox	Smallpox
Prodrome	↓	↑
Confluent lesions	↓	↑
Palmar, plantar involvement	↓	↑
Umbilicated lesion	↓	↑
Initial distribution	Centripetal (hands, face)	Centrifugal (trunk)
Stages of lesions	Variable	Uniform

↓, Low probability of being present.
↑, High probability of being present.

vaccinia vaccine. Individuals with an acute and chronic dermatologic condition—such as atopic dermatitis, psoriasis, or eczema—may experience more severe dermatologic manifestations and subsequent scarring. They are also at greater risk for vaccinia complications.

Treatment

No chemotherapeutic agent is approved to treat smallpox. Medical management is strictly supportive. Cidofovir, an antiretroviral medicine, has shown evidence of activity against orthopox viruses in *in vitro* and in animal studies. No sufficient human clinical data exist as yet to support the use of antivirals in treating smallpox, although it is an area of active research. In the event of an outbreak, cidofovir would be rapidly available from the CDC or NIH under investigational new drug (IND) protocols.

Infection Control

Respiratory isolation in negative pressure rooms with HEPA filtration and strict aerosol and contact precautions are mandatory if smallpox is suspected or diagnosed. Patients are potentially infectious until all scabs fall off. Anyone with direct contact—including household members, health care providers, and medical staff—must be offered immediate vaccination with the vaccinia vaccine. All biowaste and linens from patients with smallpox must be autoclaved and incinerated. Risk of infection is increased with direct face-to-face contact or cohabitation *after the onset of fever* and peaks with rash formation. Secondary contacts—defined as individuals who have had contact with an index case—need not be quarantined, even if infected with variola, because they are not infectious to others until the rash develops. However, secondary contacts must be monitored vigilantly for any signs of fever.

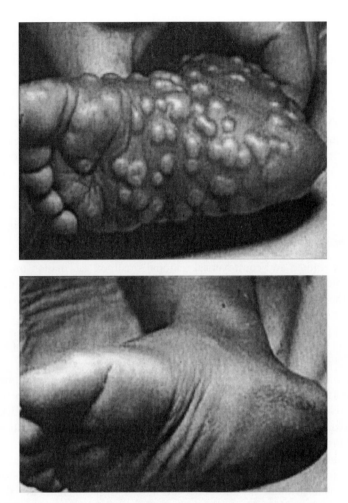

FIGURE 12–6 **Smallpox lesions of the foot (top) versus chickenpox lesions of the foot (bottom). Note the relative absence of plantar involvement in varicella.**
Courtesy of the WHO.

Onset of fever (>100.5°F; 36.2°C) 1 to 3 weeks following a potential smallpox exposure should be presumed to be smallpox and managed accordingly. The issue of inpatient isolation versus home isolation is controversial. Admitting patients to hospitals increases the risk of transmission to health care workers and other immunocompromised patients.

Differences of opinions exist as to whether quarantine of asymptomatic but known smallpox-exposed contacts is medically indicated or prudent from a public health perspective. CDC guidelines recommend that known and suspected cases, as well as febrile contacts, be admitted and isolated in predesignated hospitals, whereas asymptomatic contacts can be quarantined at home. Many experts recommend inpatient isolation only for known cases and febrile contacts be quarantined at home until diagnosis is confirmed. The conflicting recommendation is based on two facts. First, infectivity is primarily an issue only after the lesions appear. Second, in the context of a widespread outbreak, inpatient resources may be strained severely. All states allow for quarantine to be implemented and enforced by designated public health officials if the public's health is deemed to be at risk.

Nonimmune health care workers (HCWs) can work safely with smallpox patients if proper PPE is used. Isolation in a negative-pressure room and use of HEPA filter masks, gowns, gloves, and face shields are indicated. As discussed in the following section, all HCWs in any facility with a known or suspected case of smallpox should immediately be offered vaccination (Table 12–3).

Prophylaxis

Recently, a limited and controversial national smallpox vaccination program was implemented in the United States. Full consideration of this policy decision is beyond the scope of this book; however, the rationale for and experience with the largest mass vaccination effort (over twenty-five thousand people) since the 1977 Swine Flu program is worth some consideration. Further, should a case of smallpox be diagnosed anywhere in the world, mass vaccination of eligible individuals would ensue.

The smallpox vaccine is actually a live attenuated vaccinia virus, a member of the same family as Variola. It is administered intradermally, usually in the upper arm with a bifurcated needle within an area no greater than 5 mm in diameter in the deltoid for primary and secondary vaccinees. Vaccine-naïve patients receive a series of three punctures. If no blood is seen, three more punctures are made (without any additional vaccine). In previously vaccinated individuals, fifteen successive rapidly administered punctures are delivered. If no blood is seen, three more punctures are made (without any additional vaccine). If, in six to eight days, there is no evidence of a "take" (localized erythema or blistering), both groups should receive an additional fifteen successive rapidly administered punctures within one to two centimeters of the original site. If no blood is seen, three more punctures are made

 TABLE 12–3   Critical First Steps: Smallpox

Medical	Confirmatory tests, support, cidofovir (as available)
Infection Control	Aerosol isolation, negative-pressure room, full PPE including HEPA respirator, vaccination of secondary contacts
Public Health	Contact hospital infection control, security, or local law enforcement, and local and state health departments

(without any additional vaccine). A small subset of vaccinees may never have a "take." This may indicate either existing immunity, immunocompromise, or nonimmunity, although most experts believe resistance is probably present.

Within 4 days a pruritic, erythematous nodule appears at the inoculation site. This is followed by clear and then purulent vesicles that scab over and fall off, typically within three weeks (Fig. 12–7). The vaccine site and resultant scab should be kept covered to avoid any direct contact to others or self. Old dressings should be discarded by placing them in well-sealed plastic bags and may be disposed of in ordinary household trash. Partial protection against smallpox is present within two to four days of vaccination.

Although the smallpox vaccine offers 95% protection against a typical strain of Variola major, there are some risks involved. Common side effects include regional adenopathy, and low-grade fever lasting as much as 2 weeks. Less common side effects include generalized vaccinia, eczema vaccinatum, and postvaccine encephalitis. Risk of death from vaccine is estimated at one in one million. Adverse event rates from smallpox vaccination are derived from epidemiologic studies done on vaccinees during the 1960s. Recent military and civilian experience suggests lower adverse event rates, probably due to heightened prevaccination medical screening and strict adherence to infection control practices, since most vaccinees were HCWs or first responders. Cardiomyopathy and myopericarditis were added to the list of potential complications. Seriously ill vaccinees (eczema vaccinatum, progressive vaccinia, and ocular vaccinia) are offered VIG injection available under IND proto-

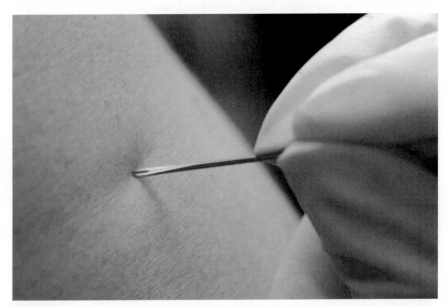

FIGURE 12–7   Vaccinia administration.
Courtesy of the CDC.

cols from the CDC (Figs. 12–8 and 12–9). VIG may be of use in postexposure prophylaxis as well.

Another issue of concern is transmission of vaccinia from the vaccinee to close contacts and household members. Vaccinia shedding does occur from the vaccination site for up to 3 weeks, or until the inoculation site scab falls off. The overall risk of inadvertent inoculation is approximately 3 per 100,000 vaccinees, and nearly all cases of generalized vaccinia occur in close contacts within a household. Again, these adverse-event rates are based largely on historical experience. Improved infection control practices and strict medical screening of potential vaccinees are associated with much lower rates of adverse events.

Several important preexisting medical conditions are contraindications to prophylactic vaccination (Table 12–4). However, with the exception of severe immunodeficiency, these contraindications are relative in a preexposure situation. There are no absolute contraindications to vaccination if exposure to a known case of smallpox has occurred. In this circumstance, the low likelihood of a vaccine-associated adverse event is outweighed easily by the risk of contracting smallpox.

Until 1972 the smallpox vaccine was given routinely to 1-year-old children in the United States with a generally good safety record. Approximately 44% of the U.S. population is unvaccinated at present. The remaining population was vaccinated at least thirty years ago. Residual protection against vaccine is uncertain. Ten years after a single-dose vaccination, antibody levels fell below quantities thought

Primary Vaccination Site Reaction

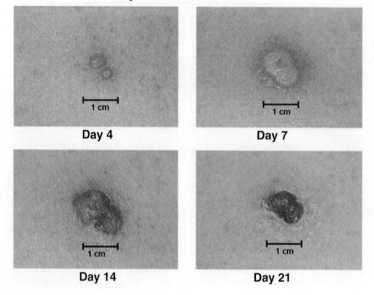

FIGURE 12–8 Progression of vaccinia site.
Courtesy of the CDC.

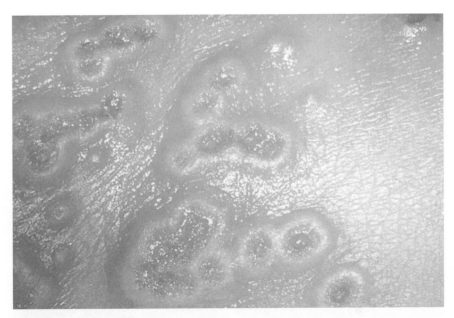

FIGURE 12-9 **Eczema vaccinatum lesions seen following vaccinia administration.**
Courtesy of the CDC.

to be protective, but recent studies have called this traditional view into doubt. Further, low antibody levels coupled with an amnestic response are likely to modify the severity of smallpox in previously vaccinated individuals. Vaccinations given at the time of a smallpox attack will offer protection or amelioration of signs and symptoms if administered up to 4 days after exposure.

As part of the national smallpox vaccination campaign, the HHS planned to vaccinate "smallpox response teams" that would provide a trained and immune workforce to care for hospitalized smallpox patients and to manage adverse events from vaccina-

TABLE 12-4 Medical Contraindications to Preevent Smallpox Vaccination

Eczema or atopic dermatitis, even in childhood

Immunosuppression (from malignancy, immunosuppressive medicines, radiation, etc.)

Immunodeficiency

Household contacts with any of the above

Pregnancy, or individuals planning to become pregnant within 30 days

Life-threatening allergy to vaccine components (glycerin, polymixin B, streptomycin, tetracycline, neomycin, phenol, latex)

Active cardiac disease, including coronary artery disease (CAD), or multiple risk factors for CAD, as well as known cardiomyopathy

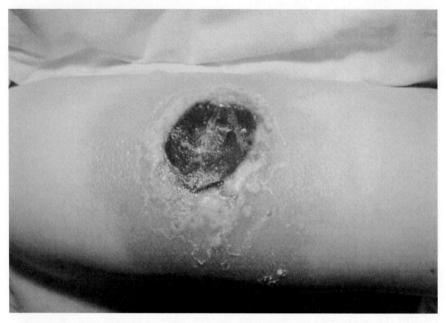

FIGURE 12-10 Progressive vaccinia, a severe complication of smallpox, is life threatening.
Courtesy of the CDC.

tion. The plan had been to vaccinate those HCWs likely to be exposed to smallpox, such as select emergency room staff, first responders, and public health officials designated for fieldwork for such an event. When the program was halted temporarily in April 2003, the total number of people vaccinated was 31,297. The HHS plan anticipated vaccinating half a million health care and public health workers. In the event of a smallpox attack, any health care worker who may come in contact with infected patients will require vaccination. At this time, there is no need for the general public to be given smallpox vaccine. There is no imminent threat, and in the event of a documented attack, there is enough vaccine to administer to every American. The CDC believes that national mass vaccination could be accomplished in less than ten days.

Clinicians need to understand what would happen in the unhappy circumstance that smallpox is once more released on the world. In a postevent situation there are no absolute contraindications to vaccination with vaccinia. Simply put, any individual who has been exposed to the smallpox virus through accidental or deliberate release should be vaccinated because the likelihood of developing smallpox far outweighs the risks associated with vaccination. Further, newer investigational drugs (e.g., cidofovir), and VIG can be used to limit adverse effects of vaccination.

Individuals who have been in contact with a confirmed, probable, or suspected case of smallpox—referred to as primary contacts—should be offered vaccination.

TABLE 12-5	Quick Reference Guide: Smallpox
Diagnostic Keys	Prodrome with high fever, followed by centrifugal, pustular rash
Risk Factors	None; one case means an attack or accidental lab release has occurred
Transmittable	Yes
Transmission	Airborne and contact high infectivity from start of rash until scabs fall off
Management	Isolation; supportive, cidofovir
Prevention	
Preexposure	Vaccinia vaccine
Postexposure	Vaccinia vaccine up to 4 days after exposure
Containment	Patient: Inpatient isolation
	Secondary contacts: Symptomatic—isolation; asymptomatic observation
Populations at Heightened Risk	The unvaccinated, immunocompromised, those with skin diseases
Long-term Issues	Scarring, hearing loss, blindness, possible teratogenic effects
Public Health Concerns	An international medical emergency
Contacts	Local, state health department, CDC, FBI

In this category, one would include household contacts of the index case(s) or persons who live in the home of the index case. The CDC recommends that these primary contacts be offered vaccination. If they refuse, they are to be isolated and observed eighteen days for the development of fever and prodromal features.

Secondary contacts—defined as contacts of primary contacts—should be vaccinated so long as they do not have documented immunosuppressive conditions, severe eczema, or atopic dermatitis. Individuals with these conditions should not be vaccinated but should be under close surveillance for the development of fever or rash. Further, they should be kept away from the primary contacts until those individuals' eighteen-day period of surveillance is over.

❋Summary

To summarize, immediate postevent vaccination is a priority for close contacts of the index case(s); those exposed to the virus but not yet ill; HCWs who have been exposed to the index case(s); along with public health, medical, emergency response, or ambulance personnel who have been involved in the transport or care of an index case. Laboratory personnel and ancillary hospital staff (e.g., laundry, housecleaning, maintenance, or dietary staff) fall into this category as well. Last, local and state law enforcement officials, military, national guard, or emergency response personnel should be offered vaccination provided they have no medical contraindications.

What was just described will be recognizable to senior clinicians and infectious disease specialists as the traditional "ring vaccination" strategy that was so effectively used in the worldwide smallpox eradication campaign. The goal of ring vaccination is to vaccinate close contacts (primary) and contacts of these close contacts (secondary). This strategy is an efficient means to creating a protective ring around the index case and the contacts of the index case and thereby disrupting transmission of the virus. It has the added benefit of being an efficient and streamlined use of resources that will likely be pushed to their limits should an actual case of smallpox be diagnosed anywhere in the world (Table 12–5).

Commonly Asked Questions From Patients Regarding Smallpox Vaccination

Q: I received a smallpox vaccine as a child and don't recall any problems with bad reactions. Has the vaccine changed?

A: Having been vaccinated without adverse reactions as a child does not ensure freedom from adverse vaccination reactions with revaccination. However, unless your personal health history has changed (e.g., newly diagnosed cancer or cancer treatment or immunodeficiency states like HIV/AIDS) a "good" experience with vaccinia vaccine is considered predictive of uncomplicated revaccination. Improved medical screening and infection control recommendations have also limited the adverse events associated with vaccination well below historical levels. Although minor side effects are common, life-threatening side effects are rare with vaccination against smallpox.

Q: Can I contract smallpox from getting the vaccine?

A: No. The smallpox vaccine does not contain smallpox virus. The vaccine does contain another virus called vaccinia, which is related to smallpox. This virus is not dangerous, but it is possible to develop an infection from vaccinia and to transmit this virus to anyone who might touch the vaccination site or anything that has come into contact with it (your hands, old bandages, etc.). Inadvertent contact inoculation is prevented by keeping the vaccination site covered at all times, by proper maintenance of the vaccination site with clean dressings and proper disposal of old dressings, and rigorous attention to hand washing.

Q: Is it possible to get vaccinia from someone who has recently been vaccinated?

A: Yes. Vaccinia is spread by touching a vaccination site before it has healed or by touching bandages or clothing that have become contaminated with live virus from the vaccination site. Vaccinia is not spread through airborne contagion. Symptoms of infection with vaccinia virus include rash, fever, and head and body aches.

Q: **If I have received a smallpox vaccine, are my small children at home at risk?**

A: Possibly. The only way to inadvertantly inoculate a child in the household or anyone else is through direct contact with the vaccination site or contact with anything (gauze, clothes, etc.) that has been in contact with the site such as clothing and bandages. Attention to the infection control practices offered at the time of vaccination should effectively eliminate the risk of both autoinoculation and inadvertent inoculation of close contacts and household members.

Q: **If I received a smallpox vaccine as a child, am I still protected?**

A: Childhood vaccination does not offer reliable protection against smallpox beyond ten years. Previous vaccination probably ameliorates the course of smallpox should one be exposed, prevents death from smallpox, and may completely prevent the disease. Insufficient real-world experience exists to be any more concrete at this time.

CHAPTER 13

Viral Hemorrhagic Fevers

Clinical Vignette

Two days after returning from a short trip to Venezuala, an otherwise healthy 48-year-old male arrives at your office complaining of fever, chills, nausea, vomiting, malaise, and arthralgias. On exam he was toxic appearing febrile, and you note abdominal tenderness, icterus, and multiple erythematous papular lesions on his lower extremities. Because of his recent tropical travel, you immediately send him by ambulance to the ER. There, staff physicians suspect a case of VHF and contact the local health department. During the course of his deteriorating clinical course in the intensive care unit, the patient developed severe coagulopathy and died. Autopsy revealed necrotic portions of the liver and histologic findings consistent with yellow fever. Yellow fever antigens were noted in blood samples, and the diagnosis was confirmed by PCR. What infection control and decontamination procedures should you have followed? In this case of a febrile individual with a rash and with recent travel abroad, what is your DDX?

Background

Extreme contagiousness, virulence, and mortality rates make VHFs among the most feared diseases of humankind (Fig. 13–1). Since the first documented case of a Marburg in 1967, VHFs have been identified all over the world and have observable

TABLE 13-1   Timeline of Ebola and Marburg

October 2000	The Ugandan Ministry of Health reports 176 cases of Ebola, including 64 deaths.
February 2001	A Canadian physician activates the national contingency plan for viral hemorrhagic fevers (VHFs) when VHF is diagnosed in a woman returning from the Congo.
December 2003	The Ministry of Health of the Republic of the Congo reported a total of 35 cases, including 29 deaths from Ebola.
March 2005	CDC confirms that VHF outbreak in Angola is Marburg. A total of 124 cases, with 117 fatalities, was reported.

OF NOTE...

September 2004: A Case of Lassa Fever in New Jersey

A businessman returning from a 4-month trip to Sierra Leone in West Africa was hospitalized in New Jersey complaining of several days of fever, chills, severe sore throat, diarrhea, and back pain. He began feeling ill just before his return to the U.S. On admission, the patient was febrile to 103.6°F (39.8°C), but alert and oriented. Due to his recent travel—which included plane rides from Africa to London to Newark, and a train ride to his home—he was placed on antimalarial and antibiotic treatments to cover the possibility of malaria or typhoid fever. Nonetheless, he developed ARDS and required mechanical ventilation. The DDx was broadened to include VHFs, specifically yellow fever and Lassa fever. Attending physicians contacted the state health department and the CDC. Under an investigational new drug protocol he was placed on IV ribavirin, but the patient died before the medication was administered. Serum antigen testing confirmed the diagnosis of Lassa fever as did additional testing, including postmortem liver immunohistochemical stains, cell culture viral isolates, and PCR testing.

Subsequent epidemiologic investigation identified 188 individuals who had contact with the patient during the period he was considered infectious. This list included family members, 139 HCWs (42 laboratorians, 32 nurses, and 11 physicians), and an additional 16 commercial laboratory workers in two additional states. Passengers on the London to Newark flight were evaluated for exposure to Lassa virus. All exposed or potentially exposed individuals were placed in an active surveillance program whereby they were checked twice daily for the development of fever >101°F (>38.3°C) for the 21-day incubation period. None contracted the disease.

MMWR/morb mortal wkly Rpt October 1, 2004; 53(38):894–897.

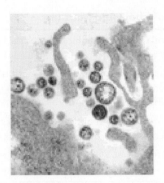

FIGURE 13-1   Electron micrograph image of Lassa virus.
Courtesy of the CDC.

naturally occurring outbreaks. No documented use of VHFs as bioweapons has yet occurred; however, it is commonly accepted that the United States, Russia, and probably other governments have successfully developed weaponized versions of VHFs. What is known about the epidemiology, transmission, clinical presentation, and prevention of VHFs is therefore based on naturally occurring epidemics and bioweapons research.

VHFs are caused by four distinct families of RNA viruses. Filoviridae, which includes Ebola and Marburg viruses; Arenaviridae, which include the etiologic agents of Argentine, Bolivian, and Venezuelan hemorrhagic fevers, Machupo and Lassa fever; Bunyaviridae, which includes the Congo-Crimean hemorrhagic fever virus (CCHFV) and the Rift Valley fever (RVF); and finally, Flaviviridae, which includes dengue and yellow fever viruses. With the important exception of filovirus, whose animal reservoirs are speculative, VHFs are zoonotic.

Epidemiology

More than twenty naturally occurring Marburg or Ebola outbreaks affecting some 2000 individuals, mainly African, have been documented since 1967 (Fig. 13–2). Infections occurred through direct contact with infected animal or human fluid or tissue, or by needle stick infection. Percutaneous exposure is associated with a higher mortality rate. Successful aerosolization and transmission of VHFs in primates has been documented. Precise information about the nature of transmission does not exist at present because outbreaks are sporadic, subside quickly, and tend to occur in areas without adequate public health involvement. Rodents and arthropods are the vectors. There are similarities and important distinctions among the various VHFs in terms of transmissibility.

FIGURE 13-2 Red Cross team disinfects a body bag of an Ebola patient in The Democratic Republic of Congo.
Courtesy of the CDC.

Indirect transmission of Filoviridae (Ebola, Marburg) from aerosolized animal feces, infected arthropod bites, or through contact with animal carcasses occurs. Person-to-person airborne transmission cannot be ruled out, though a number of these outbreaks have ended without airborne precautions ever being taken. Ebola has been found in significant amounts in human skin and sweat glands, and there is concern that casual contact could spread the virus. Primate studies indicate that the virus can be taken up through the mucosal lining. Transmission does not appear to occur during the incubation period (determined in primate studies to be several days).

Arenaviruses (Lassa or Machupo) infect humans through aerosols from infected rodent waste products or by contact of mucus membranes or open skin with the virus. Person-to-person transmission occurs secondary to direct contact with infected body fluid; transmission through airborne droplets cannot be ruled out in the case of Arenavirus.

The Bunyaviridae (RFV, CCHFV, Hartavirus) cause human infection through the bite from infected mosquitoes, inhalation of aerosolized virions from an infected animal, carcasses, and physical contact with infected animal tissue. There is also compelling evidence that infection can occur through consumption of infected animal milk. Presently, it is felt that no person-to-person transmission can occur despite the virus being found in oropharyngeal swabs.

Flaviviridae (yellow fever, dengue) are arthropodborne infections caused by mosquito or tick bites. No person-to-person transmission has been reported.

High-Risk Groups

VHFs represent a risk primarily to individuals and communities where natural reservoirs exist. Family members, HCWs, and other secondary contacts of patients with known VHF are at increased risk of acquiring the disease. Needles, syringes, and other instruments used to treat VHF patients must not be reused due to the risk of transmitting the disease to uninfected patients. Individuals involved in burial of VHF victims have acquired the disease as have those processing meat from infected livestock. HCWs, laboratorians, and individuals involved in the transport of unidentified but infected animals (especially primates) or carcasses are also at greater risk.

Pathogenesis

Basic pathophysiologic mechanisms of VHFs are not yet fully understood. As the name implies, a significant and often profound bleeding diathesis is common to all four families of VHFs. Several mechanisms, including platelet deficiency or dysfunction, direct endothelial and platelet injury, or cytokine dysregulation, have been demonstrated and vary depending on the virus family. For example, Arenaviruses cause cytokine dysregulation with resultant inhibition of platelet aggregation and thrombocytopenia, but have little or no direct cytotoxic effect. In contrast, Bunyavirus and Flavivirus are directly cytotoxic to the liver, kidney, and spleen and generate microvasculature damage.

Clinical Presentation

Naturally occurring epidemics form the basis for most knowledge of the clinical presentations of VHFs. Early clinical diagnosis is difficult due to the range and variability of signs and symptoms. The dramatic presentation of the late stages of VHFs leaves less doubt as to diagnosis. Further, distinguishing one VHF from another has less practical implication than for the other biological and chemical weapons, as treatment options are limited and infection control and public health considerations are the same. For example, ease of transmission may differ, but not to the point where the initial approach of isolation and quarantine will be affected by the form of the disease once the diagnosis of a VHF is entertained.

Following a two to twenty-one-day incubation period, VHFs present with a viral prodrome with systemic, gastrointestinal, and mucus membrane complaints. The common viral syndromes with which most primary care physicians are familiar typically pair systemic symptoms (e.g., fever or headache) with one other organ system, such as the gastrointestinal or upper respiratory tract. In contrast, VHFs present commonly with an unusually expansive range of clinical features that, early on, include systemic and multiorgan system complaints: headache, myalgias, arthralgias along with nausea, abdominal pain and nonbloody diarrhea, conjunctivitis, and pharyngitis. Rashes are common as well, but vary in their appearance depending on virus type.

The signs and symptoms of most VHFs appear rapidly. The exception is the Arenavirus whose onset is more gradual (Table 13–2). Late stages of VHFs demonstrate the vascular dysregulation and capillary leakage that is their hallmark: azotemia, oliguria, a worsening bleeding diathesis that may involve hematuria, hematemesis, petechiae, conjunctival hemorrhage, mucosal hemorrhage, disseminated intravascular coagulation (DIC), and hypovolemic shock. CNS signs such as delirium, convulsions, or coma denote a poor prognosis. When shock, multiorgan system failure, and hemorrhagic complications begin, death follows swiftly. Mortality rates for most are very high (Table 13–2). Clinical differentiation between the various VHFs is difficult without confirmatory laboratory testing.

Chronic Effects

Sixty percent of Ebola virus survivors in one study suffered long-term effects from their disease. Similar findings have been found with other VHFs. Survivors suffer from long-term sequalae such as gastrointestinal pains, weakness, alopecia, malaise, prostration, cachexia, diminished hearing or visual abilities, cerebellar dysfunction, pericarditis, and pancreatitis. Psychological impact, including depression, anxiety, impotence, and posttraumatic stress syndrome are possible. Chronic renal insufficiency, renal failure, and hypertensive renal disease have been seen in survivors from hantavirus infection.

Vulnerable Groups

Pregnant women show increased mortality during epidemics, as do the very old and young. Fetal and neonatal deaths are common.

Laboratory Tests

Initial laboratory tests are not likely to clinch a diagnosis of VHF. Common hematologic abnormalities include anemia, lymphopenia, and thrombocytopenia. However, dehydration may mask anemia or even result in hemoconcentration. Lassa fever is distinguished by an elevated white blood cell (WBC) count. Elevated liver function tests, proteinuria, or hematuria are usual. Coagulation studies are useful and should be performed whenever VHFs are suspected. Prolonged bleeding time, activated partial thromboplastin time, and prothrombin time are seen, along with signs of DIC such as fibrin degradation products and decreased fibrinogen. Viral confirmation is possible through enzyme-linked immunoabsorbent assay (ELISA), reverse transcriptase polymerase chain reaction (RT-PCR), or viral isolation. Standard viral culture

TABLE 13-2 Clinical Distinctions Among Category A VHFs

Virus	Clinical Findings	Incubation Time, Days	Contagious	Treatment	Mortality
Filoviruses Ebola	Abrupt onset of high fever, weakness. Generalized maculo-papular rash <day 5. Hemorrhaging and DIC are common.	2–21	Yes	Supportive	50–90%
Marburg	Abrupt onset of high fever, myalgias. Maculopapular rash of face, neck, trunk, and arms. Hemorrhaging and DIC are common.	2–14	Yes	Supportive	25–70%
Arenaviruses Lassa fever	Gradual onset of fever, nausea, abdominal pain, exudative pharyngitis, cough, conjunctivitis, facial flushing, diffuse lymphadenopathy. Late: edema of head and neck, pleural and pericardial effusions. Hemorrhaging less common.	5–16	Yes	Supportive, ribavirin	15–20%
New World Arenavirus	Gradual onset of fever, nausea, abdominal pain, cough, conjunctivitis, facial flushing, diffuse lymphadenopathy. Hemorrhaging less common. Petechiae possible. CNS involvement—dysarthria, fasciculations of tongue, local and generalized seizures	7–14	Yes	Supportive, ribavirin	15–30%
Bunyaviruses RFV	Fever, headache, retro-orbital pain, photophobia, jaundice. Hemorrhaging rare. Retinitis in roughly 10%.	2–6	No	Supportive, ribavirin	<1%
Flaviviruses Yellow fever	Fever, facial flushing, myalgias, conjunctival injection. Full remission occurs; or a short remission followed by fever, bradycardia, jaundice, renal failure, and hemorrhaging.	3–6	No	Supportive	20%

(Continued)

TABLE 13–2   Clinical Distinctions Among Category a VHFs (*Continued*)

Virus	Clinical Findings	Incubation Time, Days	Contagious	Treatment	Mortality
Omsk HF	Fever, cough, conjunctivitis, papulovesicular lesions on oropharynx, facial and trunk flushing, diffuse lymphadenopathy, splenomegaly. Pneumonia and CNS involvement may occur.	2–9	No	Supportive	<10%
Kyanasur Forest disease	Similar to Omsk, but biphasic. Phase 1: 6–11 days, then short remission: 9–21 days. >50% relapse and develop meningoencephalitis.	2–9	No	Supportive	<10%

Adapted from Borio L, Inglesby T, Peters CJ, et al. Hemorrhagic fever viruses as biological weapons: medical and public health management. JAMA 2002;287:2391–2405.

transport media for nasal, stool, or serum samples is adequate and will often grow virus. Clinicians are advised to get as many samples as possible from multiple sites.

Microscopy

The viruses that make up the VHFs can only be seen with electron microscopy. Ebola's classic features include a cylindrical body with a curved or rounded top. Viruses can be sampled from virtually any body fluid (Figs. 13–3 and 13–4).

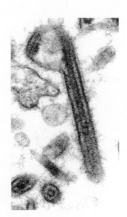

FIGURE 13–3   Electron micrograph of Ebola virus.
Courtesy of the CDC.

OF NOTE . . .

Dr. Frederick Murphy and Virus Marburg

On October 13, 1976, Dr. Frederick Murphy, DVM, PhD, was preparing a specimen transported back to the CDC from Zaire. As Dr. Murphy later recalled in an interview for the National Health Museum: "When I put the specimen in the electron microscope, I was sure it was Marburg. I had worked on Marburg in 1967 and 1968 and had done a project on experimental Marburg infection in monkeys. The specimen had come back from Zaire to the CDC in Atlanta in less than optimal condition, with the tubes in the box broken. Anyone else would have taken a look and put the whole box in the autoclave, but Dr. Patricia Webb, wearing gloves, gown, and mask, squeezed a few drops of fluid out of the cotton surrounding the broken tubes. That was the material the virus was isolated from. It was placed in tissue culture (monkey kidney cells) for a couple of days. Then I got a drop of the tissue culture fluid and prepared a specimen for the electron microscope. When I saw what I was sure was Marburg, I shut the electron microscope down and went back to the room in which I had prepared the specimen. This was in the days when hoods were a lot more primitive. I 'cloroxed the hell' out of the place where I had done the preparation and carried my discard pan with gown and gloves etc. to the autoclave and ran it. Then I went back to the microscope and called Karl Johnson and Patricia Webb to take a look. I shot a cassette of pictures and with wet negatives, not good for the enlarger, and I made prints which were available within minutes. I carried these dripping prints to the office of the Director of the CDC. It was very dramatic."

Access Excellence at the National Health Museum

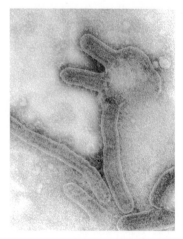

FIGURE 13–4   Electron miscroscopy of Marburg virus.
Courtesy of the CDC.

For safety purposes, testing for VHFs must be done at a BSL-4 facility. Presently, only two such facilities exist and will even accept suspect samples in the United States, the CDC in Atlanta, and the USAMRIID in Bethesda. These two labs use ELISA, PCR, and viral isolation to confirm the diagnosis with preliminary results usually available within twenty-four hours. Handling of samples may require processing in a Biosafety Level 2 (BSL-2) or Biosafety Level 3 (BSL-3) manner with laboratorians adhering scrupulously to the infection control practices described in the following sections. Plans are underway to expand to a BSL-4 capacity to select nationwide public health laboratories.

Radiographic Findings

No distinctive radiologic findings are associated with VHFs, although hemorrhagic pneumonitis is seen in two of the Flaviviruses: Omsk virus and Kayasunur Forest virus.

Differential Diagnosis

Infectious diseases that could mimic VHFs include influenza virus, typhoid fever, viral hepatitis, nontyphoidal salmonellosis, leptospirosis, rickettsial infections, shigellosis, relapsing fever, meningococcemia, and the infectious causes of DIC. Noninfectious diagnoses to consider include fulminant hepatitis, leukemia, lupus erythematosus, hemolytic uremic syndrome, idiopathic or thrombotic thrombocytopenic purpura, and the noninfectious causes of DIC.

Diagnosis

A recurring theme of this book is that clinical evaluation and astute public health awareness are needed to bring the possibility of a biological agent. As always, a high index of suspicion is needed to make the diagnosis, particularly in the context of a potential bioterrorist attack. Few physicians will have any direct knowledge or clinical experience with VHFs in the United States, and confirmatory laboratory diagnosis can only be pursued once the index of suspicion is raised. Consequently, physicians must rely on history and clinical presentation to raise the possibility of a VHF. Determining the specific virus that is causing the infection in any given individual is difficult, although certain clinical clues may help differentiate one virus from another. However, in the acute setting, differentiation does not change initial clinical management, necessary infection control steps, or public health response.

Risk factors for a naturally occurring VHF infection include travel history positive for recent visits to Africa, South America, or Asia, contact with sick animals or their corpses, sick human contact, and a recent tick bite (within 3 weeks). A bioweapons attack would reveal no risk factors other than higher risk occupational or travel histories, such as military, government officials or their employees, or recent travel to national monuments or political speeches. The WHO recommends diagnosis of an index if the series of clinical conditions are met in Table 13–3.

TABLE 13-3 Making the Diagnosis: VHFs

Pattern of Presentation	Multiple, concurrent cases of febrile illness with evidence of bleeding diathesis
Features	Febrile, temperature 101°F of less than 3 weeks duration No risk factors for hemorrhagic manifestations Two or more of the following hemorrhagic symptoms: Hemorrhagic, purpuric, or petechial rash Epistaxis Hematemesis Hemoptysis Hematochezia No established alternative diagnosis
Findings	History, presentation, and epidemiology determine the diagnosis

Modified from the WHO.

Treatment

Supportive care is the only available treatment option for VHFs at this time. Intensive fluid resuscitation, electrolyte management, dialysis, mechanical ventilation, as well as administration of pressor agents may sustain a patient through the hematologic, hemodynamic, neurologic, and pulmonary complications. Most if not all patients will require placement in critical care units, a fact that has enormous public health implications given most nations' limited critical care resources.

There are no FDA-approved drugs for VHFs, and pharmacologic treatment of VHFs is limited. Ribavirin's promise in animal studies has resulted in its approval for compassionate use for the Bunyavirus and the Arenavirus. If administered just

TABLE 13-4 VHF Treatment with Ribavirin[a]

Indication	Patients with VHF with unknown pathogen Pathogen known to be Arenavirus or Bunyavirus
Individual Case	Adult, pregnant women, children Loading dose—30 mg/kg IV (max of 2 g) × 1, then 15 mg/kg IV (max of 1 g/dose) q 6 hr × 4 days, then 7.5 mg/kg IV (max of 500 mg/dose) q 8 hr × 6 days
Mass Infection	Adult, pregnant women Loading dose—2 g po × 1, then If >75 kg, 600 mg po bid × 10 days If <75 kg, 400 q am and 600 q pm × 10 days

[a] Not FDA approved.

Adapted from Borio L, Inglesby T, Peters CJ, et al. Hemorrhagic fever viruses as biological weapons: medical and public health management. JAMA 2002;287:2391–2405.

prior to or just after infection, alpha interferon is effective in limiting viremia and liver toxicity in RVF (Table 13–4).

Public Health Considerations

There are several public health considerations with regard to VHFs. Individuals potentially exposed to VHFs—for example, lab workers working with the virus or close contacts of suspect or confirmed cases—should be monitored for fever (≥101°F) and other clinically suggestive signs for up to three weeks. The efficacy of vaccinations with heat shock proteins for VHFs is an area of active research. The only licensed VHF vaccine is for yellow fever.

OF NOTE . . .

February 3, 2001, Hamilton, Ontario

Dr. Douglas MacPherson, a tropical disease expert in Hamilton, Ontario, was called to a local hospital to evaluate a 32-year-old Congolese woman admitted the previous night. She had complained of several days of severe flulike symptoms and was bleeding from multiple sites. The patient had flown to Toronto from Ethiopia by way of New York and was apparently asymptomatic on the plane. Officials said she did not have any symptoms on the plane flight. Five minutes after beginning his examination, Dr. MacPherson was convinced that the patient's symptoms fit the picture of Ebola and he invoked immediately the national contingency plan for VHFs. The plan included strict patient isolation, use of Level A protective gear for treating medical staff, and active surveillance for fever in those known to have had contact with the patient or her body fluids. The plan also called for the implementation of a communication alert to HCWs and the general public. Although subsequent testing at the Canadian Science Centre for Human and Animal Health and the CDC did not confirm what MacPherson feared—Ebola—the speedy implementation of the national contingency plan may well have cut short what could have been an international public health crisis.

The experience also offered lessons for the future in an increasingly mobile global community. As Dr. MacPherson explained: "We have porous borders and high volumes of people moving within wide areas of the world. In every diagnosis we have to be thinking, 'Where did this person come from? What [has she] been doing since [she] got here?' This is an essential part of clinical diagnosis. . . . One of the messages to medical students should be that any fever requires these three questions: Have you travelled? When did you travel? Have you had visitors from abroad? Travel history is as important as anything else doctors ask. . . . In terms of international mobile health surveillance, it really causes you to prick your ears up. Much of this is gumshoe epidemiology, and we all have a responsibility to be alert."

Adapted from the *Canadian Medical Association Journal* 2001;164(7)

TABLE 13–5   Critical First Steps: VHFs

Medical	Supportive therapy, ± ribavirin
Infection Control	Airborne precautions, negative-pressure room, strict contact precautions,[a] N-95 or PAPR respirators
Public Health	Notify hospital infection control and Public Health Department (PHD) or CDC immediately

[a] Working Group on BT recommends enhanced body fluid precautions, including hand hygiene, double gloves, impermeable gowns, face shields, eye protection, and shoe and leg coverings.

Postexposure

There are no guidelines or even recommendations for management of postexposure prophylaxis in asymptomatic patients. If the patient begins to demonstrate characteristic signs and symptoms of VHFs, ribavirin therapy should be initiated and may limit mortality and morbidity. Ribavirin has shown no effect with Ebola or Marburg virus. Dermal and percutaneous exposures should be washed away with soap and water, whereas mucocutaneous exposures should be irrigated generously with saline (Fig. 13–5).

TABLE 13–6   Quick Reference Guide: VHFs

Diagnostic Keys	Fever for 3 weeks, bleeding diathesis (with no other risk factors)
Risk Factors	None
Transmittable	Presumed
Transmission	Airborne, body fluids
Management	Respiratory, contact isolation, supportive measures
Prevention	
Preexposure	None
Postexposure	If symptomatic, then ribavirin and supportive care
Containment	Patient respiratory and contact isolation Secondary contacts if symptomatic, same
Populations at Heightened Risk	HCWs treating VHF patients, pregnant women, the very young and old
Long-term Issues	Neurologic deficits, GI distress, weakness
Public Health Concerns	Unclear transmission, vectors, and no vaccines
Contacts	Local, state DPH, CDC, FBI

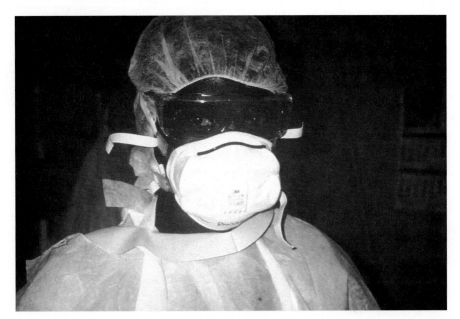

FIGURE 13-5 Nurse in PPE for ebola patient.
Courtesy of the CDC.

Infection Control

HCWs are at particularly high risk when evaluating or caring for individuals with possible or definite VHFs. Bodily fluids and secretions, as well as instruments and sharps, should be handled with extreme care, as viral loads in these fluids are impressive. The potential for aerosolization of VHFs, as has been documented for Filoviruses and Arenaviruses, mandates methicillin-resistant *Staphylococcus aureus* (MRSA)-like precautions, including negative-pressure rooms, strict contact precautions, and isolation. Enhanced contact precautions using PPE such as PAPR, N95 HEPA filter respirators, double gloving, impermeable gowns, face shields, and shoe coverings are mandatory (Tables 13–4 and 13–6).

Summary

VHFs are a family of viruses whose severe manifestations and communicability make them potential bioterror weapons. Clinically they present as a febrile syndrome characterized by widespread vascular leak syndrome and hemorrhagic signs, and are associated with high mortality rates. Treatment is supportive, although ribavirin may have some benefit. Enhanced contact precautions and airborne precautions are required to limit spreads particularly to HCWs (Table 13–6).

CHAPTER 14

Anthrax

Clinical Vignette

A large local employer for whom you do preplacement examinations calls to say that an employee in the mailroom saw an envelope with a suspicious white powder coming from its seams. Following the CDC guidelines, the company protocol for suspicious mail is followed: the envelope is left undisturbed and the room is vacated and locked. Although identification of the substance is pending, the company president has many questions, including whether his employees must be given a vaccine or antibiotics and wants your advice. What should you say?

Background

Since antiquity, anthrax has affected livestock and humans. The name anthrax derives from the Greek word for "coal" because of the black skin lesions it can cause. One of the first records of anthrax is in the Old Testament's Book of Exodus as the fifth and possibly sixth of the ten plagues (Fig. 14–1). Virgil wrote verses on the disease in 25 BC describing the toll it took on livestock and the health risks to humans from exposure to infected meats and skins. The modern history of anthrax begins in the late nineteenth century when Louis Pasteur identified the bacterium responsible for anthrax, an observation that lent early credibility to the germ theory of

FIGURE 14–1   The Fifth Plague of Egypt, 1800, oil on canvas by S.M.W. Turner is located in the Indianapolis Museum of Art.

disease. Anthrax was used as a weapon in World War I as a means to cause economic havoc through the loss of livestock.

Anthrax (*Bacillus anthracis*) is a gram-positive, nonmotile spore-forming bacterium that infects both humans and animals, particularly livestock. Anthrax is found in a worldwide distribution. The spores, most commonly found in soil, are hardy and may survive for decades. In humans, the disease presents in three forms: inhalational, cutaneous, and gastrointestinal. Mortality is greatest for the inhalational form. If diagnosed early, anthrax is a highly treatable disease. Even weaponized anthrax, such as that used in the attacks through the U.S. Postal Service, is treatable with conventional antibiotics.

Epidemiology

The incidence of anthrax has dropped significantly over the years in technologically advanced countries, in large part because of vaccination of high-risk groups—those regularly exposed to hides, wools, and other raw livestock products. Prior to the anthrax attacks in 2001, American experience with anthrax was limited. The only modern experience with the inhalational form was in 1979 with the accidental release of anthrax spores from the Soviet bioweapons facility in Sverdlosk, Russia, in which 72 died. The last case of inhalational anthrax in the United States occurred in 1972.

TABLE 14-1   Anthrax as a Bioterrorist Weapon

In 1970, the WHO estimated that 50 kg of anthrax spores upwind of a major city would result in 125,000 cases of anthrax and 95,000 deaths.

In 1979, an accidental release of weaponized anthrax from Sverdlosk, USSR, resulted in 77 cases of anthrax and 72 deaths.
Has potential for mass casualties as well as more limited covert attack (e.g., U.S. Post Office).

After the September 11 anthrax attacks, 13 cases were diagnosed, 5 died, and over 10,000 individuals were placed on post-exposure prophylaxis

Worldwide the disease is more common, with best estimates ranging between 20,000 and 100,000 cases of anthrax annually, and almost all occurring in developing nations. Most current public health policies regarding anthrax, such as vaccination strategies and medical management human experience are based on limited information and are ever evolving. The mortality rates for inhalational anthrax are greater than 80%; however, this figure was arrived at prior to modern antibiotics and the advent of critical care resources. The post-September 11 inhalational anthrax patients had a mortality rate of 40% (Table 14–1).

FIGURE 14–2   Decontamination outside Ottilie Lundgren's home in Oxford, Connecticut, November 2001.
Provided with permission from the State of Connecticut.

Occupational Risk Groups

By virtue of where they work, or their occupational exposures, postal workers, politicians and other public figures, veterinarians, and agricultural workers all represent at risk workers.

Vulnerable Groups

Immunocompromised individuals (e.g., HIVS/AIDS, patients undergoing cancer treatment), diabetics, children, the elderly, and individuals with preexisting respiratory conditions should be considered at greater risk (Fig. 14–2). Pregnant women may be treated safely, but dosing recommendations vary.

Means of Transmission

Infection occurs from direct contact with active bacteria or indirectly from contact with spores that then germinate. The spores may be in animal hair, meat, hides, or

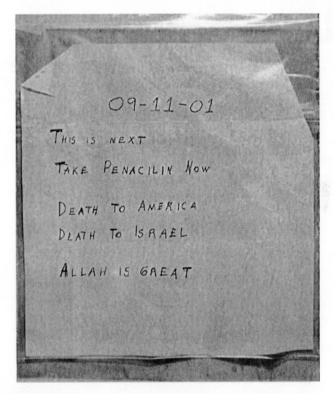

FIGURE 14–3 The letter in the anthrax-contaminated envelope sent to news anchor, Tom Brokaw.
Courtesy of the FBI.

other products. The weaponized form of anthrax is constituted as spores less than five microns in size and may be maintained as powder or aerosolized. Prior to the U.S. Postal Service attacks, the LD_{50} (see Glossary) for anthrax was thought to be between 2,500 and 55,000 inhaled spores. Weaponized anthrax is far more potent with inhalation of as few as several hundred spores sufficient to cause disease. Estimates from the U.S. Postal Service attacks suggest that the LD_{50} as well may have been an order or two below naturally occurring anthrax (Fig. 14–3, Table 14–1).

Pathogenesis

B. anthracis derives its virulence from the presence of a capsule and from three secreted proteins: protective antigen, edema factor, and lethal factor. Protective antigen facilitates the binding of the bacteria to host cell membranes and subsequent transport intracellularly of the other two toxins. Edema toxin acts to inhibit neutrophils as well as causes edema by disrupting water homeostasis. Lethal toxin causes activation and dysregulation of cytokines such as interleukin-1 and tumor necrosis factor (TNF).

Microscopy

Key identifying features include gram-positive rods measuring approximately 1 to 1.5 × 3 to 5 microns singly or in chains, with a "bamboo" or boxcar appearance (Fig. 14–4). General features include nonmotile, nonhemolytic, and encapsulated bacterium easily demonstrated with India ink culture. Anthrax spores are 1 to 1.5 microns in size and oval in shape. They grow readily on ordinary laboratory media at body temperature.

Though few laboratories specialize in diagnosis of bioweapon agents per se, diagnosis can generally be made based on the features of *B. anthracis* as a gram-positive, spore-forming bacillus. However, PCR, ELISA, and other assays are now available to confirm the presence of anthrax. Anthrax specimens may be handled

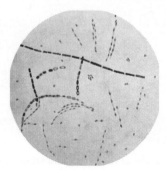

FIGURE 14–4 Anthrax microscopy. Note the boxcar-like appearance.
Courtesy of the CDC.

in a BSL-2 laboratory, that is, a standard clinical laboratory that practices "universal" precautions and has facilities for minimizing aerosols. Spore handling requires a BSL-3 laboratory.

Clinical Presentation

Although the basic pathophysiology of anthrax infections is the same whether presenting in the inhalational, cutaneous, or GI form, the clinical presentation, morbidity, and mortality of the three forms vary so that each is discussed individually.

Inhalational Anthrax

Pathophysiology

Classically, spores located in the soil or in contaminated animal products are kicked up into ambient air and inhaled directly into the respiratory tract. As a weapon, anthrax may be distributed in powdered or aerosolized forms. Anthrax spores are the ideal size for inhalation deep into the lung. Once inhaled, they easily reach the alveolar spaces where most are phagocytized and destroyed by alveolar macrophages. Surviving spores are taken up by the lymphatics and arrive at the mediastinal lymph nodes where the spores undergo germination into bacillus form. Germination may take up to six weeks longer, complicating both clinical diagnosis and, if the result of a bioterror release, the public health response. Once germination begins, both local and systemic symptoms of inhalational anthrax appear rapidly, resulting from the release of bacterial toxins. The toxins cause necrotic,

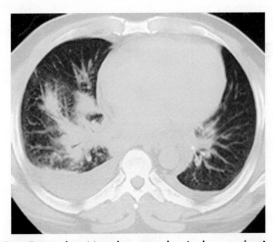

FIGURE 14–5 Post-September 11 anthrax attacks. Anthrax patient's chest x-ray at day 4. The findings include a widened mediastinum, right hilar enlargement/mass, right pleural effusion, and right perihilar airspace disease.
Courtesy of Earls JP Radiology 2002 Feb; 222(2): 305–12.

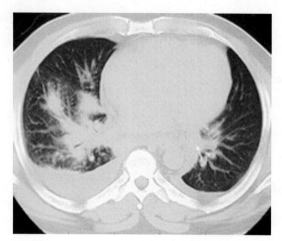

FIGURE 14–6   Post-September 11 anthrax attacks. Anthrax—chest CT scan at presentation. Note the peribronchial thickening and airspace disease.
Courtesy of Earls JP Radiology 2002 Feb; 222(2): 305–12.

edematous, and hemorrhagic thoracic lymphadenitis and a subsequent systemic inflammatory response. It is the toxin level that is associated with mortality, not the bacteria itself. This is significant because a negative blood culture does not necessarily mean the patient has reached a point of improvement. Pathologically, the bronchopneumonic processes noted are distinctive in inhalational anthrax, with a hemorrhagic mediastinitis that is virtually pathognomonic. Pleural effusions are common and often require aggressive drainage (Figs. 14–6 to 14–8).

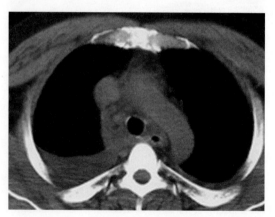

FIGURE 14–7   Post-September 11 anthrax attacks. Anthrax—chest CT scan at presentation. Note the bilateral effusion.
Courtesy of Earls JP Radiology 2002 Feb; 222(2): 305–12.

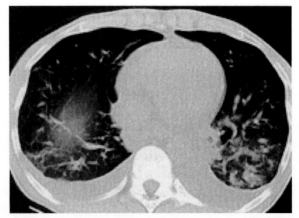

**FIGURE 14–8 Post-September 11 anthrax attacks. Anthrax—chest CT day 4.
Enlarging bilateral effusions now filled 50% of the thorax.**
Courtesy of Earls JP Radiology 2002 Feb; 222(2): 305–12.

Signs and Symptoms

Inhalational anthrax may present in a biphasic pattern (early and late) or as a continuously progressive disease. The weaponized form used in the post-September 11 anthrax attacks followed this biphasic pattern. On the whole, the victims' signs and symptoms were consistent with those seen in the classic zoonotic form. Though it was a small sample, there were two notable differences between classic anthrax and what was seen in the attacks: the inhalational rather than cutaneous form predominated, and infectivity occurred despite inhalation of spore levels below previously known pathogenic levels.

Early (Stage I)

The early stages of inhalational anthrax lasts hours to days. The prominent early clinical findings in ten of eleven of the post-September 11 anthrax victims were myalgia and fever. The constellation of signs and symptoms includes fever, myalgia, nonproductive cough, nausea, vomiting, diaphoresis, dyspnea, myalgia, chest pain, headache, and tachycardia. Other associated symptoms include chills and abdominal pain.

Late (Stage II)

Two to three days following the onset of Stage I, late signs and symptoms evolve. Stage II inhalational anthrax is presaged by respiratory paralysis, diaphoresis, cyanosis, acute fever splices, hypotension, respiratory alkalosis and terminal acidosis, massive lymphadenopathy, hemorrhagic meningitis (often with concurrent meningismus), delirium, and obtundation. This stage progresses rapidly to shock, hypothermia, and death within 24 to 36 hours. The transition from Stage I to

Stage II can occur suddenly, as a continuum, or even after a short period of improvement.

Laboratory Tests

Early

Traditional teaching was that sputum and blood cultures are an insensitive means of diagnosing anthrax in the inhalational form, particularly sputum cultures, because there is often little in the way of a pneumonic process taking place. However, in the Swerdlovsk and U.S. Postal Service outbreak, pretreatment blood cultures were positive in all instances. Once bacteremia or systemic infection is established, sputum stains and cultures may be of value. Standard laboratory testing in the early stages of anthrax is largely unhelpful. For example, there is no or minimal leukocytosis. Identification of classic "boxcar"-shaped gram-positive bacilli from sputum or blood culture in the proper setting would make anthrax a primary diagnostic possibility. Buffy coat stains may yield higher sensitivity. ELISA testing for toxin antibodies can be confirmatory. PCR is also available.

Late

Hypocalcemia, hyperkalemia, and hypoglycemia can be seen. Hemoconcentrations are seen with hematocrits often greater than 50%.

Radiographic Findings

Inhalational anthrax is characterized by a series of radiographic and tomographic changes that can aid greatly in diagnosis. Radiographs provide critical diagnostic information many hours or even days before blood or sputum cultures can turn positive. These changes, when presenting late, indicate a poor prognosis. Computed tomography (CT) taken of the post-September 11 anthrax patients also showed characteristic abnormalities that have substantial diagnostic value. Initial CTs in virtually all of the post-September 11 inhalational anthrax cases were markedly abnormal and had an unusual combination of findings of enlarged hyperattenuating mediastinal and hilar lymph nodes, diffuse mediastinal fat edema, peribronchial thickening, and pleural effusions. Given the rarity of inhalational anthrax in the modern era, this terrorist attack provided valuable information for the presentation of anthrax in the era of CT scans. The findings corroborate the classic chest radiograph findings. CT scans are now considered a standard diagnostic tool for inhalational anthrax. These key radiographic findings are summarized in Table 14–2 (see also Figs. 14–5 to 14–8).

Diagnosis

Early diagnosis requires a very high index of suspicion because of the nonspecific symptomatology. Inhalational anthrax may be hard to distinguish from ordinary

TABLE 14–2   Key Radiographic Findings: Anthrax

Chest X-ray

  Presence of a widened or abnormal mediastinum

  Hilar adenopathy

  Pleural effusions

  Peripheral airspace disease

CT Scan

  Enlarged hyperattenuating mediastinal and hilar lymph nodes

  Diffuse edema of mediastinal fat

  Peribronchial thickening

  Pleural effusions

pneumonias. Yet, early diagnosis is vital for successful management of the infected patient. The cluster of symptoms described previously—especially fever or sepsis—along with typical imaging changes in an otherwise healthy patient require that anthrax be considered in the differential. The presence of certain clinical clues may facilitate distinguishing inhalational anthrax from influenza or influenza-like illnesses. These include having an abnormal lung exam, dyspnea, nausea or vomiting, and chest pain or pleurisy. In contrast, headache, sore throat, and rhinorrhea mitigate against anthrax. Distinguishing inhalational anthrax from community-acquired pneumonia based on history and exam is more difficult and will have to rely on additional diagnostic tests, including microscopy, cultures, and radiography.

In the event of an epidemic, nasal swabs may be taken from those with possible exposure, and any positive results necessitate a prophylactic antibiotic regimen. The sensitivity and specificity of nasal swabbing for diagnosing anthrax are unknown. Certainly, a negative nasal swab should not rule out the possibility of anthrax. Because the natural course of germination and replication is occurring in the lungs, distinctive clinical features—pleural effusions, lymphadenitis, and hemorrhagic mediastinitis—are seen.

Differential Diagnosis

The differential diagnosis for anthrax includes influenza, SARS, avian influenza, and community-acquired viral pneumonias.

Prevention

Recent studies on animal models have shown that the anthrax vaccine offers protective value against the inhalational form of anthrax, whereas human studies

TABLE 14–3	Making the Diagnosis: Inhalational Anthrax
Pattern of Presentation	Multiple, concurrent cases of a rapidly progressing flulike illness
	Flulike illness seen in high-risk groups following an identified attack
Features	Febrile, pneumonia-like presentation
	Abnormal lung exam, dyspnea, chest pain, vomiting
Findings	Chest x-ray: widened mediastinum, infiltrates, pleural effusion
	CT: enlarged, hyperattenuating hilar and mediastinal nodes, mediastinal edema, peribronchial thickening
Labs	Blood cultures, buffy coat stain, PCR, ELISA

have shown protection against the cutaneous form. The vaccine is an acellular fil-trate of an attenuated form of the bacteria and is given in a series of six doses at 0, 2, and 4 weeks, then 6, 12, and 18 months, followed by yearly boosters. It has been shown that protection is attained after even two doses. Studies indicate the vaccine is relatively safe with about 1% of people developing some kind of minor reaction, headache being the most common. No long-term sequalae have been re-ported. When given with proper antibiotic therapy, there is protection against de-velopment of the disease even after exposure. Current stockpiles of the vaccine are limited, and no increases in production are expected. Vaccination is given only to those in the military. Controversy surrounds potential adverse effects from the anthrax vaccine, but these concerns have not yet been well documented. Although the post-September 11 anthrax patients were not given the vaccine, a number of people deemed to be at risk for possible exposure were given the vac-cine along with antibiotic therapy (see Appendix for full regimen and for dosing schedules). Currently, postexposure prophylaxis (PEP) for exposed individuals would include both antibiotics and vaccine. The protective effect of this is based on laboratory animal studies, but human data are not adequate to make such rec-ommendations evidence based.

Prognosis

Because there have been so few cases of anthrax—particularly inhalational—no factors are clearly predictive of morbidity or mortality. All that seems evident at this time is that early recognition and initiation of combination therapy appears to be key for survival. Among the post-September 11 anthrax patients, those who presented with fulminant anthrax before antibiotic therapy was initiated all died.

Cutaneous Anthrax

Cutaneous anthrax is the most common naturally occurring form of the disease, representing over 95% of all cases. Cutaneous anthrax usually results from direct contact with infected livestock or livestock products, such as hides or wool. This form could occur in a biological attack, particularly with a powdered form as was seen used after September 11th, where eleven of the twenty-two cases were cutaneous anthrax. Untreated, it has a mortality rate of approximately 20%. With appropriate antibiotic coverage, mortality is less than 1%.

Transmission

Infection results from spore contact with skin, particularly if any abrasions or skin breaks are present. Exposed skin surfaces are, of course, the most common sites. The incubation period appears to be short: usually less than 2 weeks after exposure (versus inhalational, which can take 6 weeks or more). Recognition of the cutaneous form may be crucial because it may be the first and best evidence that an attack with anthrax has occurred. On an historical note, it is speculated that one of the September 11 hijackers was seen by a Florida physician for what was initially diagnosed as a skin infection but was later (during the post-September 11 attack investigations) diagnosed as cutaneous anthrax. A proper initial diagnosis may have altered history, a possibility that highlights the importance of trained and vigilant clinicians.

Pathogenesis

The same three virulence factors that act in the lung in inhalational anthrax act cutaneously at the site of contact.

Signs and Symptoms

Early

Classically, there is a painless, pruritic, papular primary lesion that forms within one week of exposure to the spore. The papule is initially often mistaken for an insect bite. Within two days, 2- to 3-mm vesicles form around the papule containing serous or serosanguinous fluid these are contaminated with numerous bacilli and occasionally WBC. These vesicles enlarge, and often satellite vesicles appear. The initial site becomes highly edematous (nonpitting) secondary to the release of edema toxin by the bacteria. Once the lesion enlarges and subsequently ruptures, it becomes necrotic, forming an ulcer covered by the black eschar for which the disease is named (Figs. 14–9, 14–10 and 14–11). The eschar then dries up and falls off within 2 weeks. The lesion is often accompanied by lymphadenopathy. Secondary infection with *Staphylococcus aureus* is unusual, but if present, it manifests as painful

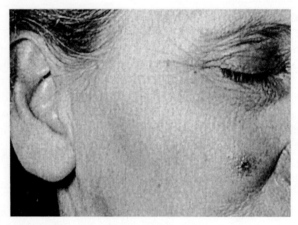

FIGURE 14–9   Cutaneous anthrax of the cheek.
Courtesy of the CDC.

lymphadenopathy, purulent discharge, and lymphangitis (Table 14–4). Concomitant low-grade fever and malaise are quite common but are more likely with extensive skin lesions.

Late

Occurring rarely, systemic disease is a late and advanced manifestation of cutaneous anthrax. It is characterized by bacteremia, renal failure, anemia, bleeding, and ecchymoses.

FIGURE 14–10   Cutaneous anthrax of the arm: day 4.
Courtesy of the CDC.

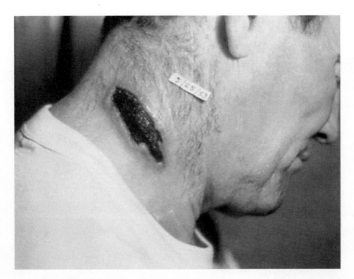

FIGURE 14–11   Cutaneous anthrax.
Courtesy of the CDC.

Laboratory Tests

Early

The early laboratory tests show significant changes in routine blood work (e.g., CBC, electrolytes, renal or liver function tests.)

Late

The late laboratory tests show Microangiopathic hemolytic anemia, coagulopathies, and hyponatremia occur as a late manifestation of cutaneous anthrax.

TABLE 14–4 Making the Diagnosis: Cutaneous Anthrax	
Pattern of Presentation	Multiple, concurrent cases of a maculopapular lesion that forms an eschar; regional adenopathy. Prodromal symptoms may be seen. Illness with maculopapular lesion that becomes an eschar seen in high-risk groups following an identified attack.
Features	Initially painless; pruritic papular lesion that drains, enlarges, ulcerates, and forms an eschar that falls off in 1–2 weeks; low-grade fever and malaise may be seen.
Findings	Skin eschar

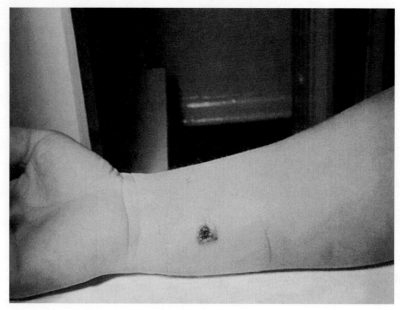

FIGURE 14–12   Cutaneous anthrax lesion at day 7.
Courtesy of the CDC.

Pathology

Any vesicular fluid should be sampled and gram stained. If the patient is on antibiotic therapy already, a punch biopsy should be taken of the lesion.

Differential Diagnosis

The differential diagnosis of cutaneous anthrax includes tularemia, plague, scrub typhus, anticoagulant necrosis, Rickettsial spotted fevers, rat bite fever, ecthyma gangrenosum, vasculitides, arachnoid bites, leprosy, lymphogranuloma venereum, and chancroid.

Diagnosis

As with the inhalational form, early diagnosis is very difficult and requires a high index of suspicion. However, the black eschar is strongly suggestive, especially if coupled with known high-risk occupation (Table 14–4).

☣Gastrointestinal Anthrax

Transmission

Transmission of this form is not clearly understood but is believed to result from either ingestion of the vegetative form of anthrax from eating undercooked, infected meat or by spore deposition in the proximal portion of the GI tract. If infected via spores, inhalational infection may also be present. No cases of GI anthrax were diagnosed in the post-September 11 anthrax patients.

Signs and Symptoms

Onset of symptoms begins classically within two to five days of ingestion. If the upper GI tract is affected, oral or esophageal ulcers form with regional lymphadenopathy, edema, and sepsis. In the lower GI tract, intestinal lesions form usually in the terminal ileum or cecum and cause nausea, vomiting, malaise, bloody diarrhea, an acute abdomen, or sepsis. Hemorrhagic mesenteric lymphadenitis can be seen as a later development, as can ascites. Systemic disease may develop and yields the same array of signs and symptoms as found in the systemic forms of inhalational and cutaneous infections.

☣ General Considerations on Anthrax

Though occurring far more commonly in inhalational anthrax, progression to systemic disease can occur in any of the forms if left unrecognized and untreated. Based on the management of the post-September 11 anthrax patients, blood cultures can become sterile after a single dose of antibiotics, making it even more critical than usual that blood cultures be drawn prior to antibiotic administration.

Diagnostic Considerations

If anthrax is suspected (or even on the list of differential diagnoses), the hospital or diagnostic laboratory should be notified and implement a protocol set forth by the CDC to ensure proper diagnosis and safe handling. In the context of biological terror, the importance of taking a thorough and relevant history cannot be emphasized enough. This includes the often-neglected travel, occupational, and social histories

TABLE 14–5 Making the Diagnosis: Gastrointestinal Anthrax

Pattern of Presentation	Multiple, concurrent cases of: a. vomiting, bloody diarrhea, acute abdomen b. oropharyngeal ulcers with adenopathy GI symptoms seen in high-risk groups following an identified attack
Features	Early—vomiting, bloody diarrhea, acute abdomen Late—sepsis; possible oropharyngeal ulcers with adenopathy
Findings	UGI series: Widened mediastinum, infiltrates, pleural effusion. CT: Mesenteric lymphadenitis, ileocecal ulcers

(see Chapter 2). As evidenced by the post-September 11 anthrax patients, certain groups are at higher risk: postal workers; mail room workers; media personnel; politicians and their associates; microbiology lab personnel; those who have had recent contact or proximity to politicians, federal, state, or local government employees; visitors to monuments or government buildings; and visitors to prominent media institutions. Obviously, clustering of cases with similar signs, symptoms, and other findings, particularly if traceable to a single focus such as a building is highly suggestive of a biological attack. In any unexplained death in which anthrax is a possible cause, it is imperative that an autopsy be performed. A finding of hemorrhagic necrotizing mediastinitis or hemorrhagic necrotizing lymphadenitis is considered pathognomonic for inhalational anthrax. Hemorrhagic meningitis is highly suggestive of a systemic anthrax infection.

Vaccination

Although a licensed anthrax vaccine exists in the United States, relatively little clinical data exist in terms of its efficacy against inhalational anthrax. There are limited data that suggest the vaccine offers some protection from cutaneous anthrax. The vaccine is derived from sterile culture fluid supernatant taken from an attenuated strain and is administered as six subcutaneous doses given at 0, 2, and 4 weeks; 6, 12, and 18 months. Yearly boosters are recommended thereafter. Mild discomfort at the injection site may be experienced by 6% of those receiving the vaccine. Pregnancy, active infection, and immunosupression are reasons to defer administration of the vaccine. Anaphylaxis to the vaccine is rare.

Treatment

Precisely because of the limited experience with anthrax, and because no clinical studies exist, treatment guidelines are far from definitive. What seems to be clear, particularly with inhalational anthrax, is the importance of early antibiotic administration. Inhalational anthrax has been fatal in almost all cases in which treatment was begun after patients were *significantly symptomatic*, regardless of the treatment regimen employed. Any person who is at high risk for possible exposure to anthrax must be put immediately on antibiotic PEP. All naturally occurring strains of anthrax tested to date have been shown to be sensitive to erythromycin, chloramphenicol, gentamicin, and ciprofloxacin. Presently, ciprofloxacin is FDA approved for treatment of inhalational anthrax. Ciprofloxacin, doxycycline, and penicillin G Procain are approved for use in PEP of inhalational anthrax. Doxycycline and penicillin G Procain are approved for use in cutaneous and GI anthrax (see Appendix for further details).

Inhalational Anthrax

Recommendations for inhalational anthrax call for a multidrug regimen of ciprofloxacin or doxycycline, plus a second antibiotic, such as rifampin, vancomycin, clindamycin, or an aminoglycoside (Table 14–6). After susceptibility testing, the regimen should be altered appropriately to include not only the most efficacious but also the least toxic antibiotics available. Recommendations are that treatment be continued for 60 days because of the risk of recurrent disease from delayed spore germination. However, once the patient is clinically improved and stable, parenteral administration can be changed to oral administration (Table 14–6). The American Academy of Pediatrics recommends the addition of intravenous clindamycin (30 mg/kg/day) to this regimin in children because it may provide additive benefit through decreased production of toxic cytokines produced by the bacillus.

Cutaneous Anthrax

Antibiotic therapy does not prevent or alter the natural history of the skin lesion but does minimize the likelihood of systemic disease occurring. Untreated, mortality from cutaneous anthrax is around 20%. Although penicillin has traditionally been used to treat cutaneous anthrax, current recommendations call for either ciprofloxacin or doxycycline to be used. If concurrent inhalation is possible, as would be the case in an anthrax attack, treatment is for 60 days (rather than for the 7 to 10 days for individuals exposed through a contaminated work environment). No topical therapy is indicated.

Gastrointestinal Anthrax

Current recommendations are to follow the guidelines for inhalational anthrax.

Postexposure Prophylaxis

There are no firm data on PEP for anthrax-exposed individuals. Asymptomatic people with likelihood of exposure to anthrax from a biological weapons attack should be treated prophylactically with oral administration of ciprofloxacin or doxycycline for 60 to 100 days to prevent infection via delayed spore germination (Table 14–7). Anthrax vaccine may also be offered along with antibiotics, but there is not an evidence-based recommendation for the vaccine at this time.

Further Treatment Considerations

- Although samples taken from the post-September 11 anthrax patients showed susceptibility to each of the three FDA-approved treatments, weaponized strains of anthrax are known to possess resistance to each of these.
- There is no clinical basis for recommending the use of multiple versus single drug regimens, but it is felt to be a reasonable therapeutic approach.
- If meningeal involvement is known or suspected, ciprofloxacin is preferred over doxycycline because of its improved CNS penetration, with additional coverage provided by penicillin, rifampin, or chloramphenicol.
- Adjunctive therapies such as steroids, anthrax IgG antisera, TNF inhibitors, and anthrax vaccine may have a role in improving the anthrax patient's condition; however, there are no clinical studies to support their use in this context.

TABLE 14–6 Recommendations for Management of Inhalational and GI Anthrax

Category	Initial Therapy	Duration
Adults	Ciprofloxacin 400 mg every 12 hr[a] or Doxycycline 100 mg every 12 hr[f] and One or two additional antimicro-bials[d]	IV treatment initially.[c] Switch to oral antimicrobial therapy when clinically appropriate: Ciprofloxacine 500 mg po BID or Doxycycline 100 mg po BID Continue for 60 days (IV and po combined)[g]
Children	Ciprofloxacin 10–15 mg/kg every 12 hr[h,i] or Doxycycline:[f,j] >8 yr and >45 kg: 100 mg every 12 hr >8 yr and ≤45 kg: 2.2 mg/kg every 12 hr ≤8 yr: 2.2 mg/kg every 12 hr and One or two additional antimicro-bials[d]	IV treatment initially.[e] Switch to oral antimicrobial therapy when clinically appropriate: Ciprofloxacin 10–15 mg/kg po every 12 hr[i] or Doxycycline:[f] >8 yr and >45 kg: 100 mg po BID >8 yr and ≤46 kg: 2.2 mg/kg po BID ≤8 yr: 2.2 mg/kg po BID Continue for 60 days (IV and po combined)[g]
Pregnant women[k]	Same for nonpregnant adults (the high death rate from the infection outweighs the risk posed by the antimicrobial agent)	IV treatment initially. Switch to oral antimicrobial therapy when clinically appropriate.[b] Oral therapy regimens same for nonpregnant adults
Immunocom-promised persons	Same for nonimmunocompro-mised persons and children	Same for nonimmunocompro-mised persons and children

[a, b] Ciprofloxacin or doxycycline should be considered an essential part of first-line therapy for inhalational anthrax.

[c] Steroids may be considered as an adjunct therapy for patients with severe edema and for meningitis based on experience with bacterial meningitis of other etiologies.

[d] Other agents with in vitro activity include rifampin, vancomycin, penicillin, ampicillin, chloramphenicol, imipenem, clindamycin, and clarithromycin. Because of concerns of constitutive and inducible beta-lactamases in *B. anthracis*, penicillin and ampicillin should not be used alone. Consultation with an infectious disease specialist is advised.

[e] Initial therapy may be altered based on clinical course of the patient: one or two antimicrobial agents (e.g., ciprofloxacin or doxycycline) may be adequate as the patient improves.

[f] If meningitis is suspected, doxycycline may be less optimal because of poor CNS penetration.

[g] Because of the potential persistence of spores after an aerosol exposure, antimicrobial therapy should be continued for 60 days.

Infection Control

Zoonotic forms of anthrax cannot be spread from person-to-person, nor does it appear that any of the post-September 11 anthrax patients became infected through person-to-person transmission. Standard barrier isolation is recommended for the inpatient setting. Neither HEPA air filtration systems nor N-95 respirators are indicated for the zoonotic forms. In the modern era in which weaponized anthrax may be presumed to be present, contact and airborne infection control practices and use of N-95 masks may be justified to prevent inadvertent exposure to health care workers and others.

The pressure on clinicians to be liberal in providing prophylaxis, administering anthrax vaccine, or prophylactic antibiotic therapy to all contacts of an index patient (including family, friends, and medical providers) is misdirected. This shotgun approach is unnecessary and has the potential to do harm. Only those who are likely to have had direct exposure to spores are candidates for PEP. Physicians should reassure those who have had contact with these patients that person-to-person transmission does not occur with anthrax and that antibiotic prophylaxis is unwarranted. This goes as well for all patient contacts, including health care workers. The risks of fluoroquinolones and doxycycline in children must also be considered when deciding who should get antibiotic prophylaxis. If exposure to spores (e.g., on infected clothing from work) is possible, PEP may be indicated. This should be decided on a case-by-case basis with adequate exposure information. Prophylaxis for the prevention of cutaneous anthrax is not indicated. In summary, then, unless there is a credible or identified risk for exposure to anthrax spores, routine PEP is not indicated.

The laboratory to be used in a suspected case must be notified so that appropriate handling measures can be in place. Local, state, and federal public health officials must be notified. Clinicians should be aware of the LRN put into place by the CDC and designed for rapid assessment of clinical specimens relating to bioterrorism.

---

[h] If intravenous ciprofloxacin is not available, oral ciprofloxacin may be acceptable because it is rapidly and well absorbed from the GI tract with no substantial loss by first-pass metabolism. Maximum serum concentrations are attained 1–2 hours after oral dosing but may not be achieved if vomiting or ileus are present.

[i] In children, ciprofloxacin dosage should not exceed 1 g/day.

[j] The American Academy of Pediatrics recommends treatment of young children with tetracyclines for serious infections (e.g., Rocky Mountain spotted fever).

[k] Although tetracyclines are not recommended during pregnancy, their use may be indicated for life-threatening illness. Adverse effects on developing teeth and bones are dose related; therefore, doxycycline might be used for a short time (7–14 days) before 6 months of gestation.

Courtesy of the CDC.

TABLE 14–7   Recommendations for Management of Cutaneous Anthrax[a]

Adults	Ciprofloxacin 500 mg bid or Doxycycline 100 mg bid	60 days[c]

Children	Ciprofloxacin 10–15 mg/kg every
	    12 hr (not to exceed 1 g/day)[b]
	or
	Doxycycline:[d]
	>8 yr and >45 kg: 100 mg every
	    12 hr
	>8 yr and ≤45 kg: 2.2 mg/kg every
	    12 hr
	≤8 yr: 2.2 mg/kg every 12 hr
		60 days[c]

Pregnant women[e]	Ciprofloxacin 500 mg bid or Doxycycline 100 mg bid	60 days[c]

Immunocompromised
    persons
	Same for nonimmunocompromised
	    persons and children
		60 days[c]

[a] Cutaneous anthrax with signs of systemic involvement, extensive edema, or lesions on the head or neck require intravenous therapy, and a multidrug approach is recommended (Table 14-6).

[b] Ciprofloxacin or doxycycline should be considered first-line therapy. Amoxicillin 500 mg po TID for adults or 80 mg/kg/day divided every 8 hours for children is an option for completion of therapy after clinical improvement. Oral amoxicillin dose is based on the need to achieve appropriate minimum inhibitory concentration levels.

[c] Previous guidelines have suggested treating cutaneous anthrax for 7-10 days, but 60 days is recommended in the setting of attack, with reasonable likelihood of exposure to aerosolized *B. anthracis*.

[d] The American Academy of Pediatrics recommends treatment of young children with tetracyclines for serious infections.

[e] Although tetracyclines and ciprofloxacin are not recommended during pregnancy, their use may be indicated for life-threatening illness. Adverse effects on developing teeth and bones are dose related; therefore, doxycycline might be used for a short time (7-14 days) before 6 months of gestation.

Courtesy of the CDC.

Sporicidal agents such as iodine (disinfectant strength must be used as antiseptic-strength iodophors are not sporicidal) may be used to clean instruments or work areas where spore contamination is a consideration. Standard antimicrobial cleansers such as hypochlorite may also be used for cleaning up any bodily fluids that may be infected; however, it must be remembered that the presence of organic material decreases the activity of hypochlorites.

Commercial tests for anthrax exist and can be used for environmental sampling. These field tests are useful as a preliminary screening of an environment for first responders and local law enforcement. Some of these are handheld devices that afford an ease of use in the field, although any positive finding should prompt more definitive testing. Sensitivity of these field-testing devices is dependent on the degree

TABLE 14–8 Interim Recommendations for Postexposure Prophylaxis for Prevention of Inhalational Anthrax After Intentional Exposure to *B. Anthracis*

Category	Initial Therapy	Duration
Adults (including pregnant women and immunocompromised persons)	Ciprofloxacin 500 mg po BID or Doxycycline 100 mg po BID	60 days
Children	Ciprofloxacin 10–15 mg/kg po q 12 hr[a] or Doxycycline: >8 yr and >45 kg: 100 mg po BID >8 yr and ≤45 kg: 2.2 mg/kg po BID ≤8 yr: 2.2 mg/kg po BID	60 days

[a] Ciprofloxacin dose should not exceed 1 g/day in children.
Courtesy of the CDC.

of contamination. Weaponized anthrax, as was used in the postal service attacks, may be present at concentrations below the detection threshold of these devices. Autonomous detection systems (ADS) are also being rapidly developed for commercial use. One such system for anthrax has been incorporated into a number of mail-sorting facilities by the U.S. Postal Service. They have the capacity for early detection and therefore can limit the spread of an epidemic and protect employees. As with all surveillance systems, they work best in conjunction with a well-designed emergency response plan and with a fully trained emergency response team.

Personal Protective Equipment

Individuals exposed or potentially exposed to anthrax spores will need to use PPE designed to limit dermal contact and inhalation. Depending on the location of work and the risk of exposure, gloves, face shields, and respirators equivalent to a NIOSH-approved N-95 should be worn in order to prevent spore exposure through airborne, cutaneous, or GI routes. Facial hair interferes with the seal of many respirators, in which case air-supplied respirators are indicated. Isolation, quarantine, or use of

Table 14–9 Critical First Steps: Inhalational Anthrax

Medical Treatment	Isolation, antibiotics, surgical masks during transport
Infection Control	Droplet precautions (surgical masks, gloves, gown, eye protection), PEP antibiotics, and respiratory protection for contacts exposed to spores
Public Health	Contact laboratory where specimens sent, LHD, SHD, CDC, local law enforcement

TABLE 14–10   Quick Reference Guide: Anthrax

Diagnostic Keys	Clustering of cases with the following symptoms in a healthy population
	Inhalational: flu-like illness, CT with dense hilar and medi astinal nodes
	Cutaneous: papular lesions that become escharic; regional adenopathy; viral-like prodrome
	Gastrointestinal: vomiting, bloody diarrhea, acute abdomen, oral ulcers
Risk Factors	Occupational risks: postal workers, public figures, veterinarians, agricultural workers
Transmittable	No person-to-person transmission known
Transmission	Contact with spores on surface or in air
Management	Fluoroquinolones
Prevention	If high risk of exposure, vaccine available
Preexposure	None
Postexposure	PEP ciprofloxacin for 60–100 days
Containment	Patient: standard precautions
	Secondary contacts: none
Populations at Heightened Risk	Postal workers, public figures, veterinarians, agricultural workers
Public Health Concerns	Determining site/source of contamination
Contacts	Local, State DPH, CDC, FBI
Medical Measures	Fluoroquinolones, supportive measures
Exposure Control	None; no person-to-person transmission
Decontamination	Washing with hypochlorite solution
Public Health Interface	Immediate notification of health departments, law enforcement

negative-pressure rooms for patients diagnosed with anthrax is unnecessary because anthrax is spread only by spores and not by the bacteria causing clinical disease. Cutaneous anthrax can be spread from surface to surface, but requires direct contact.

Anthrax Scares

There have been a spate of scares following the anthrax outbreak in the United States that have received a great deal of media attention. One of the questions that

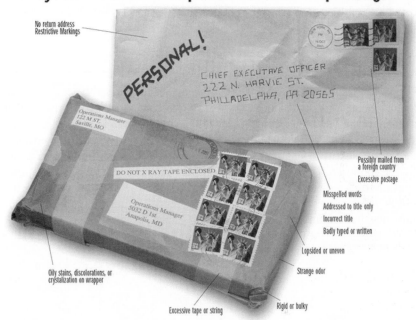

FIGURE 14–13   USPS poster on how to handle suspicious letters.
Courtesy of the USPS.

often arises is what to do should a suspicious package or envelope be noticed. Recommendations have evolved over time that have only added to the confusion among the public as to what should be done. The best advice is leave the package alone, and do not try to cover or put it into a plastic bag. Simply leave it where it is, vacate the room (Fig. 14–3), and notify local and state public health officials. They will ensure proper steps are taken to identify whether anthrax is present. If anthrax is present, risk assessment, risk communication, and PEP will be needed, and clinicians will be guided in these efforts by the expertise provided by those coordinating the public health response (Fig. 14–13, see also Chapter 3, Table 3–2).

☣Summary

Anthrax has been used as a biological weapon. It has three district clinical syndromes: inhalational, cutaneous, and GI. In the context of bioterrorism, inhalational or cutaneous forms all predominate. Inhalational anthrax presents as a flulike illness with characteristic CT findings of hemorrhagic mediastinitis. The characteristic feature of cutaneous anthrax is the eschar. Infection occurs only by contact with the spore, a fact that directly influences the public health response to anthrax. Exposure to ill individuals does not confer risk, whereas exposure to an environment in which spores are present does. Early identification and implementation of antibiotics (fluoroquinolones) is life saving.

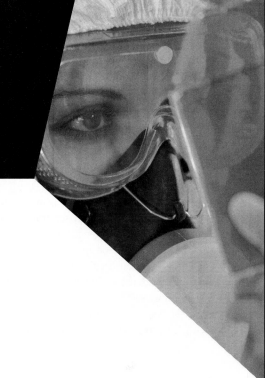

CHAPTER 15

Plague
(Yersinia pestis)

Clinical Vignette

You are a pediatrician in an affluent suburban community. The mother of a 12-year-old boy brings him in for urgent evaluation. The youngster has been ill with fever, headache, and malaise for 4 days. His family has just returned from a two-week camping vacation in New Mexico. He appears wan and fatigued but is alert and oriented. Examination reveals fever and a slightly erythematous ear drum. You diagnose otitis media and initiate antibiotics. Two days later, his distraught mother calls saying that her son's fever is worse, with relentless vomiting and severe abdominal pain. Suspecting appendicitis, the boy is referred to the ER. Surgery consultation is called for, and routine labs, including CBC, LFTs, and blood cultures are drawn. His WBC count is 25K with 60% bands, and he is taken to surgery. Intraoperatively, his appendix appears normal, but diffuse retroperitoneal and mesenteric lymphadenopathy is observed. He subsequently develops sepsis, ARDS, and DIC and is transferred to the ICU. In the ICU, he complains of severe headache and appears disoriented. His fingers are black, purpuric lesions are seen on his trunk and upper arms and Kernigs and Brudzinki's signs are positive. That night, the lab calls stating that his blood cultures are growing gram-negative rods with a "safety pin" appearance. A rapid DFA sent to the State Health Department confirms Yersinia pestis.

Background

The first documented pandemic of *Yersinia pestis* occurred in 561 AD and is often called the Justinian Plague after the Holy Roman Emperor at the time. It began in Egypt and spread along trade routes killing upwards of 60% of the populations of Europe, North Africa, and southern and central Asia. The second great Yersinia outbreak, the so-called "Black Death," occurred in 1346 and killed 25 million people, a third of the population of Europe (Fig. 15–1). Both pandemics were fueled by existing land and sea trade among countries in Europe, Asia, and Africa. The third pandemic began in China in 1855 and eventually spread worldwide. There continue to be intermittent outbreaks of plague throughout the world, although these outbreaks are isolated and better controlled primarily because of higher standards of living, improved sanitation and hygiene, and the availability of antibiotics.

    Yersinia pestis is a nonmotile, non-spore-forming, gram-negative rod. Infection with *Y. pestis* tends to take one of three forms: bubonic, septicemic, and pneumonic.

FIGURE 15–1　Boccaccio's Decameron—Depicting the Black Plague of 1348.
Courtesy of Brown University.

FIGURE 15-2 **Examination of rats suspected of carrying bubonic plague, New Orleans, 1914.**
Courtesy of United States Public Health Service/National Library of Medicine.

Historically the bubonic form of plague has been the most common. Classically, the disease is transmitted from infected rodents to humans by fleas residing on rats. As a biological weapon, pneumonic plague spread through aerosolization is expected to be the predominant form (Figs. 15–2 and 15–3).

OF NOTE . . .

Biblical Plague

Then the Lord said to Moses and Aaron, "Take for yourselves handfuls of soot from a kiln, and let Moses throw it toward the sky in the sight of Pharaoh. And it will become fine dust over all the land of Egypt, and will become boils breaking out with sores on man and beast through all the land of Egypt." So they took soot from a kiln, and stood before Pharaoh; and Moses threw it toward the sky, and it became boils breaking out with sores on man and beast. And the magicians could not stand before Moses because of the boils, for the boils were on the magicians as well as on all the Egyptians.

Exodus 8–9

FIGURE 15-3   Applying rat poison to loaves of bread to contain a plague outbreak in the United States.
Courtesy of United States Public Health Service/National Library of Medicine.

Yersinia as a Bioweapon

As Japan's use of plague in China during WWII demonstrated, the effective use of plague as a bioweapon requires little sophistication and can have dramatic effects (Fig. 15–4). Most countries engaged in biological weapons research and development programs aim to use aerosolized *Yersinia*, eliminating the need for the rat host or flea vector and enhancing transmissibility and virulence. In 1970, the WHO estimated that 50 kg (110 lb) of *Y. pestis* sprayed over a city would infect 150,000 people and kill roughly 40,000. Both North Korea and Iraq are believed to have been engaged in weaponization of plague. The former Soviet Union's Biopreparat conducted massive research on weaponized plague and large quantities were stockpiled. It is sobering to acknowledge that many scientists involved in this research, and the stores themselves, remain largely unaccounted for at present.

Distinguishing naturally occurring plague from a bioterrorist attack can be aided by epidemiologic clues, but it may not be readily apparent at the onset. Large numbers of cases occurring temporarily or spatially are highly suggestive of a biological attack. Crucial additional clues are the diagnosis of plague in areas with nonexistent, or very limited, historical experience, or in cities where no known zoonotic reservoir exists. Infection in patients with no discernible risk

FIGURE 15-4 Unit 731 bodies of soldiers subjected to experimentation by Japanese scientists.

Courtesy of Advocacy & Intelligence Index for POW-MIAs, Inc.

factors—such as exposure in research or occupational settings—or plague cases in absence of recent documentation of rodent dieoffs are useful clues as well.

The clinical form of plague might itself suggest either a naturally occurring zoonotic form or a deliberate attack. For example, naturally occurring epidemics of the pneumonic form of the plague are extremely rare and should raise suspicion of bioterrorism. Because the signs and symptoms of weaponized plague may be broad, primary care physicians must be cognizant of zoonotic and weaponized forms. Even with a plausible zoonotic source, any diagnosis of plague should raise the possibility of bioterrorism in today's society.

Epidemiology

In the past 50 years there have been fewer than 2,000 cases of plague in the United States. The clinical presentation of naturally occurring plague in the last 50 years has been dominated by the less deadly bubonic form (over 80% of all cases) with overall mortality at 14%. Another 13% of plague cases were septicemic with a mortality rate of 22%. Less than 5% of all U.S. plague cases presented in the pneumonic form, although the mortality rate is much higher (57%).

Populations at higher risk for naturally occurring plague are rural dwellers, hunters, campers, and American Indians who live in endemic areas of the Southwest. The presence of plague among prairie dogs and other domesticated animals represents a potential risk to rural dwellers, veterinarians, veterinary assistants, and others involved in animal control efforts, as well as pet store employees and pet owners.

Means of Transmission

Naturally occurring plague transmits to humans from bites of infected fleas that typically derived their infection through having bitten infected rodents (Fig. 15–5). Generally, such transmission results in the bubonic form of the plague. A small percentage of those bitten, however, will develop a septicemic form referred to as primary septicemic plague. It is important to note that neither of these two forms are transmitted from one human to another. Those infected with a pneumonic form, in contrast, can spread the disease through aerosolized droplets. In a biological attack, the most likely means of transmission would be through aerosolization of plague bacilli. Consequently, the clinical scenario would be the opposite of that seen with naturally occurring plague. That is, the pneumonic form would dominate and mortality

FIGURE 15–5   Electron microscopic image of the flea that transmits the plague.
Courtesy of the CDC.

would be far higher than with naturally-occurring plague. The success of such an attack would still be influenced by a number of variables, including climatic conditions, or the amount and type of *Y. pestis* strain used. It is possible for a biological attack to come in the form of deliberate infection of a natural animal reservoir in a major city. However, experts believe this is an inefficient means of initiating a plague epidemic.

Pathogenesis

When a flea ingests *Y. pestis* from an infected source, such as a rat, the bacterium undergoes replication in the flea's esophagus. The esophagus becomes blocked, ultimately, causing regurgitation of the bacterium when it is feeding on hosts, thus spreading the disease. In zoonotic transmission, bites from plague-carrying fleas introduce upward of a thousand organisms into the dermis (Fig. 15–6). Once present in the skin, *Y. pestis* makes its way through the lymphatics to regional lymph nodes where rapid proliferation results in lymph node necrosis. Bacteremic spread from regional lymph nodes and septicemia and shock may follow unless timely treatment is initiated.

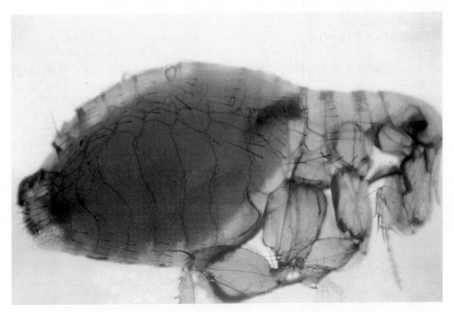

FIGURE 15–6 Multiplication of *Y. pestis* within the flea induces regurgitation and thus spread of the bacteria.
Courtesy of the CDC.

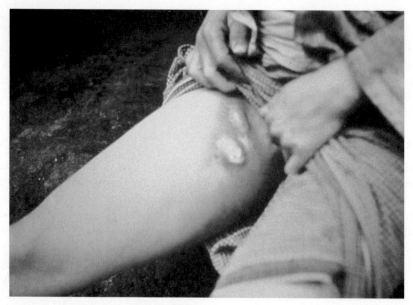

FIGURE 15–7   Plague: an inguinal bubo.
Courtesy of the CDC.

☣Bubonic Plague

Symptoms of the bubonic form of plague begin between 2 and 7 days after receiving an infected flea bite. Common clinical features include the sudden onset of fever, headache, chills, and malaise. Hepatomegaly or splenomegaly may be present as well. The pathognomonic sign of bubonic plague, of course, is the bubo— one or more painful, erythematous, warm, swollen lymph nodes usually 10 cm or less in size. Buboes are most commonly seen in the axilla, the groin, or cervical region. The pain from the bubo can be so intense that the patient may have limited movement in that area (Fig. 15–7). Patients may experience ulcerations at the bite site and secondary septicemia, the latter occurring in about 25% of cases. The mortality rate of untreated bubonic plague is 60%. With timely treatment, mortality is less than 5%.

☣Primary Septicemic Plague

Primary septicemic plague occurs when *Y. pestis* is introduced directly into the bloodstream from an infected flea. In these patients, the characteristic bubo is often not present. The clinical presentation in primary (and secondary) septicemic plague is similar to that caused by any gram-negative infection: high

FIGURE 15-8   Acral gangrene caused by septicemic plague.
Courtesy of the CDC.

fever, rigors, malaise, hypotension, nausea, vomiting, and diarrhea, and commonly, disseminated intravascular coagulation (DIC). A distinctive feature of septicemic plague is the presence of acral thrombosis, resulting in necrosis and gangrene of those regions. Black necrotic appendages and more proximal purpuric lesions caused by endotoxemia are often present (Fig. 15–8). Seeding of the CNS can occur and is accompanied by typical signs of meningitis such as meningismus and high fever (Figs. 15–8 and 15–9).

Pneumonic Plague

Primary pneumonic plague occurs when the Y. *pestis* is inhaled directly into the respiratory system. It is characterized by severe bronchopneumonia, chest pain, dyspnea, cough, and hemoptysis. Bloody sputum is one of the clinical hallmarks of pneumonic plague. The incubation period is also between one and six days. Fever occurs with dyspnea and cough that may be productive of serous, commonly bloody, or possibly purulent sputum. Pharyngitis may also be seen in these patients accompanied by cervical adenopathy. Primary pneumonic plague patients may suffer from GI distress including nausea, vomiting, pain, and diarrhea. Clinicians should be aware that a bioterrorist attack involving plague would likely present as a primary pneumonic infection. A less likely presentation is the GI or typhoidal form of the disease. Hematogenous spread of Y. *pestis* to the lungs, referred to as secondary pneumonic

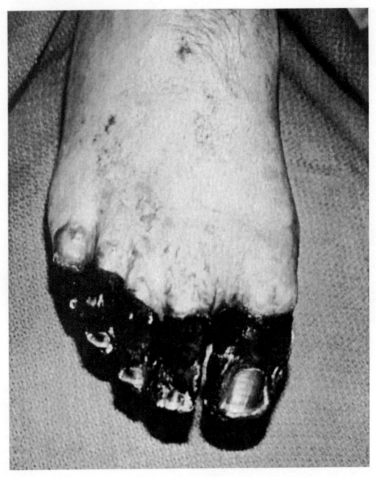

FIGURE 15-9 **Acral gangrene as part of the presentation of septicemic plague forms.**

Courtesy of the CDC.

plague, can result from either the septicemic or bubonic form of the disease. Symptoms mirror those of primary pneumonic plague with severe bronchopneumonia, chest pain, dyspnea, cough, and hemoptysis.

Distinguishing primary pneumonic plague from secondary can be difficult, but the absence of buboes is suggestive of primary plague. Unfortunately, the most definitive way to distinguish is via pathological examination.

Pneumonic plague, as with all plague forms, may progress to a septic picture. As discussed previously, complications of DIC, purpura, small vessel necrosis (leading to gangrene in the periphery), azotemia, and multiorgan failure also may evolve from pneumonic plague.

In the preantibiotic era, mortality from primary pneumonic plague was 100%. However, judicious and timely use of antibiotics has helped considerably. Timeliness of treatment is essential in reducing mortality significantly. Treatment delay of more than 18 hours is associated with very high mortality rates. Death is usually secondary to respiratory failure, circulatory collapse, multiorgan failure, or DIC.

Laboratory Tests

Routine laboratory tests are nonspecific and therefore cannot exclude or rule in plague. Common findings include a leukocytosis of up to 20,000 WBCs with greater than 80% polymorphonuclear neutrophils (PMNS) with toxic granulations. If severe, coagulopathies and elevated liver function test may be seen. A rapid DFA test for plague is now available in most state laboratories and should facilitate early diagnosis of suspected plague.

Microscopy

Yersinia is a lactose nonfermenter, urease negative, and grows best on MacConkey or blood agar plates. Detection of growth takes approximately 48 hours and yields distinctively small cultures. Gram stain will identify it as a gram-negative organism, a fact that may influence clinical management. A Wright, Giemsa, or Wayson stain reveals the *Y. pestis*'s classic "safety pin appearance" (Fig. 15–10). If plague is

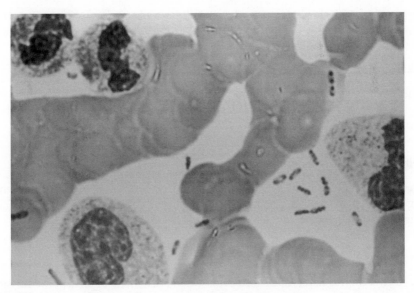

FIGURE 15-10 **Plague microscopy. Note the safety pin appearance.**
Courtesy of the CDC.

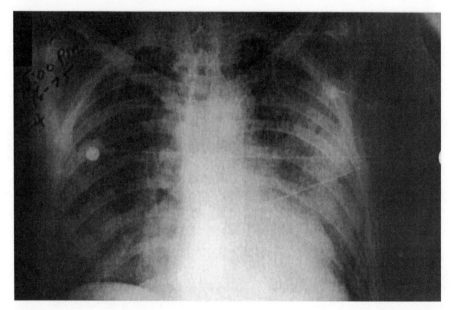

FIGURE 15-11   Chest radiograph of progressing pneumonic plague showing bilateral lung involvement.
Courtesy of the CDC.

suspected, the lab should be notified as BSL-2 protocols need to be initiated to prevent occupational infections.

Radiographic Findings

Chest radiography in the pneumonic form is likely to demonstrate a typical bronchopneumonia with bilateral infiltrates or consolidation, are of less value for other forms of plague (Fig. 15–11).

Differential Diagnosis

The DDx of plague includes ARDS, anthrax, cat scratch disease, cellulitis, necrotizing fasciitis, pneumonia, tick-born diseases, gas gangrene, and anthrax. GI anthrax mimics acute appendicitis in children.

Diagnosis

Plague infection is never an isolated event whether zoonotic or an act of terrorism. Any diagnosis of plague should signal the potential for an epidemic, and timely notification of local, state, and federal agencies is essential.

Early diagnosis of plague limits mortality in patients and protects the public's health. Reports of sudden increases in pneumonia cases complicated by sepsis may signal a bioterrorism attack. Patients who are otherwise healthy and without risk factors for pneumonia, presenting with acute onset of dyspnea, fever, and a rapidly progressing pneumonia, in the right context, should elicit consideration of plague. Hemoptysis is highly suggestive of plague, even more so if temporal case clustering occurs. If this identical scenario is seen, but without hemoptysis, consider anthrax as well.

Gram staining will identify it as a GNR and a safety pin appearance will make plague a primary diagnostic consideration. Most zoonotic strains of plague produce an F1-antigen *in vivo*, which can be detected in blood samples via a rapid DFA immunoassay. This test is available from both the CDC and state health departments. A four-fold rise in antibody titer in patient serum is diagnostic, but unfortunately offers only retrospective confirmation of the disease and treatment cannot be delayed pending this information.

Chronic Effects

Long-term complications of plague have not been well described in humans.

Vulnerable Groups

Pregnant women, children, the elderly, and immunocompromised patients are vulnerable to plague as with other bioweapons. In the preantibiotic era, maternal and fetal outcomes were dismal with markedly elevated mortality rates and fetal loss. Early antibiotic treatment has improved these outcomes dramatically, although children with septicemic forms of plague still have very high death rates.

TABLE 15–1 Making the Diagnosis: Plague	
Pattern of Presentation	Multiple, concurrent cases of fever, cough, chest pain, dyspnea; results in high morbidity and mortality
	Multiple, concurrent cases of gastrointestinal symptoms: diarrhea, nausea, vomiting, abdominal pain; with high morbidity and mortality
Features	Variable according to form of infection: cutaneous–cervical or inguinal bubo, gangrenous acral areas (late); respiratory-tachypnea, dyspnea, cyanosis, consolidations on chest exam; systemic-sepsis, shock, multiorgan system failure
Findings	Variable according to form of infection:
	Chest x-ray—consolidations, infiltrates
	Microscopy—bipolar or "safety pin" appearance of gram-negative bacilli

OF NOTE . . .

New York City, November 2002

Almost a year after the post office anthrax attacks, two cases of bubonic plague were diagnosed in New York City in a married couple. The first patient, a 55-year-old man, presented to a Manhattan hospital after 2 days of fever, chills, and unilateral inguinal tenderness and swelling. During the hospital course, blood cultures confirmed *Y. pestis*, and the patient's clinical status deteriorated. He was intubated and placed in the ICU after developing septicemia. The patient eventually recovered.

The second patient, a 47-year-old female, presented with fever, chills, myalgias, and unilateral inguinal tenderness and swelling. Because her husband had been diagnosed already, she was treated presumptively and had a better clinical course. It turned out that the cases were naturally-occurring as both patients were tourists visiting from New Mexico. It was determined that they had likely been infected at their rural home.

The NYC Department of Health and Mental Hygiene and the CDC began an investigation to determine the source whether a bioterrorist attack had occurred. In cooperation with the New Mexico Department of Health, investigations of their home revealed the presence of flea-infested rodents carrying *Y. pestis*. Laboratory studies revealed the *Y. pestis* strains carried by the fleas and the type that infected the couple were identical. Consequently, use of *Y. pestis* as a biological weapon was ruled out.

Immunosuppressed patients may be more readily infected and further may demonstrate accelerated disease progression, modified disease presentation (e.g., less prodromal features), poorer disease localization, and prolonged convalescence.

Treatment

No randomized clinical trials have been done to determine optimal treatment for plague. Current recommendations include initiating antibiotic therapy as soon as the diagnosis is suspected, because delay increases mortality significantly. The treatment of choice is 10 to 14 days of either streptomycin or gentamicin. Alternative antibiotic options include chloramphenicol, ciprofloxacin, or doxycycline. Clinical improvement is usually seen within 3 to 4 days, but the extended regimen prevents relapse. Chloramphenicol is the drug of choice for meningitis. Standard supportive therapies should be initiated when indicated clinically. Buboes will resolve with antibiotic therapy and generally require little in the way of specific management. Incision and drainage, or aspiration, of buboes may increase risk of transmission, but can be done for pain relief.

Prophylaxis with 1 week of oral doxycycline, or 1 week plus equal duration of exposure is indicated for asymptomatic persons with known exposure or in mass-casualty circumstances. Tetracycline or ciprofloxacin may be substituted (Table 15–2).

Theoretical concern with antibiotic choice in pregnant women and children are given less influence in the setting of mass casualties. For example, both ciprofloxacin

TABLE 15-2   Treatment of Plague

Category	Initial Therapy
Adults	Streptomycin, 15 mg/kg lean body mass IM every 12 hr for 10 to 14 days
	or
	Gentamicin, 5 mg/kg lean body mass IM/IV every 24 hr for 10 to 14 days. Then, 1.75 mg/kg lean body mass IV every 8 hr for 10 to 14 days
	or
	Ciprofloxacin, 400 mg IV every 12 hr
	Oral therapy may be given (750 mg orally every 12 hr) after the patient is clinically improved, for completion of a 10- to 14-day course of therapy
	or
	Doxycycline, 200 mg IV loading dose followed by 100 mg IV every q 12 hr 100 mg po q 12 hr after clinically improved, for completion of a 10- to 14-day course
Children	Streptomycin, 15 mg/kg lean body mass IV q 12 hr
	or
	Gentamicin, 2.5 mg/kg lean body mass IV q 24 hr. Then, 1.75 mg/kg lean body mass IV every 8 hr for 10 to 14 days
	or
	Ciprofloxacin, 400 mg IV every 12 hr
Pregnant Women	Gentamicin, 5 mg/kg lean body mass IM/IV q 24 hr for 10–14 days. Then, 1.75 mg/kg lean body mass IV tid hr for 10 to 14 days
	or
	Ciprofloxacin, 400 mg IV q 12 hr
	or
	Doxycycline, 200 mg IV loading q day
Widespread Outbreak and Prophylaxis	10-day course for treatment; 7-day course for prophylaxis Adults: Ciprofloxacin, 500 mg po q 12 hr Doxycycline, 100 mg po q 12 hr
Plague Meningitis	Chloramphenicol, 25 mg/kg IV loading dose, followed by 15 mg/kg IV qid. Oral therapy may be given after the patient is clinically improved, for completion of a 10- to 14-day course of therapy[a]

[a] Patient may become afebrile shortly after therapy starts, but course must be completed to prevent relapse.

Adapted from Thomas V. Inglesby, David T. Dennis, Donald A. Henderson, et al. Plague as a Biological Weapon: Medical and Public Health Management. JAMA 2000;283:2281–2290.

TABLE 15–3   Critical First Steps: Plague

Medical Measures	Initially, doxycycline 200 mg po, then 100 mg every 12 hr
Exposure Control	Uncertain; patient isolation, gown, gloves, surgical mask until treated for 48 hr Decontamination not likely needed; standard cleaning agents adequate
Public Health Interface—Local, State DPH CDC	24-hour emergency response hotline (770) 448-7100. Bioterrorism Preparedness and Response Program (404) 639-0385 FBI: Regionally, or FBI Headquarters in Washington, DC, 202-324-3000

and doxycycline are approved for use in this setting. PEP in adults is with doxycycline (100 mg bid) or ciprofloxacin (500 mg bid) for seven days. In children, the same drugs are used but calculated based on body weight (doxycycline 5 mg/kg bid; ciprofloxacin 20–30 mg/kg bid).

 TABLE 15–4   Quick Reference Guide: Plague

Diagnostic Keys	Bubos, gangrenous acral areas, signs and symptoms of pneumonic process
Risk Factors	Rural travel, government job
Transmittable	Likely with pneumonic form
Transmission	Aerosolized droplets
Management	Doxycycline supportive
Prevention	Vaccine—none
Preexposure	Doxycycline
Postexposure	Doxycycline or ciprofloxacin
Containment	Patient: isolation, droplet precautions if pneumonic form Secondary contacts: none
Populations at Heightened Risk	Immunocompromised, elderly, pregnant women
Long-term Effects	Uncertain
Public Health Concerns	One case implies a potential epidemic or bioterrorism
Contacts	Local, state DPH, CDC, FBI

Vaccine

Although vaccine research and development is underway to prevent the pneumonic form of plague, none is currently available.

Infection Control

There are little data concerning person-to-person transmission, so recommendations are based on likely means of transmission. At this time, respiratory isolation with droplet nuclei precautions are recommended for the pneumonic form, whereas for the bubonic form standard precautions are sufficient. Patient isolation is indicated until the patient has been on antibiotics for at least 48 hours, a favorable clinical response is seen, or cultures come back negative. Decontamination for plague-exposed surfaces is unnecessary because *Y. pestis* is not a particularly hardy bacterium and cannot survive for long periods outside of a host. Unlike anthrax, no spore form for *Y. pestis* exists, so normal environmental conditions do not permit prolonged survival of the organism unlike anthrax.

☣Summary

Plague has been engineered as a bioweapon by numerous countries, and clinicians must be alert to its presenting features. A deliberate release would probably occur in aerosol form and present with the pneumonic form. Pneumonic plague is characterized by chest pain, hemoptysis, and lobar consolidations. A wide range of antibiotics offered are effective for plague but must be administered early. Patient isolation, contact surveillance, and postexposure prophylaxis are important infection control and public health interventions. Vulnerable groups, particularly those with compromised immune systems, may have rapid or atypical presentations and require longer treatment and convalescence. Early antiobiotic intervention greatly improves outcomes and should be given prophylactically to contacts. Survivors demonstrate little long-term sequelae and have lifelong immunity.

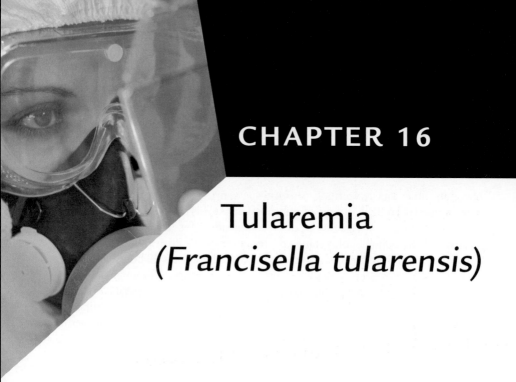

CHAPTER 16

Tularemia
(Francisella tularensis)

Case Vignette

It's mid-February when an otherwise healthy 42-year-old stockbroker presents to your office with fever, chills, and malaise. He reports no sick contacts, no exotic travel but mentions having spent the previous weekend on a fruitful hunting trip that included shooting and trapping rabbits. CXR reveals hilar adenopathy. Your differential includes influenza, tuberculosis, sarcoidosis, histoplasmosis, and tularemia. What should you do?

Background

A gram-negative coccobacillus, tularemia was first identified in 1911 by a scientist investigating an outbreak of what was initially thought to be bubonic plague in Tulare County, California. The disease has multiple means of transmission, and water-borne epidemics were seen in Europe and the Soviet Union in and around World War II (WWII). It has been suggested that the Soviet outbreak, which occurred at the Eastern front during WWII, was a result of a deliberate biological attack by the Russians. In fact, tularemia has been a favorite agent for bioweapons research since the 1930s. In 1969, the World Health Organization (WHO) estimated that 10 kg (20 lb) of aerosolized *F. tularensis* could infect 50,000 people, killing approximately 4,000. These numbers may be outdated because antibiotic and vaccine-resistant

232

FIGURE 16-1 Rabbits and other small mammals act as vectors for tularemia.
Photo provided by PDPhoto.org.

strains of weaponized tularemia—reported to have been developed at Biopreparat—would be far more deadly.

Francisella

Francisella tularensis, the causative agent of tularemia, is a small, nonmotile, facultative aerobic, intracellular gram-negative coccobacilli. Tularemia grows in aerobic environments. Despite its inability to form spores, it is nonetheless quite hardy and able to persist for several weeks in water, soil, vegetation, or in animal products. *F. tularensis* has three different species of which biovar tularensis is the most virulent form, and it is the one most commonly seen in the United States. The organism is considered to be one of the most communicable bacterial pathogens known. Disease hosts include rabbits, rats, and other small mammals that attain the infection through direct contact or by insect vectors, such as ticks and mosquitoes.

Epidemiology

F. tularensis is found worldwide, though global disease incidence has not been determined. Tularemia occurs sporadically throughout the continental United States, but predominantly in the rural areas of the South, Southwest, or Midwest. Since the

1990s, fewer than 200 cases are reported annually and occur in a bimodal seasonal pattern. The majority of cases are diagnosed in the summer/fall and are thought to be secondary to arthropod (tick) transmission. A smaller fall/winter peak coincides with hunting seasons. Most cases are secondary to direct contact of some kind, although there are infrequent cases of infection from inhalation. No significant differences in infection patterns are seen by age or by gender. Overall mortality rates prior to the advent of antibiotics ran between 5% and 15%, and for pneumonic forms between 30% and 60%. In the antibiotic era, overall mortality is less than 2%.

High-Risk Groups

Naturally occurring tularemia is a zoonosis that traditionally infects outdoorsmen, trappers, hunters, and those handling the infected carcasses of small mammals. Veterinarians, animal control workers, butchers, and farmers are occupations at risk. Laboratory workers may inadvertently infect themselves as well. Rural dwellers, particularly Native Americans, are at increased risk as well.

Means of Transmission

Most human tularemia is tick-borne. Because infection can occur through bacterial contact with skin, mucosal linings, gastrointestinal epithelium, and the respiratory tract, humans can become infected by being bitten by infected ticks, by handling contaminated animals or animal products, by ingesting contaminated food or water, or by inhaling aerosolized bacteria. Person-to-person transmission with *F. tularensis* does not occur. The greatest public health threat from weaponized tularemia is as an aerosol, although other means are possible. Aerosol dispersion in a densely populated area would likely result in widespread reports of fevers and nonspecific symptoms (see following clinical description) within five days. A large percentage of these people would develop pleuritis or pneumonitis. The abrupt onset of this constellation of symptoms in large numbers or in otherwise healthy subpopulations should alert physicians and public health officials to a possible bioweapons attack.

Pathogenesis

The factors responsible for *F. tularensis's* virulence are not well known. Once inoculation occurs, regardless of the site, an intense focal inflammatory response occurs. Immigration of neutrophils and other leukocytes results in a suppurative necrosis followed by granuloma formation. Pathology of the site shows a noncaseating, centrally necrotic region enclosed by epitheliod cells and multinucleated giant cells as well as fibroblasts in an array consistent with granulomatous pathology. *F. tularensis* survives neutrophil oxidation and is phagocytized and transported to regional lymph nodes by macrophages. Once there, the bacterium multiplies and spreads via lymphatics and blood to the kidneys, spleen, liver, and lungs. Bacteremia is detectable early in the infection. In primate studies of inhalational tularemia, peribronchial inflammation and alveolar septal inflammation occured within 72 hours of infection; later, pneumonic consolidations, granulomas, and ultimately chronic interstitial fibrosis developed. Inhalational tularemia is the bioweapon of choice and may present in the pneumonic, typhoidal, or ulceroglandular form.

Clinical Presentation

Following an incubation period of anywhere between 2 to 20 days (mean 3–5 days), the abrupt onset of the following nonspecific constellation of symptoms herald the beginning of tularemia: fevers, chills, runny nose, myalgias, dry cough, headache, and prostration; hemoptysis, pleuritic pain, or dyspnea is noted. Symptoms such as nausea, vomiting, and diarrhea may be seen. As the illness progresses, anorexia, weight loss, and sweats ensue. If left untreated, these symptoms may continue for up to 4 weeks or possibly longer.

Tularemia's clinical presentation reflects both the mode of transmission and the route of infection. The form of tularemia influences mortality, but it has little effect on treatment decisions. Regardless of the initial syndrome, infection may spread hematogenously resulting in pleuropneumonia, septicemia, and even meningitis. Tularemia sepsis is of particular concern because of its high fatality rate if untreated. Clues to septicemic tularemia include fever, diarrhea, vomiting, and abdominal pain. Further progression results in mental status changes, coma, and finally shock, DIC, ARDS, and multisystem organ failure.

Six distinct clinical syndromes have been described; however, these can exist as a continuum, and substantial syndromic overlap may present in the same individual. The six clinical syndromes are:

1. Ulceroglandular (80% of cases). This is the dominant naturally occurring form of tularemia and is seen when the means of transmission is from the bite of infected ticks or animals, or through direct contact with infected carcasses. Symptoms begin with fever and a solitary papular lesion—typically at the site of contact—that becomes centrally necrotic and tender within a few days and may be covered with an eschar; reminiscent of anthrax. Regional lymphadenopathy is noted several days after the papule appears (Figs. 16–2 and 16–3). Despite initiation of antibiotic therapy, lymph nodes may become progressively more fluctuant and may rupture. Pneumonia develops in 10% to 15% of patients (see discussion of pneumonic tularemia).

2. Glandular (10% of cases). This form (considered by some to be a subset of ulceroglandular) presents with one or more enlarged regional lymph nodes but without any cutaneous lesions or ulcerations (see Fig. 16–4).

3. Typhoidal (10% of cases). Generally occurring after an inhalational exposure (though it can occur from skin or mucosal contact), patients present with fever and chills but without any clear focus of infection. Patients may develop rhabdomyolysis and subsequent renal failure, sepsis, and DIC (even with apparently negative blood cultures). Pneumonia may develop, as well, and can be quite severe. Pneumonia can occur in all the forms but is most commonly seen as

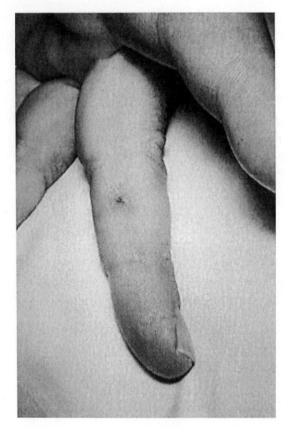

FIGURE 16-2   Finger ulceration typical of infection with tularemia.
Courtesy of the CDC.

part of typhoidal tularemia (see later for discussion of pneumonic tularemia). The typhoidal form is distinctive in that, unlike the other forms, lymphadenopathy is absent. In the event of a deliberate attack using aerosolized *F. tularensis,* typhoidal tularemia would likely be a dominant form. Untreated, typhoidal tularemia has a mortality rate of roughly 35%.

4. Pneumonic (less than 17% of cases). Occurring rarely in nature, the primary pneumonia form results from inhaling *F. tularensis.* It is seen mostly in laboratory workers. Secondary pneumonic tularemia from hematogenous spread of an existing infection, is more common. The mortality rate for pneumonic tularemia is high, regardless of the source. Infection or inflammation of any part of the respiratory tract occurs. If the pneumonia is secondary to hematogenous spread, then these symptoms are superimposed on the preexisting systemic symptoms. Conversely, inhalational exposure may cause typhoidal symptoms without any major respiratory symptoms. It is important to note that the aggressiveness of

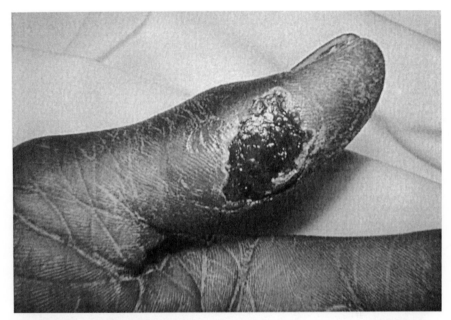

FIGURE 16-3   Ulceroglandular tularemia.
Courtesy of the CDC.

the pneumonic process is highly variable sometimes leading to respiratory failure and at other times does not even develop into full-fledged pneumonia. Because aerosol distribution is the probable form of a biological attack, it is important to be familiar with the pneumonic form.

5. Oropharyngeal (fewer than 5% of cases). Ingestion of contaminated tissue or water, or possibly through inhalation of aerosols, is the means of infection in this form. The presentation is marked by exudative pharyngitis. Impressive cervical or retropharyngeal lymphadenopathy may be noted on exam, and occasionally oropharyngeal ulcerations. Diagnosis is sometimes made after treatment for more common causes of pharyngitis fails.

6. Oculoglandular (fewer than 1% of cases). This form results from ocular contact with *F. tularensis.* Unilateral, painful, purulent, or ulcerating conjunctivitis are the presenting symptoms. Other findings include preauricular lymph node enlargement and periorbital edema.

Chronic Effects

Little is known about the long-term sequelae of tularemia. Survivors of pneumonic tularemia may demonstrate chronic granulomatous changes and interstitial fibrosis. Ocular disease may result in permanent conjunctival scarring. Individuals who survive tularemic sepsis or meningitis may demonstrate multiorgan system effects,

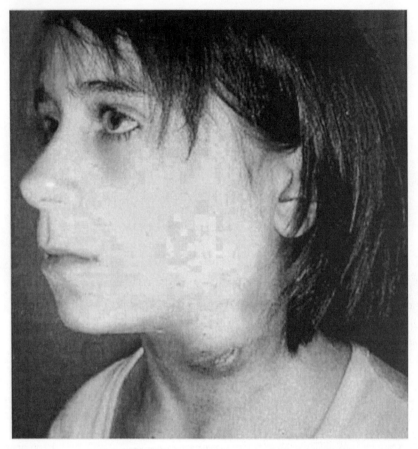

FIGURE 16–4   A young girl from Kosovo with ulcerating lymphadenitis.
Courtesy of the CDC.

such as renal insufficiency. There is no evidence that tularemia causes cancer, or that it is teratogenic, or mutagenic.

Vulnerable Groups

Tularemia is an intracellular organism (as are all viruses and most mycobacteria), subsequently, hosts with impaired cellular immunity are at heightened risk. Cell-mediated immunity is impaired in the following circumstances: human immunodeficiency virus (HIV) infection, extremes of age, immunosuppression (especially in association with solid organ transplantation), pregnancy, malignancy, diabetes, renal failure, alcohol abuse, and protein calorie malnutrition. Immunocompromised patients may take longer to respond to antibiotics as well. Individuals with preexisting respiratory diseases are likely to be at greater risk from the inhalational form of tularemia.

Laboratory Tests

Findings are generally nonspecific. Leukocytosis may be normal or exceed 20 K cells/ml. Other blood work is generally normal, especially early on. Up to 50% of patients may have mildly elevated transaminases, alkaline phosphatase, and lactate dehydrogenase. There may be evidence of creatinine kinase elevation as a result of rhabdomyolysis.

Microscopy

Microscopy reveals a small, gram-negative, coccobacillus, appearing as distinct, single cells. Its small size (0.2 × 0.2–0.7 microns) helps distinguish it from anthrax, and unlike Yersinia, it has no bipolar staining features. It is a fastidious organism with relatively slow growth, and most strains require special nutritionally supplemented media for growth (Fig. 16–5).

At present there is no readily available rapid diagnostic laboratory test for *F. tularensis*. Routine microbiological screens of sputum, blood, and other cultures can be helpful but often fail to identify the organism. It is unusual for the bacteria

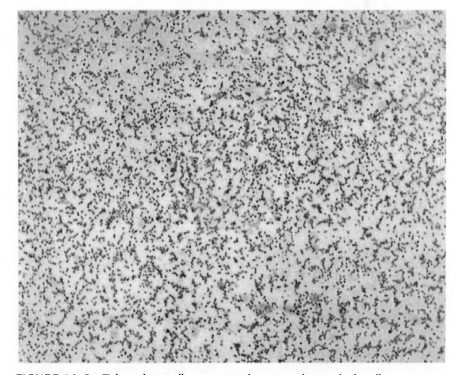

FIGURE 16–5 Tularemia: small, gram-negative, appearing as single cells.
Courtesy of the CDC.

to grow from blood cultures. If a case is suspected, the lab should be notified so special screening methods as well as precautions can be taken. Samples should include respiratory secretions and blood. Preliminary diagnosis may be made using PCR, fluorescent antibody testing, enzyme-linked immunoassay (ELISA), and pulsed field gel electrophoresis. Most of these diagnostic tests are available at designated state public health labs, and results can be back in a matter of hours. Rising serum titers take 2 weeks to develop, so that serologic testing is not of immediate clinical utility.

Radiographic Findings

Half of all cases of tularemia, regardless of presenting syndrome, show abnormalities in chest radiographs, including infiltrates (50%) or pleural effusions (15%). Often early pneumonic tularemia shows only mild infiltrates that soon evolve into a more typical bronchopneumonic pattern. Pleural effusions may also be seen as may hilar adenopathy. Additional pulmonary findings with the pneumonic form include interstitial patterns, abscesses, cavitary lesions, hilar adenopathy, and calcifications. The typhoidal form would likely reveal mediastinal lymphadenopathy or pneumonia. In general, radiological changes are not highly sensitive findings in tularemia.

Differential Diagnosis

It follows from the broad and nonspecific nature of tularemia's symptoms that the differential is large. Diagnosing tularemia requires thorough exposure and travel history. The differential diagnosis for pneumonic tularemia includes: atypical pneumoniac, influenza, plague, anthrax (both of which would progress faster and have higher fatalities than tularemia), Q fever, or brucellosis. The diagnosis for the typhoidal form includes: salmonella, rickettsia, malaria, other "typhoidal" illnesses. Lastly, the differential diagnosis for glandular tularemia includes: mycobacterial infections, cat-scratch disease (Bartonella infection), lymphogranuloma venereum, streptococcal or staphylococcal lymphadenitis, malignancy or lymphoma, fungal infection, and plague.

Diagnosis

In recent decades, tularemia has become so rare that it warranted almost no diagnostic consideration. Unfortunately, the emergence of a plausible bioweapons

 TABLE 16–1 Critical First Steps: Tularemia

Medical Treatment	Antibiotics, supportive care
Infection Control	Isolation, surgical masks during transport
Public Health Interface	Contact local, state public health officials

TABLE 16–2	Making the Diagnosis: Tularemia
Pattern of Presentation	Multiple, concurrent cases of severe respiratory, febrile illness
Features	Early: febrile illness; otherwise, variable according to form of infection: Pharyngitis, pneumonitis: cutaneous—ulcerations; lymphadenitis; late-sepsis, ARDS, inguinal bubo, gangrenous acral areas Late: respiratory—tachypnea, dyspnea, cyanosis, consolidations on chest exam; systemic-sepsis, shock, multiorgan system failure
Findings	Variable according to form of infection: Chest x-ray—bronchopneumonic findings in at least one lobe, pleural effusion. Less common: diffuse granulomatous lesions, discrete infiltrates, enlarged hilar lymph nodes. Microscopy—small, gram-negative coccobacilli in respiratory secretions

threat requires consideration of tularemia in the differential diagnosis of acute febrile illnesses presenting as a temporal or geographic cluster, particularly if associated with a pulse-temperature differential, atypical pneumonia, or hilar lymphadenopathy. These findings in association with an exposure history will facilitate earlier diagnosis. For syndromic recognition of biological weapons, tularemia is best characterized as one of the flulike syndromes, along with such Category B agents as brucellosis and Q-fever (see Chapter 5).

Treatment

Individual patients diagnosed with tularemia should be given parenteral streptomycin for fourteen days. Alternative choices include gentamicin (particularly for immunocompromised hosts where bacteriocidal antibiotics are preferred), tetracycline, or chloramphenicol. Rates of failure and relapse are higher with the latter two because they are bacteriostatic drugs. Fluoroquinolones, such as ciprofloxacin, are efficacious based on in vitro studies, but are not FDA approved for tularemia. Macrolides and beta lactams show little effect against *F. tularensis* and should be avoided. Parenteral administration can be switched to oral when clinical improvement is seen. In large-scale outbreaks, oral administration is the rule. Doxycycline and ciprofloxacin are the preferred antibiotics. A biological attack with a modified strain of *F. tularensis* resistant to current antibiotics is possible. Antibiotic susceptibility testing is mandatory in order to guide clinical and public health decision making in regards to treatment and prophylaxis (Table 16–3).

TABLE 16–3   Tularemia Treatment Guidelines

Individual Case

Adults including pregnant women:

Streptomycin, 1 g IM bid for 10–14 days,

or

Gentamycin, 3–5 mg/kg IM/IV bid

Or

Ciprofloxacin, 400 mg IV bid—can finish course with 500 mg po q 12 hr when clinically better

Children:

Streptomycin, 15 mg/kg lean body mass IV q 12 hr

or

Gentamicin, 2.5 mg/kg lean body mass IV q 24 hr, then, 1.75 mg/kg lean body mass IV every 8 hr for 10 to 14 days

Mass Infection and Prophylaxis

Adults including pregnant women:

Doxycyline, 100 mg po bid for 14 days

or

Ciprofloxacin, 500 mg po bid

Children:

Ciprofloxacin, 15 mg/kg orally twice daily for 14 days

Adapted from David T. Dennis, Thomas V. Inglesby, Donald A. Henderson, et al. Tularemia as a Biological Weapon: Medical and Public Health Management. JAMA 2001;285:2763–2773.

Preexposure Prophylaxis

A live attenuated vaccine for individuals working with *F. tularensis* is available under investigational new drug (IND) protocol. At present, the FDA is assessing its efficacy and safety. Small retrospective studies of USAMRIID employees showed that the vaccine offered significant protection against inhalational tularemia and moderated the morbidity associated with ulceroglandular disease. Current recommendations are that the general public not be vaccinated.

Postexposure Prophylaxis

Protection from the vaccine takes two weeks, and given tularemia's typically short incubation period, postexposure vaccination is not recommended. Antibiotic therapy with either doxycycline or ciprofloxacin (not FDA approved), if started within twenty- four hours of exposure and continued for the full fourteen-day course, offers excellent protection against inhalational tularemia. If more than twenty-four hours has passed since exposure, active surveillance for fever is required. Exposed individuals who become febrile or evidence prodromal symptoms within a two-week period should be presumed to have tularemia and treated accordingly.

TABLE 16-4   Quick Reference Guide

Diagnostic Keys	Variable according to form, ulcers, glandular enlargement, respiratory symptoms with fever, chills
Risk Factors	Animal, carcass handling
Transmittable	No
Transmission	Tick bites or direct skin contact
Management	Aminoglycosides, macrolides, fluoroquinolones
Prevention	
Preexposure	None
Postexposure	Doxycycline, cipro if symptomatic
Containment	Patient: none
	Secondary contacts: none
Populations at Heightened Risk	Immunosuppressed
Long-term Issues	Fibrotic changes, renal insufficiency
Public Health Concerns	No person-to-person transmission
Contacts	Local, state DPH, CDC, FBI

Close contacts of infected patients need not be prophylaxed since person-to-person transmission is not known to occur.

Infection Control

Quarantine and strict isolation is not justified given that tularemia is not transmitted person-to-person. Standard infection control practices are sufficient. Because *F. tularensis* is hardier than many biological weapons, patients' rooms should be cleaned thoroughly with standard disinfecting protocols. The impact of tularemia's innate hardiness coupled with modified strains that might be used in a biologic attack, has unclear implications from a public health perspective, although secondary dispersal through aerosol droplets is considered unlikely. Adequate decontamination of contaminated surfaces and objects using a 10% chlorine bleach solution (1 part household chlorine for 9 parts water) for ten minutes is sufficient. A 70% alcohol solution can then be used to for additional disinfection and for bleach removal. Individuals with direct contact with contaminated fluid or objects should wash the appropriate body parts and clothes in soap and water.

☣ Summary

Tularemia has been the subject of bioweapons research since the 1930s. It is a Category A agent due to its virulence, communicability, and its potential for genetic manipulation and weaponization as an aerosol bioagent. Tularemia presents with an acute flulike picture, accompanied by fever. Hematogenous spread can result in sepsis, meningitis, and multiorgan system failure. The pneumonic form is considered the most probable form in which tularemia would be used as a bioweapon. Long-term sequelae are poorly characterized. Vulnerable groups include individuals with impaired cellular immunity (e.g., HIV, malnutrition, and alcoholism) and preexisting respiratory conditions. Diagnosis is clinical with confirmation by advanced microbiologic testing available in BSL-2 or BSL-3 labs. Treatment with antibiotics is effective, although weaponized tularemia may be genetically altered to confer widespread antibiotic resistance. Research into vaccines is ongoing. Postexposure prophylaxis with ciprofloxacin or doxycycline is indicated if begun within 24 hr of direct exposure to the bacterium, but person-to-person transmission does not occur. Standard isolation precautions are indicated for individuals hospitalized with tularemia.

SECTION III

Biotoxins and Category B and C Agents

CHAPTER 17

Introduction to Biotoxins

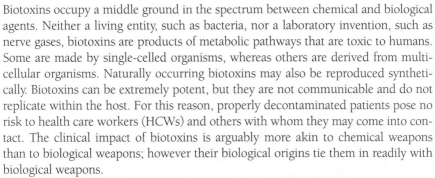

Biotoxins occupy a middle ground in the spectrum between chemical and biological agents. Neither a living entity, such as bacteria, nor a laboratory invention, such as nerve gases, biotoxins are products of metabolic pathways that are toxic to humans. Some are made by single-celled organisms, whereas others are derived from multicellular organisms. Naturally occurring biotoxins may also be reproduced synthetically. Biotoxins can be extremely potent, but they are not communicable and do not replicate within the host. For this reason, properly decontaminated patients pose no risk to health care workers (HCWs) and others with whom they may come into contact. The clinical impact of biotoxins is arguably more akin to chemical weapons than to biological weapons; however their biological origins tie them in readily with biological weapons.

Exposure to biotoxins occurs through ingestion, dermal absorbtion, or as an aerosol. Militarily or as a weapon of terror, aerosol forms of biotoxins pose the greatest risk not only in terms of numbers of people exposed, but the rapidity of symptom onset. With some of the biotoxins, inhalational effects are understood only from animal studies, leaving many open questions about how an inhalational exposure might present. Not all biotoxins are likely candidates for use as weapons of mass destruction or terrorist weapons. Due to their biological and physiochemical properties, Army biowarfare experts consider botulinum toxin and Staphylococcal Enterotoxin B (SEB) to be of most concern from the point of view of battlefield exposures. SEB, a so-called superantigen, causes hyperactivation of the immune system and prompt incapacitation (see Chapter 19). As a Category A agent, botulinum is discussed in its own chapter (Chapter 18).

TABLE 17–1 Quick Facts: Botulism as a Bioweapon

During WWII, the Japanese fed Chinese POWs cultures of *C. botulinum* as part of its research into the military use of botulinum toxin.

The United States ceased its research into botulinum toxin as a bioweapon in 1972, although the USSR and Iraq continued their efforts.

Iraq, Iran, North Korea, and Syria are believed to have active research and development efforts on botulinum.

In 1990, the Aum Shinrikyo cult twice failed in its attempt to attack U.S. naval bases in Japan using botulinum toxin.

Since terrorists may have more limited goals—for example, sowing fear, as was seen with anthrax—wider options in regard to potential biotoxin weapons, including ricin, abrin and mycotoxins, exist. This, too, is discussed briefly.

The CDC classifies the biotoxins as biological weapons. Others make biotoxins a category unto itself, or classify them as chemical agents. For the purposes of this text, we have chosen to follow the CDC categorization, thereby providing discussion of particular biotoxins according to their Category A and Category B biological weapons designation. The only exception is ricin, which although listed with Category B agents by the CDC is included among the Category A agents because of a number of features, including its use already, indicating it to be a threat comparable to other Category A agents.

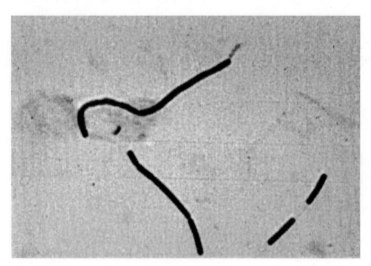

FIGURE 17–1 Gram stain of *C. botulinum*, note the characteristic thin, narrow rods.
Courtesy of the University of Arizona/Glenn Songer, Ph.D.

The other deviation from the CDC categories is our having included four biotoxins that the CDC leaves off of their lists: abrin, a toxin made from the seeds of the commonly found rosary pea, aflatoxin, a mycotoxin usually produced by species of fungus that have a proclivity for peanuts; T-2 trichothecenes, another mycotoxin, but one that is believed to have been used previously as a weapon; and saxitoxin, a potent neurotoxin produced by a commonly found algae species. These agents were selected for inclusion because each possesses various features that make them legitimate threats as biological weapons. They are listed at the end of the section on biological weapons, because though legitimate, they are less likely to be employed than are the biological agents that precede them.

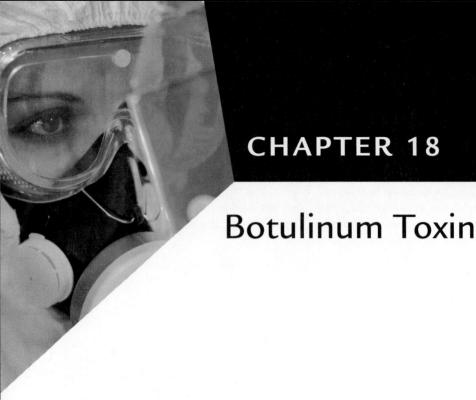

CHAPTER 18

Botulinum Toxin

Clinical Vignette

A frantic mother calls your office stating that her sister had just told her that feeding honey to children causes botulism. She had given a quarter teaspoon of a locally grown honey to her infant daughter three days earlier and was now in a panic. The mother states that the child appears normal and has shown no evidence of illness over the past three days, specifically no constipation, lethargy, or swallowing difficulty. You ask her to bring the child to your office where you examine her. The child appears healthy and in no respiratory distress. She is breathing normally and is not pooling secretions. The girl readily drinks a baby bottle of water without any difficulty. Her eyes are wide open with no proptosis, and her pupils are normal and react swiftly to a penlight. No cranial nerve findings or muscle hypotonia are present. You tell the mother that although botulism has been associated with honey in infants, none of the disease's features of descending paralysis, or odynophagia, are present and reassure her that her child does not have the disease. The mother asks whether there is "a test" to prove it definitively and whether there is an antidote for botulism. What do you tell her?

Background

Botulinum is the first of the eight biotoxins in this book to be discussed. It is the only biotoxin included among the Category A agents, largely because of the ubiquity of the bacteria that produce it, its toxicity, the relative ease of production and dissemination, as well as the precedence of its use in warfare. Botulism is a neurologic syndrome caused by a toxic proteolytic enzyme produced by the bacteria *Clostridia botulinum*. Botulism derives its name from the Latin word for sausage because contaminated sausage was the source for several of the earliest described outbreaks of the disease. *C. botulinum* is a spore-forming, obligate anaerobe found most commonly in soil. The neurotoxin produced by the bacterium is responsible for the clinical disease known as botulism. Botulinum toxin includes seven different proteins (identified as A through G) that are secreted by four distinct but closely related types of Clostridia bacteria.

Botulinum toxin (botulinum) is the most potent toxin in existence: If dispersed ideally, 1 g could kill over a million people. Botulinum is colorless, odorless, and said to be without taste. Botulism is a medical emergency since proper treatment must be implemented quickly, including antitoxin and life support systems, to prevent death. Botulinum toxin as a weapon of bioterrorism possesses several distinctive features compared to other Category A agents, the most obvious being that it is the product of the microbe and not the microbe itself that causes the disease.

There are three naturally occurring forms of botulism: foodborne, wound, and intestinal. An additional form, resulting from the weaponization of botulism, is inhalational. Historically, many of the deaths associated with botulism resulted from exposure to improperly prepared and canned foods. It has found medical application in treating such conditions as tetanus, blepharospasm, strabismus, and most recently in cosmetic surgery. Further, although not FDA approved for these purposes, botulinum has been used to treat migraines, chronic back pain, achalaisa, and other conditions.

Botulism's potential as a biowarfare agent can be traced to WWII when it was fed to Chinese prisoners of war by their Japanese captors. The U.S. military feared its possible use by Germany and so vaccinated soldiers just prior to the D-Day invasion. In accordance with President Nixon's directive, botulism research and development ended in 1972, although other countries, notably the USSR and Iraq, continued their research and development efforts throughout the 1970s and early 1980s.

Acts of terrorism involving botulinum have occurred, with three separate attacks in Japan, including one at a U.S. military base. Fortunately, none of the attempts were successful. Of note, the botulinum used in these attacks was made from clostridium cultures grown from local soil samples. Governments believed to presently possess or have begun production of botulinum included Iran, North Korea, and Syria.

As a bioweapon, botulinum is far more potent per equivalent weight than any synthesized toxins, and is considered one of the most toxic substances known to humanity. Through automated processes, large quantities of botulinum can be produced and introduced in an attack as an aerosol or through contamination of the food or water supplies. Given that sporadic botulism is a foodborne illness, deliberate contamination of food sources pose a likely threat. No naturally occurring waterborne illnesses have ever been reported with botulinum. Inhalation results in a similar presentation to foodborne infection albet with a faster onset. This is possibly because of its slower absorption through the intestinal mucosa. Fortunately, the toxin has limitations as a weapon because aerosolization affects its stability and it cannot withstand standard water treatment methods.

Sources

C. botulinum is found throughout the world in soil. It may contaminate improperly cooked or prepared foods, particularly honey and home-canned foods with low acid content. Although botulinum toxin is stable in water, it is rapidly inactivated by standard water treatments and contamination of drinking water supplies requires enormous quantities of the biotoxin. For these reasons, waterborne botulism is impractical and unlikely as a means of terrorism. Consumers of contaminated foods, rather than food preparers, have become ill in the sporadic nondeliberate outbreaks of botulism reported in the medical literature.

Epidemiology

Natural outbreaks of botulism are uncommon but can be seen in agricultural areas. Based on patterns from natural occurrences, there are fewer than 200 cases (including all forms) each year in the United States with no obvious age or gender discrimination. Foodborne botulism results in roughly 9 cases per year and 2.5 cases per outbreak of disease. The largest foodborne outbreak was in 1977 at a restaurant in Michigan where 59 cases occurred. Three of the seven isomers of botulinum (A, B, and E) are the most commonly seen in humans. F rarely occurs in humans, whereas C and D are more common in non-humans, and G does not appear pathogenic.

Means of Transmission

All three naturally occurring forms of botulism toxin (foodborne, wound, and intestinal) result from Clostridium synthesis of botulinum either in vivo or prior to entering the bloodstream. Botulinum enters the body via absorption through the mucosal linings of the gastrointestinal or respiratory tracts, or from a contaminated wound site. Importantly, botulinum cannot enter the body through intact skin. A bioterrorist attack with botulinum would likely be from an aerosolized form, although it is possible that it could occur through contamination of food or water. Some protection is offered by simply covering one's mouth with thick or folded cloths such as a

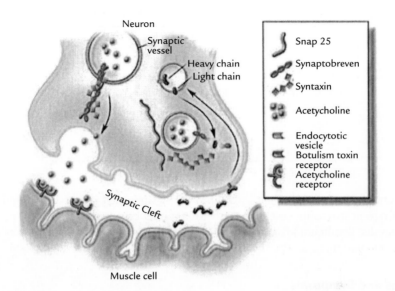

FIGURE 18-1   Mechanism of botulinum.
Reprinted from Baumann L. "The Cosmetic Uses of Botulinum Toxin." In Baumann, L: Cosmetic Dermatology: Principles and Practice. McGraw-Hill, April 2002;140, with permission.

handkerchief. Fortunately, sunlight denatures the toxin rendering it harmless within one to three hours of exposure, and the chlorine content of most municipal water supplies inactivates approximately 85% of botulinum toxin. Foodborne outbreaks, natural or deliberate, require foods that are uncooked or are poorly cooked. Natural infections tend to occur from contaminated vegetables—particularly those with a relatively high pH, such as beans, corn, carrots, and peppers. Life support systems and botulinum antitoxin have decreased the mortality rate to less than 5%.

Naturally occurring botulism or botulinum toxin is not contagious and therefore is not passed from person-to-person. Unfortunately, advances in genetic engineering pose a theoretical risk for contagion. Reports that the former Soviet Union experimented with splicing the Clostridium gene that codes for botulinum toxin into an infectious bacterial agent remain unconfirmed. However, if successful, this would effectively create an infectious form of botulinum toxin.

Pathogenesis

Botulinum toxin is a protease that binds to neuromuscular junctions outside the central nervous system. The biological effects of botulinum toxin are composed of two polypeptide chains. When the B subunit binds (irreversibly) to the pre-synaptic motor neuron, the toxin is endocytosed into the terminal end of the axon. Once inside, subunit A cleaves the proteins needed for neuronal vesicles to release the

TABLE 18–1 Cranial Neuropathies: Botulinum Toxin

Mydriasis
Diplopia
Ptosis
Photophobia
Dysarthria
Dysphonia
Dysphagia

neurotransmitter acetylcholine (ACh) into the synaptic cleft. Without the binding of ACh to post-synaptic receptors the neuron cannot be activated and motor paralysis results. Botulinum toxin does cross the blood-brain barrier so that its effects are limited to peripheral and autonomic nerves outside the central nervous system.

Signs and Symptoms

All forms of botulism present with the same general signs and symptoms. Foodborne botulism begins with nausea, vomiting, cramping, or diarrhea, although this is believed to be the result of other clostridial metabolites rather than the toxin itself. This distinction is noteworthy because even if a bioterrorist attack involved contaminated food or water, only the neurological effects will be seen because purified toxin rather than bacteria will be used. All presentations of botulism regardless of exposure route will be neurological.

Onset of symptoms is seen between twelve and thirty-six hours after exposure depending on the extent and amount of exposure. In animal studies, higher doses have symptom onset well under twelve hours whereas low doses have shown delays as long as days. Although there can be extreme variation in the scope and timing of symptoms, all cases include cranial nerve (CN) paralysis because the toxin always affects bulbar musculature (see Table 18–2 for clinical features).

Early Stages

CN palsies, especially those affecting the eyes and the oral pharynx, are pathognomonic findings for botulism. Patients seek medical care for one or more of the following complaints: difficulty seeing, speaking, and swallowing. Next affected are the skeletal muscles which become weak, and are accompanied by diminished deep tendon reflexes and flaccid paralysis. This occurs in a symmetrical, descending, and progressive pattern (Fig. 18–2).

Late Stages

Collapse of the airway due to oropharyngeal muscle paralysis signals advanced botulism. Two other late features are loss of the gag reflex and pupil dilatation (possibly fixed). Weakening of the diaphragm and accessory muscles of respiration

TABLE 18–2   Differentating Botulinum Toxin from SEB, and Nerve Agents[a]

Features	Botulinum Toxin	SEB	Nerve Agents
Time of onset	12–24 hr	1–6 hr	Minutes
Nervous system	Descending flaccid paralysis	Myalgia, headaches	Convulsions, fasiculations
Cardiovascular	No effect	Mild elevation	Bradycardia
Respiratory	Early—normal Late—paralysis	Cough—nonproductive, chest pain	Dyspnea, airway constriction
Gastrointestinal	Ileus	Nausea, vomiting, diarrhea	Painful diarrhea
Ocular	Mydriasis Ptosis	Conjunctival injection	Miosis
Salivary	Decreased	Slight increase	Copious
Death	2–3 days	Rare	Minutes
Response to atropine	None	Improves GI symptoms	Improvement

Adapted from USAMRIID's Medical Management of Biological Casualties Handbook.

cause cyanosis or CO_2 retention and narcosis. Not surprisingly, the mortality of botulism is secondary to paralysis of the respiratory muscles. Without ventilatory support, overt respiratory failure occurs within twenty-four hours of symptom onset. Classic signs of botulism underscore the loss of synaptic ACH with anticholinergic symptoms and impaired autonomic function. These include dry mouth, ileus, constipation, and urinary retention.

Pertinent negative findings also assist in the diagnosis. Patients are afebrile (unless there is a secondary infection), lack sensory nerve impairment, and have a clear sensorium. The paralyzing effects from botulism can last for weeks or months even with proper management (see section under Treatment).

Laboratory Tests

Routine laboratory studies and CSF show no abnormalities. The mouse bioassay (this entails injecting serum or stool samples into mice and waiting for the telltale signs of botulism) is the current gold standard for diagnosing botulism. A serum enzyme-linked immunoassay (ELISA) test for botulinum toxin is also available commercially. *C. botulinum* may be grown from gastric aspirates (useful in setting of potential aerosol exposure), serum and stool cultures at LRN or CDC laboratories. Electromyography shows characteristic findings of axonal denervation, but this is not pathognomonic for botulinum toxin. Urinary excretion of hydroxy indole acetic acid (5'HIAA) may increase in botulism. In the future, a rapid PCR that detects DNA segments from the toxin is likely to be forthcoming.

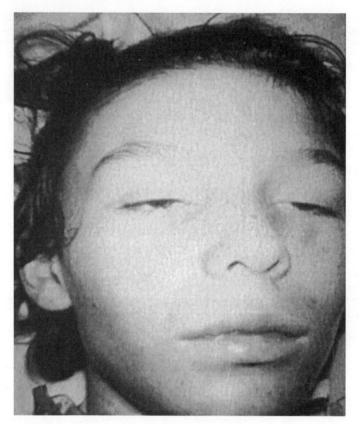

FIGURE 18-2 Botulinum affects with the classic features of ptosis.
Courtesy of Department of Health and Human Services/Office of Public Health Emergency Preparedness.

Microscopy

Because most laboratories do not have the appropriate tests for diagnosing botulism, physicians should send samples—including serum (>30 mL), stool, gastric aspirate, and if possible, vomitus that includes the suspect food—for testing. Specimen collection should be done prior to antitoxin administration. However, if the patient is symptomatic, antitoxin should be provided immediately regardless of whether samples have been taken (see section under Treatment). A toxicology screen and current patient medication list should be sent with the samples because some of these may interfere with the mouse assay.

Radiographic Findings

No distinctive radiological findings are associated with botulinum exposure.

Differential Diagnosis

The following illnesses may present with similar findings: Guillain–Barré, stroke, myasthenia gravis, or tick paralysis enteroviral myelitis (which would have CSF consistent with viral infection and usually an antecedent fever); inflammatory myopathies (noted by elevated creatinine kinase), viral encephalitis, atropine poisoning (which demonstrates mental status changes), chemical nerve agents (marked by copious respiratory secretions and miotic pupils), and SEB (see Table 18–2). Among the possibilities included in the differential diagnosis, only the bioweapons—nerve agents and SEB—would likely occur as a cluster. Other possibilities include medications and toxic ingestions with depressant effects—such as ethanol, methanol, organophosphates, carbon monoxide, thallium poisoning, multiple sceloris. Diagnostic tests to exclude diseases such as Guillain–Barré syndrome, stroke, and myasthenia gravis may help narrow the differential. These tests include CT or MRI scanning, lumbar puncture, electromyography, and a tensilon test for myasthenia gravis.

Prominent cranial nerve findings distinguish botulism from other causes of flaccid paralysis especially in its earliest stages. This contrasts with SEB or nerve agents where peripheral motor paralysis dominates the acute clinical presentation. Supporting the diagnosis in an afebrile patient with no sensory impairment or mental status changes with a descending symmetrical flaccid paralysis with prominent cranial nerve involvement, are the so-called four D's of diplopia, dysarthria, dysphonia, and dysphagia.

Diagnosis

Few laboratories have the capacity to test for botulinum toxin, and a lag time of one to two days is likely before confirmation of the diagnosis.

Distinguishing sporadic or naturally occurring botulism from a bioterrorist attack is difficult. An outbreak of acute flaccid paralysis with prominent cranial nerve involvement is a significant clue. Clinicians must take a scrupulous history, including the often neglected social history inquiring as to recent travel, diet, and occupation. Inquiring about recently eaten foods and other people presenting with similar symptoms is important. A clustering of signs, symptoms, and other findings in a cohort of patients with a common geographical link, but no dietary link is highly suggestive of

TABLE 18–3 Clinical Diagnosis: Botulism	
Pattern of Presentation	Multiple, concurrent cases of a symmetrical, descending progressive motor paralysis, without common food source
Features	Symmetrical, descending, and progressive flaccid paralysis with cranial nerve palsies but no mental status changes, afebrile
Findings	No findings found on imaging or in blood work

a deliberate biotoxin attack. If no common dietary source can be detected and geographic clustering is present, a deliberate attack, possibly inhalational, should be entertained and the appropriate public health and infection control steps instituted. Also suggestive of a deliberate release of botulinum is the identification of one of the less common forms of toxins, (A, B, or E) that are typical of foodborne botulism. C, D, and G isomers are considered possible options for use in an aerosol attack. Multiple reports of botulism occurring in dispersed geographic areas is also suggestive and highlight the importance of national and state surveillance programs.

Vulnerable Groups

Given its neurotoxicity, individuals with preexisting peripheral neuropathies may be susceptible at lower levels of exposure. Lab workers and military recruits represent at-risk occupational groups. Individuals who do home canning are at greater risk of foodborne botulism. Drug users are at higher risk from infected injection sites. The rapid expansion of the medical use of botulinum toxin brings with it potential for risk to consumers, physicians, and other HCWs, but currently its use has an enviable safety record.

Chronic Effects

Advances in advanced life support and intensive care medicine have improved outcomes in botulism poisoning. Survivors often complain of chronic fatigue and dyspnea, though the mechanisms for these long-term sequelae are unknown.

Treatment

Treatment is a twofold approach: antitoxin and supportive measures. Botulinum antitoxin is available from the CDC and should be administered as early as possible. The antitoxin's neutralizing effect acts on circulating botulinum in patients whose symptoms are still progressing, but does not reverse symptoms that are already present. Antitoxin administration should not be delayed for the sake of sample collection. Once the symptoms have reached a plateau, there is no more botulinum in the bloodstream and the antitoxin offers no additional therapeutic value. In foodborne botulism, in which the toxin is thought to be absorbed slowly and steadily by the intestinal mucosa, the antitoxin is most efficacious. The usefulness of antitoxins in the event of an aerosolized bioterrorist attack is uncertain, primarily because it has never been tested on inhalational botulism. Animal studies are reassuring, however, suggesting that antitoxin remains highly effective if given before onset of symptoms in inhalation exposure. If given after onset of symptoms, respiratory failure still results (see Table 18–4). With the recommended dose, the amount of neutralizing antibody provided far exceeds the toxin levels found in naturally occurring botulism patients, so further administrations are not needed. In contrast, a deliberate attack with botulinum may result in dramatically higher serum toxin levels. Repeated

TABLE 18–4   Botulinum Treatment

Dosing for botulinum antitoxin for A, B, E

Available only through CDC and SHD

A single 10-ml vial is used per patient, diluted to 1:10 in 0.9% saline solution
   given IV infused over 20 minutes.

administration of antitoxin—with or without testing of serum for the presence of toxin—may be justified. Should antitoxin be unavailable or delivery delayed, and foodborne botulism is suspected, standard detoxification measures such as activated charcoal may be administered; however, no data exist to indicate the efficacy of this approach. Currently only the trivalent form of antitoxin is available for use against the A, B, and E forms of botulinum. Identifying the toxin type is important. Whereas A, B, and E are the most common naturally occurring forms, a biologic attack may well use a different form. A heptavalent (equine-derived) antitoxin for use on all seven botulinum toxin forms (A, B, C, D, E, F, and G) is available from USAMRIID under investigational new drug (IND) status. It has been used successfully

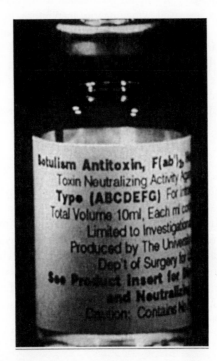

FIGURE 18–3   Botulism antitoxin for all six forms.

Courtesy of the Department of Health and Human Services/Office of Public Health Emergency Preparedness.

and safely in experimental lab workers. This antitoxin has been "despeciated" by cleaving the F_c fragments from the IgG molecules and should theoretically result in a smaller risk of anaphylaxis and serum sickness. In contrast, botulinum antitoxin is an equine-derived antidote and may cause hypersensitivity reactions in as many as 9% of recipients. Current recommended dosing is lower than that used historically, so the prevalence of severe allergic reactions is probably less. Nevertheless, patients should be monitored for urticaria, wheezing, or frank anaphylaxis. Skin testing with antitoxin may help predict those who are at risk for hypersensitivity responses. If the skin test is positive after twenty minutes, then desensitization is carried out by giving 0.01 to 0.1 ml of antitoxin subcutaneously, doubling the dose every twenty minutes until the patient tolerates 1.0 to 2.0 ml. Challenge testing with antitoxin may help predict those who are at risk for hypersensitivity responses. Human-derived anti-toxin is being investigated but is not yet approved for clinical use excepting as an IND for infantile botulism. At present, there is no evidence that antitoxin is a unique risk to children, pregnant women, or the immunosuppressed.

For the most part, supportive measures include the use of enteral or parenteral feedings, admission to the intensive care unit, and mechanical ventilation. Any patient in whom botulism is suspected, or who is diagnosed but is having progressing symptoms, should be monitored carefully for progression to respiratory failure, as well as for the loss of the gag reflex, swallowing integrity, inspiratory strength, and vital capacity. Twenty percent of foodborne botulism patients will require mechanical ventilation. A large bioterrorist attack could quickly saturate available advance life support systems especially given the high hospitalization rate and the long recovery period of the disease. Secondary infections may occur, but antibiotic choice must be decided by the recognition that certain antibiotics (e.g., aminoglycosides and clindamycin) may exacerbate neuromuscular blockade and are therefore contraindicated in botulism.

Vaccine

A vaccine is available from the CDC under IND status. It is a pentavalent toxoid, given in a series of four shots, and offers preexposure prophylaxis against types A, B, C, D, and E. Use is suggested only for individuals and populations at high risk to inhalational botulinum, such as CDC lab workers and military recruits. The use of a heptavalent toxoid vaccine for postexposure prophylaxis has shown positive results in animal studies; no human studies have been done. Preexposure prophylaxis is not recommended and is not available to the general public. Research into a recombinant vaccine is well underway, and an experimental vaccine has been used successfully in lab workers. Recent research into an inhaled form of botulinum vaccine is very promising.

Infection Control

Botulinum is destroyed readily. Temperatures greater than 85°C for more than five minutes decontaminates food or beverages. Any clothes or skin that come in

TABLE 18–5	Critical First Steps: Botulism
Medical	Supportive care, antitoxin
Infection Control	Hospital epidemiologist or infection control team. Consider prophylactic antitoxin for asymptomatic but exposed individuals or emergency response workers.
Public Health	Notify LHD, SHD, or the CDC

contact with botulinum toxin can simply be washed in soap and water. Objects or surfaces may be cleaned in 0.1% hypochlorite solution. Natural degradation occurs within two days. Physicians and their staff need only standard precautions when dealing with botulinum exposure. Isolation is not required. Botulinum is not transmitted person-to-person.

Public Health Considerations

Diagnosis of botulism is considered a medical and possibly public health emergency (regardless of the source) and public health notification must begin at the

TABLE 18–6	Quick Reference Guide: Botulism
Diagnostic Keys	Descending progressive motor paralysis, clear mental status, afebrile
Risk Factors	None
Transmittable	No
Transmission	No
Management	Antitoxin, supportive
Prevention	
Preexposure	None
Postexposure	Antitoxin, supportive
Containment	Patient: none
	Secondary contacts: none
Populations at Heightened Risk	Unclear
Long-term Issues	Unclear
Public Health Concerns	Diagnosis signals a medical and possibly public health emergency
Contacts	Local, state DPH, CDC, FBI

local and state levels. If bioterrorism is suspected, then federal notification of both public health and law enforcement must be initiated.

✷Summary

Botulism is a neurologic syndrome characterized by a progressive, symmetric, descending paralysis with prominent cranial nerve findings. Clustering of multiple cases with the neurologic syndrome is a clue to a potential deliberate release. Treatment includes early use of antitoxin and supportive measures. Aerosolized botulinum represents the greatest threat from a bioterrorist perspective, although evidence that an engineered aerosol form exists is lacking. Survivors often complain of chronic fatigue and dyspnea. A diagnosis of botulism should prompt notification of public health and law enforcement authorities.

CHAPTER 19

Ricin

Clinical Vignette

A 22-year-old with a history of allergic rhinitis and childhood asthma presents to your office for evaluation. He noticed a flulike illness with low-grade fever, congestion, watery itchy eyes, and dry cough 1 week earlier soon after he arrived at work. The symptoms resolved by midweek and he felt fine through the weekend. On Monday morning he noticed the same constellation of symptoms, but this time he developed hives and by midafternoon he felt quite short of breath. He works as a maintenance worker at a local manufacturer of castor oil for the automotive industry. His job involves removing the mash residual following processing of the castor plants. He has no significant cardiac history and is a nonsmoker. Because he was frankly wheezing and very dyspneic, you referred him to the local emergency room. You inform the ED that you suspect he may be having an allergic reaction to or even frank ricin intoxication from his work. As neither you nor the ED is familiar with this syndrome, you recommend that the attending contact the Poison Control Center regarding diagnostic testing and treatment.

Background

Ricin is a protein derived from Ricinus communis, a plant whose bean produces a natural laxative "castor oil" long known for its medicinal value and the bane of

American school children through much of the 20th century. Ricin's toxic properties were first identified in the 19th century. After castor oil is extracted from the bean, ricin remains in the residual mash and can easily be separated out. Ricin is not found in castor oil itself. Each year, approximately a million tons of castor beans are involved in industrial uses primarily to extract the oil for use as a lubricant in aerospace machinery. Such a figure represents an annual availability of 50,000 tons of industrial ricin byproduct and making it a viable BCN threat.

Ricin has played valuable roles both in the development of medical knowledge and as a therapeutic agent. At the end of the 19th century, experimentation

FIGURE 19–1   The castor bean plant, *Ricinus communis.*
Courtesy of Cornell University.

with ricin laid the groundwork for understanding the mechanisms of immune response in regards to antibody stimulus and specificity. Ricin, owing to its anti-tumor effect, has been used as a chemotherapeutic agent. In recent years it has been of interest in the development of monoclonal antibodies in cancer treatment, as well as for treatment of chronic pain syndromes. Processing of castor plants industrially is also associated with occupational illness and disease. Castor bean dust or castor oil are organic compounds and may induce a typical allergic syndrome consisting of upper airway irritation, urticaria or rashes, and even reactive airways or frank asthma. Occupational exposure to plant mash also puts workers at risk for the toxic syndromes to be summarized in the following sections.

Ricin has a notorious history as a weapon of war and espionage as well. Ricin can be manufactured cheaply and easily, and it is a highly potent phytotoxin. Ricin first became of interest at the end of WWI when the American and British military recognized its potential as a military weapon. Efforts to create a ricin bomb were never completed. In 1978, it is believed that Soviet agents used a ricin pellet shot from an umbrella to assassinate Bulgarian dissident Georgi Markov in London. Ricin remains a concern as a terrorist weapon. In January 2003, the antiterrorist branch of Scotland Yard arrested six men of North African descent after small amounts of ricin were found in their apartment. In 2004, several federal buildings in Washington DC were evacuated when ricin powder was found in a letter to Senator Bill Frist. Ricin has also been reportedly found in some Al Qaeda and Ansar Al Islam caves in Afghanistan. These facts underscore the importance of understanding ricin and other biotoxins.

Properties

Ricin is a globular protein that comprises roughly 3% to 5% of the castor plant. Ricin is at its most potent if inhaled, but is also dangerous if ingested; however, skin contact appears to offer little to no risk. Chewing or swallowing as few as three to six castor beans may be fatal. Ricin can be made as a liquid, crystallized into a solid form or even into powder. Large-scale dissemination requires aerosolization of particles less than 5 microns. This is technically difficult, and for the present limits ricin's usefulness as a mass terrorist threat or military weapon. Smaller scale use would likely involve contamination of food or water, points that could help in clinical decision-making.

Sources

Castor oil is used in a wide variety of industrial processes and is arguably one of the most important vegetable oils in terms of versatility and use. Castor oil is used in paints and varnishes, as a water-resistant treatment for fabrics, plastics, and metals. Its viscous and lubricating properties make it useful in the automotive industry, particularly as a high-performance engine motor oil and fuel additive. Castor oil is a

FIGURE 19–2 The natural source of ricin is the bean of the castor plant.
Courtesy of M. Welienger.

raw ingredient used in the production of nylon and other synthetic fibers and resins. Its use for medicinal purposes is well known; less well known is its use in the perfume, flavoring, and odorant industries where it is incorporated into soaps, and synthetic fruit and flower scents. It is grown worldwide. Workers involved in the harvesting and processing of castor plants are at risk for toxicity if appropriate work practices are not followed. Workers involved in fertilizer manufacturing where pomace is used are also exposed to ricin, and dermatologic and allergic reactions have been described in this industry as well. Phytotoxins are an active area for medical research, a fact that increases potential sources for ricin and risk of injury to lab workers and scientists. Modified biotoxins that retain their immunogenicity with less neurotoxicity may prove useful in medical treatment for a wide variety of conditions, including cancers, and ricin inhibitors are being actively investigated for their potential use as an antidote.

Pathogenesis

Ricin's toxicity is a result of its capacity to inhibit protein synthesis by stripping ribosomes of their purine bases, thereby slowing cellular protein production. It is

FIGURE 19-3 **A small vial containing ricin sent to Senator Frist.**
Courtesy of the FBI.

classified as a ribosome inhibiting protein (RIP). One of its two polypeptide chains binds to a receptor on cell membranes and is brought into the cell via endocytosis. Once in the cell, the second polypeptide chain enzymatically cleaves RNA subunits thereby impairing protein synthesis and inducing cell death. A single ricin molecule is capable of incapacitating 1,500 ribosomes per minute making it one of the most potent cytotoxins known. The route and degree of ricin contamination determine where the toxin is likely to act and how it impacts the victim. With higher doses, particularly from ingestion of castor beans, splenic, hepatic, gastrointestinal, and renal necrosis can occur. This is thought to result from agglutination and hemolysis of red blood cells, rather than from direct insult to these organs.

Signs and Symptoms

The clinical presentation, including lab findings, varies according to the route and dose of exposure. In general, ricin causes direct endothelial toxicity leading to a vascular leak syndrome that accounts for most of its clinical effects.

Inhalational

There are very little data regarding the effects of inhalation exposure of ricin in humans. What is known comes from a limited number of case reports and from animal studies. Symptoms caused by the inhalation of ricin are dose dependent and the result of direct toxicity to the respiratory epithelium, rather than through systemic absorbtion. Possibly this explains why inhalational exposures are associated with a delay of clinical features for about 6 hours. At that time, nonspecific findings such as fever, itchy eyes, cough, congestion, chest tightness, shortness of breath,

OF NOTE...

Ricin Letter

On October 15, 2003, a letter addressed to the Department of Transportation (DOT) and containing a small vial of ricin was identified in a South Carolina postal facility. The author claimed to be a tanker fleet owner angry about changes in DoT trucking regulations. The identification was made easier as the sender wrote on the envelope that the letter contained ricin. The FBI released a version of the text:

I have easy access to castor pulp. If my demand is dismissed I'm capable of making Ricin. My demand is simple, January 4 2004 starts the new hours of service for trucks which include a ridiculous ten hours in the sleeper berth. Keep at eight or I will start dumping.
You have been warned this is the only letter that will be sent by me.
Fallen Angel

The other letter, processed through a Tennessee postal facility on October 17, 2003, and containing ricin was sent to the White House, but also directed at the DoT, echoing the same sentiment toward the regulation and including a threat: "If you change the hours of service on January 4, 2004, I will turn D.C. into a ghost town. The powder in the letter is ricin. Have a nice day."

There is a $100,000 reward posted by the FBI for information leading to the mailer's conviction.

On February 2, 2004, a letter to Senator Bill Frist was found to contain ricin. No connection between this and the letters from the previous October have yet been made.

arthalgias, and nausea occur. Symptoms may last several hours and may then resolve. The start of symptom resolution may be marked by the patient becoming diaphoretic. No terminal cases of inhalational exposures have been noted in humans, but animal studies suggest development of necrosis and inflammatory changes leading to pulmonary edema and even ARDS. If these features are severe enough, death usually occurs within two to three days. Less intense exposures may result in reactive airways disease, or persistent asthma.

Ingestion

The harmful effects of ricin are moderated when it is ingested, due in part to enzymatic destruction of the protein or modification resulting from environmental factors such as pH. Still, ingestion of ricin results in localized gastrointestinal signs and symptoms such as fever, nausea, vomiting, diarrhea, or abdominal pain. Necrosis of the gastrointestinal tract, liver, spleen, and kidneys is possible with higher exposures and hemodynamic instability and shock may ensue. Autopsies from case reports of fatal ingestion indicate that the necrosis tends to occur in the reticuloendothelial system—thought to be vulnerable because of an elevated number of receptors available for the toxin to bind among resident cells.

Parenteral

Parenteral exposure through injection is uncommon but deadly. With the exception of clinical trials, parenteral exposures have been associated with self-experimentation, suicides, or homicides. Symptoms include myonecrosis and lymphonecrosis followed by multiorgan failure and death.

Laboratory Tests

There are no pathognomonic laboratory findings for ricin exposure at present, although efforts to develop a diagnostic test for ricin or ricinine, a component of the castor bean plant, are underway. If available, enzyme-linked immunoassay (ELISA) testing can confirm the presence of ricin as it tends to evoke a potent antibody response. Polymerase chain reaction (PCR) can also be used to detect castor bean DNA. Ricin's toxic effects are mirrored in the lab findings and follow the route of exposure. Renal and liver function test abnormalities are noted with kidney or liver necrosis. Similarly, severe inhalational exposure can cause extensive fluid extravasation in the lungs resulting in hypoxia and acidosis. Limited clinical data indicate that ricin may cause a significant leukocytosis, with a predominant neutrophilia. Hypomagnesemia is a reliable indicator of ricin exposure. Chest radiographs show bilateral infiltration in the lung fields of patients who had severe inhalational exposure to ricin. If ricin poisoning is suspected, environmental samples should, if possible, be sent to the Laboratory Response Network (LRN) or Centers for Disease Control and Prevention (CDC) for confirmatory testing using standard gas or liquid chromatography. New techniques for diagnosing ricin are not yet available for clinical use, although radioassays for abrin and ricin can detect these biotoxins at extremely low concentrations.

Diagnosis

Diagnosing ricin poisoning on clinical parameters alone is difficult because of the nonspecificity of tissue injury. However, signs and symptoms do follow from the route and intensity of exposure. For example, inhalation of ricin causes predominantly respiratory signs ranging from mild or severe. Similarly, ingestion will present with a range of gastrointestinal complaints depending largely on dose. As a practical clinical matter, small doses of ricin might cause an airway syndrome indistinguishable from a common respiratory virus, or a gastrointestinal syndrome easily misdiagnosed as gastroenteritis.

Differential Diagnosis

The differential diagnosis for ricin poisoning is extensive and runs from the mundane to the obscure, including acute respiratory distress syndrome (ARDS), infectious agents, and toxic exposures. Careful attention to the context of clinical presentation, including epidemiologic clues to be discussed, is essential. Confirmation requires matching an appropriate clinical presentation with environmental testing or biological

TABLE 19–1   Differential Diagnosis of Ricin Poisoning

Inhalation	Infectious: ARDS, tularemia, influenza, anthrax, Q fever, pneumonic plague
	Toxic: SEB, phosgene, oxides of nitrogen, pyrolysis products, especially organoflourine compounds (e.g., Kevlar, Teflon)
Ingestion	Infectious: enteric pathogens (e.g., salmonella, shigella)
Toxic	Mushrooms, arsenic, lye, metals, colchicine

fluid assays. Clinical experience is limited, but case reports and animal studies suggest that following a brief delay, toxicity progresses relatively rapidly over twenty-four hours. Table 19–1 indicates key differential diagnoses for the two likely forms of ricin poisoning.

Chronic Effects

Survival from the acute effects in humans in general will result in full recovery. Little evidence exists on chronic effects and what does exist is largely extrapolated from lab animal studies. Cardiotoxicity and hemolytic effects have been described in animals. As mentioned earlier, any acute lung injury may lead to persistent reactive airways or frank asthma. Individuals with preexisting lung, renal, hepatic disease may be more vulnerable to sublethal exposures. Workers who develop sensitivity— as manifest by urticaria, rashes, and asthma—to ricin will require reassignment to limit exposure to the antigen. Literature is sparse on the mutagenic or fetotoxic effects of ricin, although lab animal data support maternal and fetal toxicity. Both ricin and abrin are cancerostatic, a fact that has been used by molecular biologists to create hybridomas with potential use in cancer treatment.

Vulnerable Groups

HCWs and first responders may be the first to respond to a ricin attack. Given the potential for environmental contamination, proper field decontamination and personal protective equipment must be used. Atopic individuals, particularly those with allergic rhinitis and asthma, are likely to be more sensitive to the allergenic dust produced in castor bean processing, or other industrial processes.

Decontamination

Ricin is extremely stable under ambient conditions but denatures at temperatures of 80°C (176°F) after ten minutes. Ricin is nonvolatile and is not readily absorbed through the skin. Dermal absorption is not associated with illness unless there is a break in the skin or mucus membrane barriers allowing systemic absorption to then occur. Primary routes of exposure leading to toxicity are parenteral (injection),

TABLE 19-2	Critical First Steps: Ricin
Medical	Observation and early respiratory support (e.g., oxygen, positive end-expirating pressure), gastric lavage, fluid support.
Decontamination	Remove contaminated materials, soap and water shower. HCWs and first responders must use full hazmat PPE.
Public Health	Contact the regional poison control center or medical toxicologist, if terrorist event suspected notify local or state health department.

ingestion, or inhalation. Consequently, and unlike virtually every other chemical agent threat, delay in treatment to decontaminate is less crucial. Decontamination is best achieved in the field before transport, presuming life-threatening complications are not present. Once all clothes, accessories, and jewelry are removed, washing down the patient with soap and water is effective. Alternatively, a 0.1% hypochlorite solution may be used for skin or clothing decontamination, as well as for surface environmental decontamination. Decontamination can be done with concentrated chlorine as well.

Provided that gross decontamination has taken place, standard precautions are sufficient to protect HCWs who might come in contact with a ricin-exposed individual. This would include using personal protective equipment, such as disposable gowns, nitrile gloves, and surgical mask and safety goggles or face shields. Hand hygiene should always be practiced. However, if patients have not been decontaminated in the field as recommended, full hazmat protective gear is necessary in order to achieve decontamination (see Chapters 2 and 3) and limit secondary exposures to first responders and medical personnel.

TABLE 19-3	Making the Diagnosis: Ricin
Pattern Identification	Multiple, concurrent cases of a fever, dyspnea, cough if inhaled. Multiple, concurrent cases of a nausea, vomiting, diarrhea if ingested.
Presentation	Inhaled—cough, chest tightness, dyspnea, wheeze, worsening respiratory distress
	GI—nausea, vomiting, diarrhea
Findings	Neutrophilic leukocytosis
	Inhalation—chest x-ray may show pulmonary edema, hypoxia, acidosis.

TABLE 19–4	Quick Reference Guide: Ricin
Diagnostic Keys	Variable according to means of exposure
Risk Factors	None
Transmittable	No
Transmission	No
Management	Supportive
Prevention	
Preexposure	None
Postexposure	None
Containment	Patient: decontamination
	Secondary contacts: decontamination
Populations at Heightened Risk	Occupational workers, first responders, HCW in bioterrorism event
Long-term Issues	Cardiotoxicity, reactive airways
Contacts	Local, state DPH, CDC, FBI

Treatment

There is no specific medical therapy for ricin poisoning. Unlike botulinum, ricin has no presently available antidote and is not dialyzable. Interventions are strictly supportive in nature and determined according to the clinical presentation. Severe cases of inhalational exposure may require treatment of pulmonary edema, intubation, and assisted ventilation, including positive pressure airway support. If ingestion is identified early enough, activated charcoal and lavage may be useful followed by cathartics to minimize absorption. Severe cases of ingestion may require aggressive fluid and electrolyte repletion.

☣Summary

Ricin is a plant biotoxin that inhibits intracellular protein synthesis and causes profound respiratory and vascular effects. Wide-scale commercial availability makes ricin a viable candidate for biological terrorism. It has limited use as a weapon of mass destruction. The acute syndromes appear as an ARDS-like picture with fever and hemodynamic instability. Chronic effects are poorly characterized but may include respiratory sequelae and possibly developmental or teratogenic effects. Currently, treatment is supportive.

CHAPTER 20

Category B and C Additional Agents

Category B Agents

The agents in this section are typically diseases of animals; with the exception of psittacosis, the animals are usually livestock, although other animals may be carriers. Psittacosis is commonly viewed as an avian disease (it is also known as ornithosis). Typhus fever is the only one of the group carried and passed by an invertebrate. As is characteristic of Category B diseases, these agents do not carry a high mortality rate. In fact, many infections with Category B agents will resolve without medical intervention. Typhus fever has the highest mortality rate of the group, roughly 50% if untreated, but death is rare with treatment. With Category B agents the health concerns are more with debility and psychological consequences, rather than with death or permanent or serious injury. Clinicians and public health officials are likely to be familiar with many of these agents in their natural or sporadic form, but they have potential as low-tech biological weapons and are therefore included in the *Sourcebook*.

Infectious Agents: Q Fever

Q fever (*Coxiella burnetii*).

Background

The "Q" in Q fever stands for "query" because investigators looking into one of the earliest documented outbreaks of the disease in the 1930s could not identify the causative agent. Later it was determined that the etiologic agent of Q fever was *Coxiella burnetii*, a rickettsial bacterium. Livestock, such as cattle and sheep, as well as other animals (e.g., cats and dogs) serve as natural reservoirs for the disease. Infected animals are typically asymptomatic.

There are reports that in the past the Soviet Union attempted to weaponize *C. burnetii*; whether this was successful is unknown. As a biological weapon, *C. burnetii* is most likely to be aerosolized, but it has potential as a food or waterborne contaminant. Q-fever is a self-limited illness, its primary impact as a bioterror weapon is to cause morbidity and spread fear.

Properties

C. burnetii has a hardy spore form that is capable of surviving for long periods outside the host, even with exposure to heat or dehydration. The spore form also tolerates many ordinary disinfectants. This agent is not highly pathogenic, is rarely fatal, and usually resolves even without treatment of any kind. It is highly infective, however. Indeed, some experts believe that even a single organism is sufficient to cause the illness. With infection, *C. burnetii* is found in higher concentrations in the products of gestation including placenta, amniotic fluid, as well as in the animal's waste products.

Sources

C. burnetii is found throughout the globe and is considered a potential bioweapon primarily because of its impressive infectivity. Livestock and other animals are the primary hosts and therefore the source for cross-infection of humans living in close proximity to these animals, or having direct contact with the animals.

Epidemiology

Although Q fever is found worldwide, its exact incidence is difficult to establish because so many cases go unreported due to the relatively mild clinical syndrome it causes. Mortality rates are very low; no more than 2% of those infected die as a result of the disease. This probably reflects underlying host factors, such as co-morbidities, immunosuppression, or the extremes of age.

Means of Transmission

Infection with Q fever occurs following inhalation of contaminated dust, consumption of infected animal products, such as milk or meat, or bites from ticks carrying the disease. As a biological weapon, dissemination as an aerosol is the most probable method for spreading the disease. It follows that inhalation is the primary route of exposure.

Pathogenesis

C. burnetii is an obligate intracellular parasite. Once inside the body, it is ingested by phagocytic cells where, in turn, the organism takes up residence in intracellular lysosomes. At this point, the bacterium does not cause injury to the cell or host organism. Once bacterial replication within the lysosome reaches a maximum load cell, cell lysis occurs. At this point, the infection triggers symptoms secondary to the host's immune response. *C. burnetii* induces a biphasic antigenic response. Acutely, IgM antibodies predominate. In the chronic form, both IgG and IgM antibodies are demonstrable.

Signs and Symptoms

Following an incubation period of fourteen to twenty-one days, approximately half of all people infected with *C. burnetii* evince signs of clinical illness. Signs and symptoms tend to be nonspecific and, for the most part, benign. Most commonly, illness begins with sudden onset of fevers (up to 104–105°F) lasting up to fourteen days. This febrile phase is associated with severe headache, general malaise, myalgia, confusion, sore throat, chills, sweats, nonproductive cough, nausea, vomiting, diarrhea, and abdominal pain. Up to half of all patients will develop chest x-ray findings of pneumonia—with or without cough. Cough when present can be productive or dry, and pleuritic chest pain may occur as well. Approximately one-third of patients will develop acute hepatitis or signs of hepatic injury.

In most cases, signs and symptoms of infection resolve fully within a few weeks, even without treatment. Resolution may be delayed for as long as several months in some circumstances. Infection confers lifelong immunity.

Chronic Effects

Chronic forms of Q fever are rare but do exist. Infection may reemerge even after twenty years following an initial infection. Clinical manifestations of chronic or recurrent Q fever include chronic hepatitis, bacterial-negative endocarditis (seen most commonly in the presence of synthetic valves), encephalitis, and osteomyelitis.

TABLE 20–1 Q Fever: Clinical Features

Sudden onset of high fever

Headache, malaise, confusions

Nonproductive cough, possible progression to pneumonia

Nausea, vomiting, diarrhea, abdominal pain

Chronicity carries a significantly higher mortality rate, upward of 70% in some clinical case series.

Laboratory Tests

Mild elevations of liver enzymes, leukocytosis, and thrombocytopenia are laboratory findings seen with Q fever infection. In acute infection, IgM followed by IgG antibodies are present. In chronic disease, IgG antibodies predominate.

Microscopy

C. burnetii bacteria are pleomorphic bacilli or cocci that are difficult to discern on standard microscopy because they are intracellular. They are more readily identified with special stains such as Giemsa or using electron microscopy (Fig. 20–1).

Radiographic Findings

Chest radiographic findings are not consistently seen even with respiratory symptoms such as cough and pleuritic pain. Conversely, radiographic abnormalities may be noted even in the absence of respiratory symptoms. When present, findings are

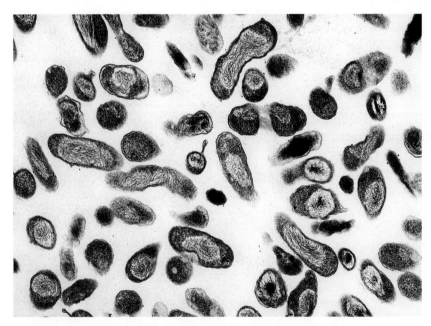

FIGURE 20–1   Electron micrograph of *Coxiella burnetii*, the bacteria responsible for causing Q fever.
Courtesy of Rocky Mountain Laboratories, NIAID, NIH.

consistent with pneumonia: a unilateral or occasionally multilobar infiltrate. Consolidation and effusions are less common.

Differential Diagnosis

Q fever is included in the differential diagnosis of flulike bioterror syndromes, including as well as the Category A agents *Yersinia pestis* and *Franciscella tularensis*. The differential diagnosis is expansive, including atypical bacterial pneumonias (e.g., Legionella or Mycoplasma) or influenza-like illnesses. See Table 20–2 for a list of other possible causes.

Treatment

Nearly all untreated patients will recover; however, antibiotic therapy appears to serve a dual function: shortening the duration of the natural course as well as reducing the likelihood of recurrent disease. Drugs of choice for Q fever are tetracyclines, specifically doxycycline. Macrolides have been used empirically with good success and quinolones have shown efficacy in vitro as well. The optimal duration of therapy is less clearly defined but has the greatest impact on morbidity if started within 3 days of symptoms. The Centers for Disease Control and Prevention (CDC) recommends doxycycline 100 mg po twice a day for 14 to 21 days, whereas military sources cite a shorter 2 to 7 day course as adequate. Chronic or recurrent Q fever is treated with prolonged combination treatment with doxycycline and a quinolone. The duration of treatment for chronic infections varies depending on the source: the CDC recommends four years; U.S. military texts recommend two years.

Vaccine and Postexposure Prophylaxis

A fully protective vaccine for Q fever exists. Presently, it is made in Australia and is not commercially available in the United States. Postexposure prophylaxis with doxycycline 100 mg orally twice a day for 5 to 7 days is appropriate if the patient demonstrates clinical symptoms or signs. Starting antibiotic prophylaxis during the incubation period, that is, prior to the onset of signs and symptoms is thought only to prolong the incubation period.

TABLE 20–2   Q Fever Differential Diagnosis

Legionella	Pneumonia, viral
Mycoplasma	Ehrlichiosis
Chlamydial pneumonia	Lyme disease
Endocarditis	Relapsing fever
Hepatitis	Rocky Mountain Spotted Fever
Legionnaires disease	Plague
Typhoid	Tularemia
Mononucleosis	

Infection Control

All contaminated clothing and surfaces must be handled carefully as standard cleaning agents and laundering do not destroy the spores. Autoclaving should be done before disposal. Direct person-to-person transmission does not occur, so that standard precautions are sufficient.

Brucellosis (*Brucella* species)

Background

One of the world's most significant agents of animal disease, *Brucella* species are found naturally in a wide range of animal reservoirs including livestock, farm animals, and dogs. Brucellosis—also known as undulant fever due to its characteristic biphasic fever curve—is caused by a number of different species within the genus. Four of those are known to be infectious to humans. Because it is both an animal and a human disease, *Brucella* species can be wielded to inflict both human health and economic injuries. As an agent of bioterror it would most likely be used in aerosol form.

Properties

In its natural form, Brucellosis is highly infective, but as with Q fever it is not very virulent: it is debilitating but rarely fatal to those infected. *Brucella* species do not sporulate; nonetheless, they are quite hardy. The bacterium is able to survive for weeks in water or soil and resists environmental stresses such as cold temperatures.

OF NOTE . . .

Brucella in New Hampshire—A Forme Fruste Terrorist Attack?

In early 1999, a 38-year-old New Hampshire woman was admitted to a community hospital and eventually ended up on mechanical ventilation. Three weeks into her hospital stay she was diagnosed with brucellosis. She had no risk factors for the disease, such as animal contacts, camping, or recent travel history. Her family came forward with petri dishes, flasks, growth media and *Brucella* cultures belonging to her live-in boyfriend. The man was a foreign national and claimed to be a marine biologist who had been working at a local university but had recently returned to his country of origin. The woman eventually died of adult respiratory distress syndrome. The FBI and CDC investigated the matter, and no explanation of how she acquired the infection was ever found. Bioterrorism was not ruled out as having played a role.

It does not survive desiccation caused by hot weather or direct exposure to sunlight.

Sources

Brucella species can be found indigenously in much of the world. Infected animals serve as naturally occurring sources via contaminated animal products, such as dairy or meat products, or by inhaling contaminated dust. As a result of diligent veterinary practices—specifically vaccination—most U.S. livestock are free of the disease. The primary natural reservoirs are bison and elk. Since the 1940s, several nations have sought to weaponize *Brucella* species, but whether these efforts were successful is unknown. There are no confirmed examples where *Brucella* has been used as a biological weapon.

Epidemiology

Brucellosis is found worldwide. Roughly 0.04 cases per 100,000 population are reported in the United States. Outside the United States, the incidence is considerably higher, according to some estimates as high as 88 cases per 100,000 population in certain regions of the world.

FIGURE 20-2 **In the United States, elk are one of the only remaining natural hosts of *Brucella*.**
Courtesy of the Agricultural Research Service, USDA/Keith Weller.

Means of Transmission

Transmission occurs as a result of ingestion of contaminated animal products such as unpasteurized milk or cheese, or from meat. It can also occur as a result of inhalation of contaminated dust or soil particulates as well as through direct contact of contaminated matter with skin openings, such as skin abrasions. Entry into the body occurs through epidermal breaks, the respiratory tract, the gastrointestinal tract, or the conjunctiva. Person-to-person transmission is extremely rare and documented only through breast-feeding and unsafe sexual practices. As a biological weapon, aerosolization of brucellosis and subsequent respiratory disease due to inhalation of the bacterium is the primary concern.

Pathogenesis

After finding its way into the body, *Brucella* is endocytosed by phagocytic macrophages. *Brucella* seems able to inhibit normal lysosomal activity within the macrophage and so is not destroyed by the intrinsic enzymatic pathways. Macrophages deliver the bacteria to local and regional lymph nodes where intracellular proliferation occurs until the lymphatic cells burst. The infection then spreads hematogenously throughout the body, being taken up preferentially by the reticuloendothelial system (liver, spleen, bone marrow, and lymph nodes). Less commonly, infection of the central nervous, cardiovascular, and renal systems may result from hematogenous dissemination. Granuloma formation is noted characteristically at tissue sites where infection is found, particularly in the liver and large lower axial joints (Fig. 20–3).

Signs and Symptoms

The clinical manifestations of brucellosis in the acute and chronic forms are nonspecific. Onset of signs and symptoms in the acute form may be immediate or insidious and occur following a 2 to 4 week incubation period. Documentation of incubation periods as long as two months underscores the potential difficulty in identifying deliberate biological attack with *Brucella*. Because of its predilection for the reticuloendothelial and lymphatic systems, symptoms of brucellosis may be localized or systemic. Symptoms include malaise, fatigue, myalgias, and headache. CNS symptoms such as depression and difficulty concentrating are reported commonly. Signs are few, with fever and chills being the most consistent clinical findings. *Brucella* gets its moniker of "undulating fever" from the diurnal pattern of the disease's fever curve. Hepatosplenomegaly and lymphadenopathy are also found with some frequency. Complications of brucellosis include pancytopenia, axial joint arthritis, and tenosynovitis (commonly sacroiliitis), as well as orchiditis and epididymitis. Within the lung, brucellosis may cause interstitial pneumonia and pleural effusions, including empyema. Symptoms may last up to a month before resolving. Rare, but serious complications of brucellosis occur in fewer than 5% of cases,

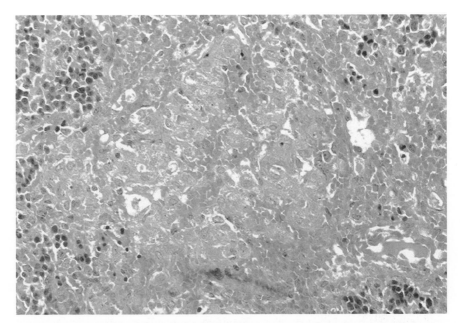

FIGURE 20-3   Spleen biopsy from *Brucella* infection showing the characteristic granuloma formation.
Courtesy of the CDC.

including encephalitis, meningitis, or endocarditis. These are responsible for the majority of brucellosis fatalities.

Chronic Effects

There is a chronic form of brucellosis characterized by symptoms that relapse every 2 to 3 weeks for upwards of a year and presents similarly to chronic fatigue syndrome. A deep or walled-off source of infection that seeds the body intermittently is thought to be the cause. Sequelae after infection are also quite variable and include granulomatous hepatitis, distal arthritis, anemia, leukopenia, thrombocytopenia, meningitis, uveitis, optic neuritis, papilledema, and endocarditis.

TABLE 20-3   Brucellosis: Clinical Features

Fevers in a diurnal pattern
Malaise, fatigue, myalgias, arthritic pain
CNS findings include depression, inability to concentrate; encephalitis or meningitis in more advanced cases
Possible pneumonia

Laboratory Tests

Pancytopenia may be noted, or individual hematological values may be low in isolation. White cell counts are usually within the normal range but may be depressed.

Microscopy

Brucella species are small, aerobic, encapsulated, nonmotile, slow-growing facultative, intracellular gram-negative coccobacilli. Tissue samples of infected sites will show pronounced granuloma structure and cytology.

Radiographic Findings

The chest x-ray may be normal, but abnormal findings when present include lung abscesses, single or diffuse nodules, pneumonia, and effusions.

Differential Diagnosis

The nonspecific clinical manifestations of brucellosis make the differential diagnoses quite broad. Any chronic febrile illness, particularly one in which the fever relapses and remits, must include brucellosis. Syndromic surveillance categorization would places it in the flu-like category, along with the Category A agents plague, tularemia, and the Category B agent, Q fever. The differential diagnoses for brucellosis is found in Table 20–4.

Diagnosis

Blood cultures offer a reasonable yield for diagnosis. Bone marrow cultures, however, are even more useful, with yields over 90%. Serum agglutination tests are available to detect both IgM and IgG antibodies; a titer of 1:160 or greater is diagnostic for brucellosis.

Treatment

Even without medical intervention, the majority of patients will recover from *Brucella* infection. Antibiotics reduce the course and severity, as well as the likelihood of recurrence and sequelae. Doxycycline 200 mg orally each day plus Rifampin 600 mg orally per day for 6 weeks is the CDC recommended regimen (See Appendix).

TABLE 20–4 Differential Diagnosis: Brucellosis

Viral infections	Mycobacteria
Mycoplasm infections	Familial Mediterranean fever
Lymphoma	Tularemia
Drug fever	Typhoid
Yersinia enterocolitica	

Vaccine

No vaccine is currently available for preexposure protection, nor has antibiotic intervention been shown to be beneficial for *Brucella*. For postexposure prophylaxis in an attack, the Doxycycline/Rifampin regimen for 3 to 6 weeks is recommended.

Infection Control

Person-to-person transmission does not occur, so isolation is not necessary; standard precautions will suffice. Contaminated objects are easily sterilized or disinfected by common disinfecting agents, such as phenol- or formalin-based solutions. Pasteurization is effective for treatment of contaminated dairy products.

☣ Glanders (*Burkholderia mallei*)

Background

Glanders is a disease primarily affecting members of the equid family, including horses, mules, and donkeys. Other animals—humans included—are also susceptible to the disease. Human cases of glanders occur sporadically. In all likelihood, glanders has been used as a biological weapon for nearly a century. In World War I, the German army was suspected of having intentionally infected a large number of horses of the Russian army along the eastern front with this zoonotic disease. The Japanese infected horses, civilians, and prisoners of war with glanders (and other biological warfare agents) during its occupation of China. During World War II, unconfirmed reports suggest that the Soviet army used glanders during its occupation of Afghanistan in the 1980s. Excepting the biowarfare examples noted, there are no documented naturally occurring epidemics of the disease. As a biological weapon, aerosolized glanders represents the most concerning possibility.

Sources

Horses, mules, donkeys, goats, and domesticated animals, such as dogs and cats, are the natural reservoirs for *Burkholderia mallei*. Unlike its close relative *Burkholderia pseudomallei* (see next section), it is not found in environmental reservoirs such as water, soil, or plants. Laboratorians, veterinarians, outdoorsmen, and others with close physical contact with animal reservoirs, animal products, or microbial sources for research are at greatest risk of acquiring the disease in its sporadic form.

Properties

B. mallei is an aerobic, nonmotile, non-spore-forming, gram-negative bacillus. Four glanders species are capable of infecting animals. Humans, although not immune, do not acquire the disease as readily.

FIGURE 20–4   **Horses serve as the primary reservoir for *Burkholderia mallei*.**
Courtesy of Pacific Northwest National Laboratory.

Epidemiology

There has been one documented case of glanders infection in the United States since 1945. The infected individual was a laboratorian who was working with the bacteria.

Means of Transmission

How the bacteria spreads is not well understood. Epidermal breaks due to lacerations or other lesions are believed to be one method of transmission following direct contact with infected animals: entry through mucosal surfaces of the upper airway or on the conjunctiva appear to be another way. Infections among laboratorians working with *B. mallei* have demonstrated that aerosol infection is possible as well. In the context of biological weapons, the latter is the most likely means of transmission.

Pathogenesis

The virulence of *B. mallei* is due to compounds synthesized by the bacteria that are toxic to the host's metabolism. The toxins tend to disrupt enzyme activity at key metabolic pathways. For example, the blue-green pigment pyocyanin, disrupts oxidative phosphorylation. Lecithinase, another of the toxins elaborated by the bacterium, causes cell lysis by degrading lecithin contained within cell membranes, collagenases, lipases, and hemolysins. If the production of these compounds reaches a certain threshold, clinical symptoms ensue. Because the symptoms result from toxins, antibiotics are not especially helpful in treating this disease.

Signs and Symptoms

There are three acute forms of glanders: focal, pulmonary, and septic. They may occur in isolation or in combination. A chronic form of the disease, known as "farcy," presents with lymphangitis and lymphadenopathy. The focal form is seen following exposure through an epidermal break and direct bacterial contact after an incubation period of about five days. Following implantation, a nodular lesion with local lymphangitis and lymphadenopathy results. Similarly, infection of an upper respiratory site or ocular infection causes a pustular, ulcerative lesion. In the former instance, production of bloody and/or purulent sputum results. If systemic seeding occurs from these sites, abscesses in a variety of organs (or a diffuse maculopapular rash that may mimic smallpox) may arise. The pulmonary form of glanders occurs after inhalation of the bacterium or as a result of hematogenous spread from a focal infection. The incubation period in the respiratory tract is up to fourteen days. Clinical manifestations include fever, myalgias, headache, and pleuritic chest pain. The septicemic form of glanders can occur as a primary infection, or it may result from hematogenous spread from pulmonary or focal infections. Symptoms include flu-like features such as fever, myalgias, cervical adenopathy, and diarrhea, as well as cutaneous findings such as necrotizing, pustular lesions, generalized erythroderma, and jaundice. CNS findings of septicemic glanders include headache and photophobia. This form of the disease is distinctly uncommon but carries with it a high mortality rate with death occurring within 2 weeks of onset.

A bioterrorist attack with *B. mallei* would likely come in the form of an aerosol, and pulmonary forms would no doubt predominate. Further, signs and symptoms might occur faster with perhaps a higher likelihood of progression to the septicemic form. For the purposes of syndromic surveillance, glanders is best grouped with the flulike bioterrorism syndromes.

Chronic Effects

Farcy, the chronic form of glanders, presents years or decades following the acute infection. Abscesses involving abdominal organs (e.g., liver and spleen), skeletal muscles, bone, or the skin may develop. These finding may be associated with lymphangitis and lymphadenopathy in the affected area. CNS involvement may occur with meningitis, encephalitis, or brain abscesses.

Laboratory Tests

Blood cultures are usually negative until very late in the illness and may require supplementation of growth media with nutrients such as meat and sugar. White blood cell counts are normal or slightly high.

TABLE 20-5   Glanders: Clinical Features

Three acute forms: focal, pulmonary, and septic

Focal—nodular lesion with adenopathy; possible progression to pulmonary or septic forms

Pulmonary—myalgias, headache, and pleuritic chest pain

Septic—typical sepsis features as well as meningeal features. This form is associated with a high mortality.

Any of these forms can include a maculopapular rash secondary to bacterial seeding.

Microscopy

B. mallei are a nonmotile, gram-negative rods. Staining with methylene blue of contaminated fluid may reveal scant small bacilli, whereas standard gram staining reveals small gram-negative, bipolar rods.

Radiographic Findings

Chest radiographs may show consolidation, a bilateral miliary pattern, or multiple lung abscesses.

Differential Diagnosis

Because of the highly nonspecific nature of the symptoms caused by glanders, a broad differential must be assembled. The differential diagnosis must obviously be altered to the clinical presentation. Among others, tuberculosis must be considered with the pulmonary form, and smallpox must be considered if the dermatological features are present. A partial list is included in Table 20–6.

Diagnosis

The most specific diagnostic test available for glanders is complement fixation, and it is considered positive if the titer is greater than or equal to 1:20. However, this is not as sensitive as agglutination titers which have poor specificity. PCR would be needed to distinguish between glanders and melioidosis, the latter caused by the related microbe, *Burkholderia pseudomallei*.

TABLE 20-6   Differential Diagnosis: Glanders

Viral pneumonia	Typhoid anthrax
Legionella	Smallpox
Mycoplasma	Tularemia

Treatment

There is surprisingly little clinical experience treating glanders infection. The U.S. military's recommended management for the focal form and pulmonary form are amoxicillin/clavulanate (20 mg/kg tid), tetracycline (40 mg/kg po tid) or trimethoprim-sulfamethoxazole (TMP 2 mg/kg, sulfa at 10 mg/kg bid). Regardless of which regimen is selected, the recommended course is 2 to 5 months. If the clinical picture is more severe, oral treatment should include two of the three standard regimens for 1 month followed by monotherapy for an additional 2 to 5 months. Treatment is prolonged 10 to 12 months further if pulmonary glanders is present or if abcesses have formed. The U.S. military's recommended management for the septicemic form is 2 weeks of intravenous therapy with ceftazidime (40 mg/kg/day every 8 hours) plus TMP (2 mg/kg/day) and sulfa (10 mg/kg/day) every 6 hours. After 2 weeks, parenteral administration may be changed to oral administration and continued for an additional six months.

Vaccine

At present, there is no vaccine available. Postexposure prophylaxis (PEP) has not been fully evaluated, but use of trimethoprim/sulfamethoxazole has been suggested.

Infection Control

Glanders is not transmitted from person to person, and so standard infection control precautions are sufficient. If there is skin compromise—laceration or open sores—contact precautions are indicated. Adequate decontamination of surfaces can be achieved with standard bleach or disinfectants. Clothes and laundry that come into contact with cutaneous sites infected with glanders should also be cleaned.

Melioidosis (*Burkholderia pseudomallei*)

Background

Though closely related to *Burkholderia mallei* genetically and clinically, *Burkholderia pseudomallei* is distinctive in a number of important ways. Melioidosis, also known as Whitmore's disease, is endemic to much of Southeast Asia and is not uncommon in other tropical regions. The disease takes multiple forms in humans including a septic form that is almost always fatal. A chronic form of melioidosis may activate years after the initial exposure, and is itself potentially life-threatening. Both the United States and the former Soviet Union attempted to weaponize *B. pseudomallei*, but the outcome of this research and development effort is unknown. It is generally agreed that as a biological weapon, *B. pseudomallei* will be used in aerosolized form.

Sources

B. pseudomallei bacteria are found in worldwide distribution in soil and water, particularly in rice paddies. Like with glanders, horses are a natural reservoir for B. pseudomallei, along with goats, monkeys, and rodents. The microbe's presence in soil and water is thought to represent the primary route of exposure for humans. Australian outbreaks have been associated with potable water sources.

Properties

B. pseudomallei is a motile, flagellated facultative intracellular pathogen. It is resistant to dry heat and survives for long periods under arid conditions. The bacteria are saprophytic, surviving by attaching to organic debris in soil, streams, pools, stagnant water, rice paddies, and even vegetables. Land and marine animal reservoirs excrete and shed the bacteria, and this represents the primary means by which the bacterium is spread.

Epidemiology

Unlike glanders, melioidosis has caused epidemics in humans and continues to do so in parts of Southeast Asia.

Means of Transmission

Infection occurs as a result of either direct contact with contaminated objects or via material that enters the body through epidermal breaks, such as lacerations, scratches, or abrasions. The other common means are ingestion of contaminated water or inhalation of contaminated dust or particles. As a biological weapon, most experts feel that exposure would occur via an aerosolized form of the bacteria. Based on the natural history of sporadic melioidosis, the majority of those infected from a deliberate biological attack should not become ill. They would, however, be at a significant risk for relapse—particularly if immunosuppression or other co-morbid conditions exist or develop in the infected host.

Pathogenesis

The pathophysiology of melioidosis is poorly understood. As with Q fever, brucellosis, and glanders, it is an intracellular pathogen that thrives inside the lysosomes of macrophages. Similar to glanders, the bacteria resists phagocytic destruction, but unlike glanders, no exotoxins have yet been identified to explain this effect. Speculation that the virulence of melioidosis is related to the presence of a particular surface lipopolysaccharide has not been confirmed scientifically. Epitheliod granuloma formation is a common finding at infection sites. Bacteremic seeding often results in deep-tissue abscess formations.

Signs and Symptoms

Melioidosis is sometimes referred to as a "mimicker" because of its wide clinical manifestations. Like glanders, melioidosis occurs in three forms: focal, pulmonary, and septic. These may occur distinctly or in combination. Additionally, there is a chronic form that occurs following an incubation period that runs from a few days to a few years. The focal form usually results from direct contact between a contaminant and a break in the host's epidermis. Nodular or pustular regional lymphadenopathy, myalgias, malaise, and fevers are often associated with dermal melioidosis. The lesion is painful, unlike cutaneous anthrax which is painless. The most common naturally occurring form of melioidosis is pulmonary. This results from inhalation of the microbe or hematogenous spread from a focal infection. Pneumonia with the formation of solitary or multiple lung abscesses are seen, usually with caseating necrosis. Involvement of the upper lobes of the lung is not uncommon, putting tuberculosis into the differential diagnosis. Cough is common, but is not invariably productive and if productive, it is rarely purulent. Myalgias, fevers, headaches, and weakness round out the clinical symptoms associated with pulmonary melioidosis. Septicemic melioidosis may occur as a primary infection, or again following pulmonary or focal infections. As with glanders, it is associated with a mortality rate greater than 90%. The clinical picture is that of septic shock, with hypotension and multiorgan system failure. Septicemic melioidosis is more likely in a host with significant co-morbidities, including diabetes, HIV, or chronic obstructive pulmonary disease (COPD). Pustular cutaneous lesions, myalgias, and deep tissue abscesses can occur anywhere throughout the body, including the central nervous system.

Vulnerable Groups

Individuals with exposure to water in endemic areas of Southeast Asia and Northern Australia are more likely to develop melioidosis. Individuals with diabetes, liver failure, or renal disease, alcoholism, COPD, and other immunocompromised states are at increased risk for contracting the disease. These groups experience increased case fatality rates as well.

Chronic Effects

Chronic melioidosis occurs in infected persons who remained asymptomatic—and therefore untreated—during the acute period. It is a multisystem illness affecting about 3% of those infected with the bacterium. Any organ of the body—joints, viscera, muscle, lymph nodes, skin, brain, liver, lung, bones, and spleen—may be involved. Lymphadenopathy and lymphangitis may be seen proximal to the abscess site.

Laboratory Tests

Complete blood counts may demonstrate a mild leucocytosis with band forms. Alternatively, a modest leukopenia can be seen. Complement fixation tests are

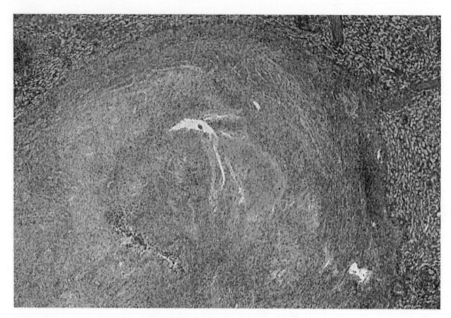

FIGURE 20-5   Splenic abscess found in a man with Melioidosis.
Courtesy of Fujita Health Univ. School of Medicine/Yutaki Tsutsumi, MD.

highly specific, and a single titer above 1:160 in the setting of suspected infection supports the diagnosis of meliodosis.

Microscopy

B. pseudomallei is a non-spore-forming, gram-negative, bipolar (safety-pin), flagellate bacillus. Isolation of the bacteria in culture is key to making the diagnosis. Purulent sputum or discharge offers the highest yield, but negative cultures are common. Meat and sugar supplementation of the culture medium may stimulate growth of the bacterium.

Radiographic Findings

Chest radiographs may show single or multiple nodular lesions that gradually enlarge, coalesce, and cavitate. Patients are often more ill systemically than radiographic findings might suggest. Pleural effusions are infrequent.

Differential Diagnosis

The nonspecific signs and symptoms of melioidosis once again present the clinician with a broad differential diagnosis, including Category A agents: plague, anthrax, and tularemia; and several Category B agents including glanders.

OF NOTE . . .

Singapore, 2004: An Investigation of a Melioidosis Mini-epidemic

In the first 8 months of 2004, 79 people in Singapore became ill with melioidosis, leading to the death of 24. Typically, there are about 45 cases each year in the island nation. The increased numbers aroused suspicion among public health officials there, as did the unusually high mortality rate of nearly 30%, making it nearly twice as deadly as SARS. Although the increased number of infections and the increased case fatality rate were worrisome to local officials, this clustered outbreak prompted a government investigation to determine whether the infections were actually a result of deliberate exposures or bioterrorism. The Singapore government was particularly concerned because melioidosis is classified as a Category B agent. Results of the investigation showed no evidence of *B. pseudomallei* having been used as a biological weapon. The increased mortality was found to be a result of infection of people already bearing the burden of co-morbid conditions, such as diabetes mellitus, hypertension, or renal insufficiency. Although the increased mortality rate may thus be explained, the reasons for the increased incidence are still unclear.

—*MMWR*

Diagnosis

A complement fixation titer above 1:160 in the setting of suspected infection confirms a diagnosis of melioidosis with excellent specificity.

Treatment

Antibiotic therapies for melioidosis follow the same regimes as for those of glanders. The U.S. military's recommended management for the focal form and pulmonary form are amoxicillin/clavulanate (20 mg/kg/day every eight hours) or tetracycline (40 mg/kg/day in three divided oral doses) or TMP (2 mg/kg/day) with sulfa (10 mg/kg/day every 12 hours). Regardless of the regimen selected, the recommended

TABLE 20-7 Melioidosis: Clinical Features

Wide ranging clinical manifestations—fever, myalgias, arthralgias

Three forms—focal, pulmonary, and septic

Focal—painful nodular/pustular lesions, fever, myalgia

Pulmonary—pneumonia, possible pulmonary abscess in any part of lung, cough, fever

Septic—typical sepsis picture along with deep tissue abscesses

TABLE 20–8   Differential Diagnosis: Melioidosis

Mycoplasma	Glanders
Atypical pneumonias	Typhoid
Viral pneumonias	Anthrax
Tuberculosis	Tularemia
Plague	

course is two to five months. If the clinical picture is more grave, oral treatment should include two out of the three drugs for one month followed by monotherapy for an additional two to five months. An additional ten to twelve months is needed if, as with the pulmonic form, abscesses are present with drainage. The U.S. military's recommended management for the septicemic form is two weeks of intravenous therapy with ceftazidime (40 mg/kg/day every 8 hours) plus TMP (2 mg/kg/day); and sulfa, (10 mg/kg/day every 6 hours). After 2 weeks, administration may be changed to oral forms and continued for an additional six months. In the septicemic form, empiric evidence supports the addition of granulocyte colony stimulating factor (GCSF).

Vaccine

There is no vaccine for melioidosis.

Infection Control

Person-to-person transmission is possible through contaminated fluids. Contact precautions are therefore required. If sputum is culture positive, or while cultures are pending, respiratory isolation is suggested until confirmation of melioidosis, whereupon respiratory isolation is not necessary. Surface decontamination with standard cleaning agents is adequate. Medical equipment in contact with the patient should be autoclaved.

☣ Psittacosis (*Chlamydia psittaci*)

Background

Chlamydia psittaci is a member of the *Chlamydia* genus which is currently under taxonomical scrutiny because of its peculiar microbiological features. *C. psittaci* infection results in acute respiratory illness and may reactivate many years following acute infection. Seventy-five percent of all naturally occurring cases or outbreaks have been linked directly back to contact with or near birds carrying the bacteria. There are documented cases of *C. psittaci* infection having resulted from humans kissing birds and other cases from doing mouth-to-mouth resuscitation on birds. However, the other 25% appear to have had no exposure to birds. The largest outbreak in the Unites States occurred in 1930 when over 800 individuals became infected with *psittacosis*. As a biological weapon, *C. psittaci* has two potentials: the first as an aerosolization targetting human populations and the second as a deliberate agroterrorist attack on the poultry industry.

Sources

Classically, and most commonly, *C. psittaci* is found in bird droppings. Infection occurs from exposure to the droppings of healthy specimens serving as a reservoir. If there were widespread use of *C. psittaci* as a biological weapon, birds could sustain an epidemic by serving as natural reservoirs. The term psittacosis is somewhat of a misnomer as it implicates parrots as the primary source, whereas "ornithosis" would be a more inclusive and accurate term. To be sure, parrots tend to carry one of the more virulent forms of the disease, as do turkeys. Other animal reservoirs include livestock, for example, cattle and sheep; and domesticated animals, including cats and dogs.

Properties

C. psittaci is a gram-negative bacillus that functions as an obligate intracellular parasite. The bacteria survives unprotected for several days in bird excrement and can be spread as a wind-blown aerosol from dried feces. *C. psittaci* is a hardy organism,

FIGURE 20–6   Birds, including parrots, are the primary reservoirs of *C. psittacii*.
Courtesy of National Parks Service/Jerry Bauer.

resistant to drying, and remaining viable for months under common environmental conditions.

Epidemiology

C. psittaci occurs in worldwide distribution. There are roughly 200 cases diagnosed annually in the United States. This number is probably an underestimate given the diagnostic difficulty of identifying a disease whose presentation is often subtle and mild. Birds function as the primary reservoir, but other animal reservoirs provide opportunity to cross species to infect humans. It has been reported that samplings of bird populations show a baseline infectivity rate of nearly 8%; this can rise 10-fold during outbreaks at commercial poultry settings.

Means of Transmission

Infection most commonly occurs as a result of inhalation of contaminated bird feces in the form of dust. Inhalation of contaminated respiratory or urinary products can lead to infection, as well. Cases of person-to-person transmission are unusual, but have been documented, often presenting with a more severe illness than those acquired from avian or animal sources.

Pathogenesis

The precise mechanism of *C. psittaci* is not understood, but the general actions are described. Chlamydia takes on two distinct morphologies: elementary and reticulate. The elementary form, or elementary body, is the infective form. The reticulate body is the proliferative form. Once in the airways, the elementary form of the bacterium adheres to the epithelial cells lining the respiratory tract and are endocytosed. As with the other obligate intracellular pathogens described earlier in this chapter, *C. psittaci* inhibits phagocytic mechanisms once inside the host cell. Within the endocytotic vesicle, the elementary body transforms into the reticulate body. This form of the microbe is responsible for bacterial reproduction, capable of producing hundreds of new bacteria inside the host cell. The bacteria cannot synthesize its own energy sources, so this replication consumes the host's stores of energy. When bacterial replication reaches a maximum load, the cell ruptures and proximal cells are infected with the newly released bacteria. Cell lysis also results in hematogenous dissemination to the host's reticuloendothelial system.

Signs and Symptoms

C. psittaci's incubation period is between seven and fourteen days, but it can take as long as five weeks. Symptoms begin rapidly, gradually, or may remain subclinical and range from mild to severe, including death. Once again, the signs and

TABLE 20–9   Psittacosis: Clinical Features

Fever, chills, sweats, myalgia, malaise, sore throat

Nonproductive cough, rales

Hepatomegaly and/or splenomegaly

Maculopapular rash

symptoms are typically nonspecific, including fever, chills, sweats, myalgia, malaise, sore throat, headache, photophobia, and gastrointestinal distress. The textbook presentation of psittacosis is that of an atypical pneumonia—fever, nonproductive cough (often late appearing), malaise, dyspnea, and commonly a maculopapular rash. Physical examination may reveal rales, hepatomegaly, or splenomegaly.

Chronic Effects

In the more severe cases, respiratory failure, hepatitis, renal injury, and CNS effects may be present. Proteinuria, reactive arthritides, endopericarditis, myocarditis, and pericarditis may also be found. Many of these complications linger long after the acute infection has resolved.

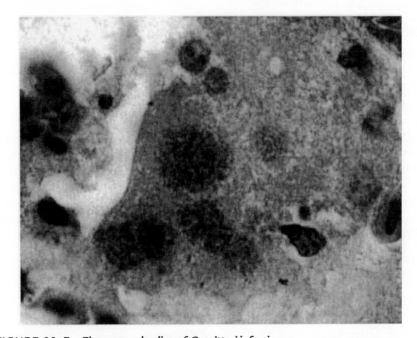

FIGURE 20–7   Elementary bodies of _C. psittaci_ infection.
Courtesy of the Department of Small Animal Medicine, College of Veterinary Medicine, University of Georgia.

Laboratory Tests

The leukocyte count is often within normal; however, an increase in band forms and neutrophilia is common. Liver function studies show transaminase elevations (commonly aspartate aminotransferase, or AST) and a concomitant drop in albumin levels. Serum chemistries may demonstrate elevation in blood urea nitrogen and creatinine levels and moderate hyponatremia.

Microscopy

Culturing *C. psittaci* is too hazardous for BSL-1 labs and should be deferred to BSL-2 or BSL-3 laboratories that are specially equipped to handle the specimens.

Radiographic Findings

Chest radiographs in *C. psittaci* infection are usually normal. When abnormal, the CXR classically reveals lower lobe consolidation, often disproportionate to the clinical syndrome. That is, the patient may appear mildly ill, but have substantial lung involvement based on CXR findings. Less common radiographic findings include segmental consolidations or miliary patterns. As is common with other pneumonic processes, radiographic clearing often occurs weeks after clinical resolution.

Differential Diagnosis

The differential diagnosis for *C. psittaci* is vast. It is probably best included in the bioterror syndrome of flulike illnesses. Characteristically, it is a febrile respiratory illness and ranks among the short list of classic atypical pneumonias, such as Legionella and mycoplasma pneumonia. The differential also includes the panoply of bacterial, viral, and fungal pneumonias. Viral illnesses such as influenza and respiratory viruses, as well as rickettsial diseases and mycobacterial diseases (e.g., tuberculosis), are also included in the differential. The key to diagnosing psittacosis is careful social and environmental history taking. Close contact to animals—particularly birds—is a useful clue that can increase clinical suspicion for the disease. This questioning is even more essential if a possible bioterror attack is being investigated. A partial differential diagnosis for *C. psittaci* is included in the following table.

TABLE 20-10 Differential Diagnosis: *C. psittaci*

Influenza	Viral pneumonia
Typhoid	Q fever
Legionella	Tuberculosis
Tularemia	Plague
Mycoplasma infections	Rickettsial illness

Diagnosis

Diagnosis of *C. psittaci* using complement fixation testing is a the gold standard. A fourfold rise in *C. psittaci* antibodies over a 2-week period confirms the diagnosis. Antibody testing does not distinguish among different chlamydial species, however. Definitive diagnosis of *C. psittaci* requires microimmunofluorescence of IgM titers greater than 1:16, or a fourfold increase in titers over a 2-week period.

Treatment

Tetracyclines are first-line agents in chlamydial infections. A 2-week treatment with doxycycline (100 mg orally every 12 hours) is the standard regimen. Critically ill patients should receive intravenous doxycycline (4.4 mg/kg every 12 hours). Typically, clinical improvement occurs in the first 48 hours after initiating antibiotic therapy.

Prophylaxis

There is no vaccine for *C. psittaci,* and postexposure prophylaxis is not recommended. Instead, exposed individuals should be placed under medical surveillance. Development of fever or other symptoms should lead to implementation of antibiotic treatment. Prior infection with *C. psittaci* does not confer lifetime immunity.

Infection Control

Respiratory isolation is not recommended because person-to-person transmission occurs so infrequently. Contact isolation and adequate decontamination are recommended. The organism is susceptible to standard disinfectants, including sodium hypochlorite solution.

Typhus fever (*Rickettsia prowazekii*)

Background

Infection with *Rickettsia prowazekii* leads to a clinical presentation often referred to as "typhus fever" or "epidemic typhus," or the reactivation of a prior infection termed Brill–Zinsser disease. In its natural form, the bacteria is passed on in the feces of body lice (*Pediculus humanus corporis*) to individuals on whom the lice reside. Epidemics of typhus have been described through much of human history and occur whenever crowding and poor sanitary conditions exist. Natural outbreaks are more likely to occur in less industrialized nations, often as a result of the dislocation and social disruption incident to natural disasters or war. During WWI, tens of millions of typhus infections occurred in combatants with roughly 3 million deaths.

From a bioweapons perspective, intentional induction of a typhus outbreak using lice vectors is logistically difficult. However, *R. prowazekii* has the potential to be

FIGURE 20-8 *Pediculus humanus corporis,* **the primary vector for typhus.**
Courtesy of the CDC.

aerosolized and spread though dispersal and subsequent inhalation. This strategy would be most effective in densely populated areas.

Sources

Typhus is a global infectious disease. It is an arthropod-borne disease, being carried in body lice who have themselves become infected by feeding on infected hosts. Hosts include humans and animals, such as flying squirrels.

After the acute illness resolves, infection with *R. prowazekii* may become subclinical as the microte remains hidden in undetermined sites in the body, and may reactivate at some later point, causing Brill-Zinsser disease. Patients with latent typhus infection may then serve as a resevoir of infection, capable of passing on, via the louse vector, to other individuals.

Properties

The mortality rate in untreated typhus is approximately 50%, whereas fatal cases are rare with treatment. Rickettsia are obligate intracellular bacteria. They survive for long periods in lice feces. The ID_{50}—the dose required to cause infection in 50% of those exposed—is remarkably low. Just a few organisms are sufficient to cause disease. The hardiness of the microbe and its low ID_{50} make it a viable candidate for use as a biological weapon.

Epidemiology

Disease caused by R. *prowazekii* is found worldwide, although it is extremely uncommon in the United States. The last case in the United States was in 2001 in New Mexico, but no source could be identified. Outbreaks of typhus follow natural disasters such as floods and famines, and manmade disasters such as war.

Means of Transmission

Person-to-person transmission does not occur; rather body lice serve as the vector for the disease. Lice acquire the bacteria after biting an infected person during the first five days of symptoms. The bacteria take up residence and proliferate within the louse gastrointestinal tract and are then passed out in its abundant feces. The louse bites the human host causing a local inflammatory response with histamine release. The resulting pruritis elicits scratching which brings contaminated feces or contaminated louse body parts into the lesion effectively inoculating the host. Mucus membrane contact with contaminated lice feces also results in infection. It is speculated that inhalation of dust contaminated with louse feces can cause infection as well. Typhus epidemics, however, generally require crowded conditions, direct contact, or shared clothing; otherwise spread is difficult because infected lice tend to live only about 5 days before dying from the infection. An outbreak could also result from previously uninfected lice becoming infected by biting someone with a reactivation of a prior infection (Brill–Zinsser disease).

Pathogenesis

After inoculation from a louse bite, the bacteria are taken up into the capillaries where they then adhere to and invade the endothelial cells lining of vessels in which they are carried. Once inside, the bacteria proliferate until the bacterial load reaches maximum capacity and the cell bursts. Subsequent injury causes a local vasculitis and thrombogenesis with fibrin and platelet deposition at the rupture site. Thrombosis may continue from this point, with potential for thrombotic clots and damage to vital organs, the heart, or the kidneys.

Signs and Symptoms

After an incubation period of approximately 10 days, patients experience the sudden onset of signs and symptoms that last as long as 28 days if untreated. Findings include high fever, chills, weakness, cough, myalgias, abdominal pain, and headache. A rash beginning centrally and spreading centrifugally starts at about Day 4 with discrete macular lesions that evolve into petechial or necrotic lesions. These skin changes spare palmar and plantar surfaces. Severe CNS features can occur after 5 to 7 days of illness including mental status changes and meningitis.

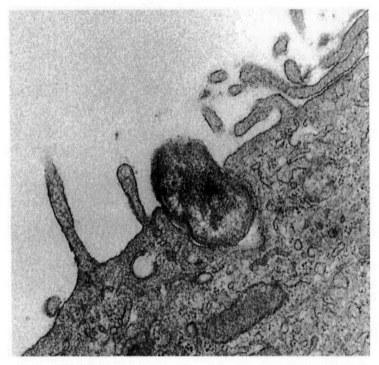

FIGURE 20-9   **Electron micrograph of** *R. prowazekii,* **an rickettsial parasite, being taken up into the host cell.**
Courtesy of the CDC.

When present, gangrene occurs cutaneously or in the extremities depending on the size and location of thrombi formation.

Full recovery is slow, often taking 2 to 3 months before full strength is restored.

Chronic Effects

Brill–Zinsser disease is a reactivation of *R. prowazekii* often years after the initial infection. Reactivation usually correlates with an immunosuppressed or immunocompromised state. The signs and symptoms are the same as the initial phase but generally take a milder form. The diagnosis is easily missed unless the prior history of typhus fever is known.

Microscopy

R. prowazekii are gram-negative coccobacilli that survive as obligate intracellular parasites. Microscopic identification of the bacteria within host cells using Giemsa staining is difficult.

TABLE 20-11   Typhus Fever: Clinical Features

Abrupt onset of high fever, chills, weakness, cough, myalgias, abdominal pain, and headache

Discrete macular lesions that begin truncally and spread peripherally evolving into petechial or necrotic lesions

CNS features can occur with meningeal signs and symptoms and/or mental status changes

Thrombi formation occurs; signs and symptoms result depending on the size and location of the thrombi. Gangrene may be seen in peripheral or cutaneous thrombi formation

Radiographic Findings

Chest radiographs are either normal, or infrequently reveal subtle pulmonary infiltrates.

Differential Diagnosis

The differential for *R. prowazekii* infections is quite large, including other rickettsial diseases, bacterial, and other febrile viral illnesses. A partial list is included in Table 20–12.

Diagnosis

Clustering is a key historical feature for both naturally occurring forms of typhus and a potential bioterrorist attack. Identifying rickettsial infections can be done routinely using a Weil–Felix reaction; however, identifying *R. prowazekii* specifically is difficult and requires PCR. Antibodies appear after about 2 weeks of infection and can be diagnosed with immunofluorescent staining.

Treatment

Tetracyclines are first-line agents against virtually all rickettsial infections. A one-time dose of 200 mg of doxycycline is recommended, although some sources suggest

TABLE 20-12   Differential Diagnosis: Typhus Fever

Meningococcemia	Smallpox
Gonnococcemia	Mononucleosis
Staphylococcemia	Vasculitides
Measles	Subacute bacterial endocarditis
Rubella	Erythema multiforme
Varicella	Secondary syphilis

100 mg orally twice a day for 2 to 3 days. Symptoms should begin resolving within 48 to 72 hours of treatment. Relapse is rare but is responsive to antibiotic therapy should it occur.

Prophylaxis

An inactivated vaccine offers some degree of protection is available, but it is not recommended for the general population. Should an attack (or epidemic) occur, resources and the clinical setting will dictate whether prophylaxis will be given or if those exposed should be watched carefully for 2 weeks. Prophylaxis consists of a one-time dose of doxycycline following exposure. Some believe that antibiotics taken within the first 48 hours are less effective and should be avoided.

Infection Control

In the event of an attack with typhus, person-to-person transmission is not a concern. To prevent epidemic propagation after an attack, or in the event of a natural outbreak, delousing is the mainstay of infection control. Generic lindane treatments along with disposal of contaminated clothing are the most effective measures, though there are increasing reports of lindane resistance. Use of permethrin (1% powder, 40 mg) as a first-line management for bedding and clothing is effective and has long-term activity against lice.

☣Bacterial Toxins

Staphylococcal Enterotoxin B (SEB) and epsilon toxin of *C. perfringens* are two bacterial toxins notable in the context of bioterrorism because they are the metabolic products of ubiquitous bacteria that are relatively easy to mass produce. In their naturally occurring forms, the toxins are ingested typically. It is widely held that in a weaponized form they are more likely to be aerosolized, and thus affect the respiratory tract. Inhalation can additionally result in gastrointestinal features as is discussed later in this section.

Background

Biotoxins are organic compounds excreted as products of cellular metabolism from bacteria or fungi. *Staphylococcus aureus* produces a number of such exotoxins that are commonly associated with food poisoning. They are included among other bacterial exotoxins that affect the GI tract and are referred to generically as enterotoxins. SEB is one of the most common of the enterotoxins. Although the GI effects are the basis for classification, SEB affects a wide array of metabolic and physiologic functions. As a biological weapon, SEB would most likely be used in aerosol form exerting its harmful effects through the respiratory tract. The toxin could also be used to contaminate food or water supplies though, only on a relatively small scale.

Sources

S. aureus is a ubiquitous bacterium. A variety of foods may be contaminated with its preformed toxin, including meat products, poultry, baked goods, and dairy and egg products. *S. aureus* also colonizes the nares and axillae. It is precisely *S. aureus's* ubiquity that makes SEB a potential bioweapon. It is relatively easy to culture and certain strains of *S. aureus* yield copious amounts of the toxin. However, mass-scale production requires the use of industrial microbiology equipment. As with all the biotoxins, technological advances in molecular biology make using the bacteria itself unnecessary. Instead, insertion of the gene responsible for toxin production into other types of bacteria could increase toxin output dramatically.

SEB is reported to have been included among several countries' stockpiles of biological weapons.

Properties

SEB is the most studied of the many antigenically different exotoxins produced by *S. aureus*. It consists of 239 amino acid residues and has a molecular weight of 28. SEB is classified as a "super antigen" because of its effect on the immune system (see pathophysiology later). A heat-stable molecule, the toxin is readily soluble in aqueous solutions and is denatured only with prolonged boiling.

SEB is less of a concern in terms of mortality, but its morbidity is substantial. Fully 80% of those exposed become debilitated for up to 2 weeks. In aerosolized form, the ED_{50} for SEB is 0.0004 µg/kg, whereas the LD_{50} is 0.02 µg/kg, nearly fifty times greater.

Epidemiology

The incidence of naturally occurring SEB illness is unclear for several reasons: There are countless illnesses that give a similar gastroenteritis-like picture, and all of these are generally treated empirically. Diagnosing SEB specifically is difficult, as symptoms often are mild enough that no medical treatment is sought.

Means of Transmission

There are several potential portals of entry for SEB. As a biological weapon, the most likely means of transmission is in an aerosolized form with inhalation as the primary route of exposure. However, small-scale contamination of food sources or water supplies with the preformed toxin is possible, resulting in the more classical gastrointestinal effects. Person-to-person transmission does not occur with SEB.

Pathogenesis

SEB, like other so-called superantigens, binds to the major histocompatibility complex (MHC) directly outside of the peptide binding cleft rather than inside the cleft, where antigens are typically presented by the host cell. By binding outside the cleft, superantigens nonspecifically stimulate a large number of T cells (Fig. 20–10).

In the case of SEB, the toxin binds outside of the MHC class II molecules of macrophages resulting in activation of an inappropriately large number of CD4 T cells. The activation of such of a large number of T-helper cells results in a cytokine storm, with the release of large amounts of tumor necrosis factor (TNF), interferon (INF), interleukin (IL)-1, and other cytokines. This highly inflammatory milieu causes cellular injury and the clinical feature resultant of SEB exposure.

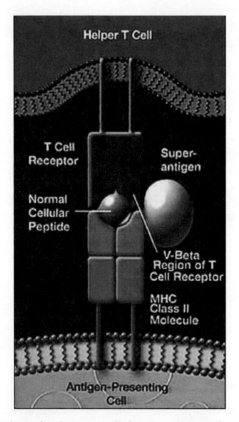

FIGURE 20-10 Schematic of a prototypical superantigen, such as SEB.
Courtesy of Eade Creative Services, Inc./George Eade.

Signs and Symptoms

Ingestion

When SEB is taken into the gastrointestinal tract, either through contaminated food or water, GI symptoms predominate. Onset of symptoms can occur within 8 hours. Classically, symptoms seen with ingestion include abrupt onset of intense nausea, vomiting, abdominal cramps, and diarrhea. Most cases are self-limited and resolve within a day. Physical examination reveals little, although patients will appear to be in discomfort secondary to abdominal cramping, nausea, and diarrhea.

Inhalation

After a period of up to twelve hours, inhalation of SEB into the respiratory tract presents with abrupt onset of headache, high fever lasting up to 5 days, chills, myalgia, prostration, and shortness of breath. A cough, sometimes productive, may be present and persist for up to 4 weeks. Noncardiac substernal chest pain has been reported. Physical exam is generally unremarkable except in severe cases where pulmonary findings such as crackles may be noted.

Inhalational forms may be accompanied by GI signs and symptoms from incidental ingestion.

Regardless of the route of exposure, debilitation for upwards of 2 weeks is common with SEB and if severe infection occurs, it may cause shock and even death.

Chronic Effects

Although full resolution of symptoms may take up to a month, no long-term sequalae are documented.

Laboratory Tests

An elevation in white cell count with neutrophilic predominance occurs within 24 hours of symptoms.

Microscopy

Because the illness is not a result of *S. aureus,* but rather of its exotoxin, no bacteria are likely to be found in routine microscopy, and the toxin itself is too small to be detected with light microscopy.

TABLE 20–13   SEB: Typical Clinical Features

Ingestion—self-limited nausea, vomiting, diarrhea, discomfort
Inhalation—abrupt onset of fever, chills, dyspnea, nonproductive cough, crackles, headache, noncardiac chest pain

Radiographic Findings

Chest radiographs are generally normal except in the most severe cases where signs of pulmonary congestion such as infiltrates, Kerley lines, and increased interstitial markings may also be seen. Incipient respiratory failure may present as ARDS radiographically, but this, again, is rare. Likewise, abdominal radiographs are unhelpful.

Differential Diagnosis

A broad differential diagnosis exists for SEB because of the wide list of gastroenterites. A gastroenteritis epidemic from SEB would still likely be the result of unintentional food poisoning, rather than an attack, though such use has occurred (see section on enteric bacteria in this chapter). Likewise, a broad differential exists because of the vast number of causes of acute febrile respiratory illnesses (see Chapter 5). Category A infectious agents (Tularemia, VHFs, plague, anthrax), chemical agents (mustard, nerve agents, chlorine gas), selected biotoxins (ricin), and category B agents (Q fever) that present primarily as pulmonary bioterror syndromes are among the diagnoses in the differential diagnosis of SEB. A partial differential diagnosis is included in Table 20–14.

Diagnosis

Though not widely available, enzyme-linked immunosorbent assays (ELISA) of body fluids or nasal swabs can detect the presence of SEB. Diagnosis is rarely done for naturally occurring cases, because diagnosis and treatment are empiric. With such a short time until onset, clustering of cases will be pronounced, so epidemiologic surveillance would be a helpful means of detecting an outbreak.

Treatment

Treatment consists of supportive measures dictated by the dominant clinical findings. Pulmonary edema from over-aggressive fluid resuscitation can occur if substantial

TABLE 20–14   Differential Diagnosis: Staphylococcal Enterotoxin B

Ricin	Pulmonary anthrax
Cholecystitis	Hantavirus pulmonary syndrome (HPS)
Pancreatitis	Tularemia pneumonia
Gastroenteritis	Pneumonic plague
Giardiasis	Q Fever
Influenza	Chemical agents: cyanides, chlorine gas
Chlamydial pneumoniae	Phosgene
Adenovirus infection	Nerve agents
Mycoplasma infection	Mustard

lung injury is present. The role of steroids is uncertain. Most symptoms will resolve shortly, and while it may take up to 2 weeks for patients to be fully functional, nearly all patients make a full recovery.

Vaccine

Currently, no vaccine exists for SEB. Postexposure prophylaxis (PEP) is not recommended as the disease is caused by an enterotoxin, for which no antidotes exist.

Infection Control

There is no person-to-person transmission of SEB; thus, standard precautions suffice. Decontamination is sufficient using hypochlorite solution or soap and water for 15 minutes. Clothing is decontaminated with conventional laundering.

Epsilon Toxin of *Clostridium perfringens*

Background

Clostridium perfringens is a gram-positive, spore-forming bacterium living in soil and the gastrointestinal tract of animals throughout the world. Like other clostridial species, *C. perfringens* are saprophytes that produce a number of different toxins. Epsilon toxin is produced by Species B and D and is responsible for a fatal livestock disease. Little is known about the effects of epsilon toxin on humans with most available data extrapolated from animal studies.

As a biological weapon, the toxin is most likely to be used in aerosolized form resulting from inhalational exposure. Exposure is possible through contamination of food or water supplies as well.

Sources

C. perfringens species are ubiquitous and easily cultured, and the toxin is readily harvested. With current techniques in molecular biology, the gene responsible for producing the toxin could be inserted into other bacterial hosts resulting in amplified production of the exotoxin.

Properties

C. perfringens types B and D release a preformed enterotoxin that is highly soluble in water. It has an LD_{50} of 100 mg/kg, making it one of the most potent toxins known. Only tetanus and botulinum are more potent.

Epidemiology

The incidence of illness from epsilon toxin is hard to assess. Epsilon toxin primarily affects livestock, though human cases have occurred.

Means of Transmission

There is no person-to-person transmission of epsilon toxin. Exposure results from overgrowth of clostridia species in the gastrointestinal tract or from direct ingestion of preformed toxin as would be the case in a deliberate attack involving contamination of food or water. An aerosolized attack could also occur with illness resulting from inhalation. Animal studies indicate that mortality is potentially very high in the aerosolized form.

Pathogenesis

Epsilon toxin is inactive toxin until it enters the body and host enzymes inadvertently activate it. The toxins generate cellular damage by forming pores that penetrate the host cell membrane, disrupting homeostatic flow of ions. It is postulated that the toxin itself is pore forming. The result is edema fluid at the target site, as well as electrolyte imbalances. Following this logic, inhalation exposures will probably cause massive pulmonary edema, whereas ingestion will likely yield a gastroenteritis syndrome. Because the toxin is absorbed systemically, all organ systems are susceptible. Most significantly, epsilon toxin crosses the blood brain barrier and is a potent neurotoxin, causing neuronal death and vacuole formation. Epsilon toxin also appears to damage bone marrow, impeding blood cell line production. It is speculated that the toxins may also function as superantigens (see pathophysiology of SEB).

Signs and Symptoms

Gastrointestinal

Ingestion of *C. perfringens* toxins results in impairment of the intestinal mucosa. Clinical features include watery diarrhea, abdominal cramping, and discomfort. Necrosis of the bowel is possible resulting in blood-tinged or bloody diarrhea. Onset of symptoms occurs between 8 and 24 hours after ingestion. Symptoms typically resolve within 3 to 4 days without sequelae.

Inhalational

If aerosolized, respiratory tract damage would likely result in inflammatory cytokine release with cough and bronchospasm, and potentially increased vascular permeability with pulmonary edema and acute respiratory distress syndrome (ARDS).

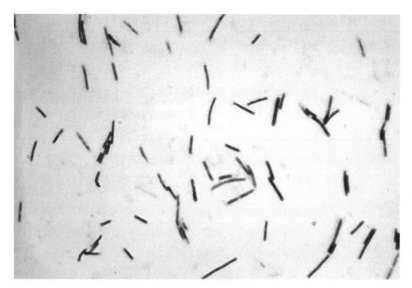

FIGURE 20–11 *Clostridia perfringens* **gram-positive rods.**
Courtesy of the CDC.

Systemic

Systemic dissemination following inhalational or ingestion of the toxin is possible, with impact on multiple body organ systems. There are two important systemic effects worth noting. First, epsilon toxin crosses the blood brain barrier, causing neurotoxicity. This is manifest as prostration, cerebellar dysfunction, impaired consciousness, and even coma. The other serious consequence of systemic absorption is hemodynamic instability. Epsilon toxin poisons the vascular endothelium causing massive fluid shifts and subsequent hypotension and tachycardia. Bleeding may also be seen in late stages. Animal studies indicate that necrosis of the kidneys with subsequent kidney failure also occurs with systemic absorption.

Laboratory Tests

Laboratory values may be normal although pancytopenia may be noted secondary to toxic effects on hematopoietic stem cells.

Microscopy

C. perfringens is a gram-positive, anaerobic spore-forming rod. Because the illness is not a result of *C. perfringens*, but rather of its exotoxins, bacteria are unlikely to be identified. The toxins are too small to be detected with light microscopy.

TABLE 20–15   Epsilon Toxin of *C. perfringens*: Clinical Features

Exposure can result in three clinical forms: gastrointestinal, inhalational, and systemic.

Gastrointestinal—self-limited, watery diarrhea that can become blood tinged, cramping

Inhalational—cough, wheeze, pulmonary edema—can develop into ARDS

Systemic—CNS features including, weakness, cerebellar dysfunction, impaired
  consciousness, and even coma. Hemodynamic instability including hypotension,
  tachycardia; progression to shock and bleeding can occur in severe cases

Radiographic Findings

Gastrointestinal

Abdominal imaging yields little diagnostic information.

Inhalational

With advanced cases, infiltrates may be seen; rarely, diffuse patchy infiltrates or frank ARDS are present.

Differential Diagnosis

A broad differential diagnosis exists because—depending on the route of exposure—*C. perfringens* has gastrointestinal, respiratory, vascular, and systemic effects. The presence of neurological findings and lack of fever help distinguish this from other causes of gastroenteritis. A partial list of the differential diagnosis is found in Table 20–16.

Diagnosis

Though not widely available, ELISA of stool samples or serum can identify the presence of the exotoxin.

Treatment

Treatment consists of supportive measures based on the clinical picture. As always, care should be given when giving fluid resuscitation. Pulmonary edema may occur due to the vascular leak caused by the toxin's endothelial damage, and can be made worse through iatrogenic interventions, such as in fluids.

Table 20–16   Differential Diagnosis: *C. perfringens* Epsilon Toxin

SEB exposure	Nerve gas
Alpha viruses	VHFs
Nipah virus	Hanta virus
Botulinum toxin	Viral or bacterial gastroenteritis

Vaccine

No vaccine exists, and postexposure antibiotic prophylaxis is not offered since the disease is caused by a toxin and not a microbe.

Infection Control

There is no person-to-person transmission of *C. perfringens* toxin, making isolation unnecessary. Standard precautions will suffice. Decontamination can be achieved with hypochlorite solution or soap and water for fifteen minutes. Clothing or linens are safely decontaminated with conventional laundering.

☣Food-Borne Pathogens

Salmonella species
Escherichia coli O157:H7
Shigella species

Background

The food-borne pathogens (FBP) include three naturally occurring bacteria that can contaminate food causing infection in the gastrointestinal tract infections. Unlike SEB, or episilon toxin, pathogenicity results from the presence of the bacteria in the GI tract rather than from toxins which form prior to ingestion. The ill effects of infection with any of the three are commonplace. These bacteria have been well studied, and their clinical features are well known to most clinicians. Because of their pathophysiology, clinical presentation, and other features are covered fully in traditional references, readers are referred to these resources for more detailed discussion. From a biological weapons perspective, the similarities of use, presentation, and management justify collective discussion.

One of the dubious distinctions of the FBPs is that, other than anthrax, they are the only other category of biological weapons that has ever been used in an attack in the U.S. In 1984, *Salmonella* was intentionally spread in an Oregon town by members of a local religious sect resulting in 750 cases of gastroenteritis. Outside the United States, these agents are believed to have been used in more than one instance. The biological weapons division of the Japanese army Unit 731 is reported to have contaminated a Manchurian river with large amounts of *Salmonella typhi* during WWII. No cases or outbreaks are associated with that action; however, *S. typhi*, along with *paratyphi*, cause a distinct clinical scenario referred to as "typhoid fever." Typhoid fever has a particular place in American history, forever associated with the infamous "Typhoid Mary" (see "of note . . ." on page 312).

OF NOTE . . .

Bhagwan Rajneesh

In 1984, the small rural town of The Dalles, Oregon, experienced the largest biological weapons attack in modern U.S. history, although it received little national attention at the time. Followers of the Indian guru Bhagwan Shree Rajneesh, known as Rajneeshees, contaminated salad bars at a number of restaurants with salmonella and infected 752 people. The group apparently had intentions of running its members for elected positions in the local government as a way of advancing their agenda. According to some reports, some of the sect's members conspired to impede voter turnout by debilitating the town via food poisoning. One of these members was a nurse who acquired an incubator and salmonella sample. A later CDC investigation never identified the actual source; however, an FBI investigation ultimately confirmed that the group had acquired the pathogen. No charges were brought, and the Bhagwan Shree Rajneesh denied knowledge of the event. He was, however, found guilty of violating immigration laws and ultimately deported. Soon thereafter, the Rajneeshees left the area.

Sources

The bacteria included among food-borne pathogens are ubiquitous, associated with water, soil, insects, animal GI tracts and feces, uncooked meats, poultry, and seafood. Fecal contaminations are arguably the single largest common source of infection. In the case of *S. typhi*, chronic carriers serve as a source of bacteria and can spread infection.

Epidemiology

The incidence of food poisoning is hard to estimate as numerous small-scale outbreaks are missed, but sporadic identified outbreaks occur worldwide. In the United States, outbreaks are most commonly associated with poor safety practices of restaurants or food manufacturers.

Means of Transmission

The primary means that food preparers transmit food-borne pathogens is fecal–oral cycling secondary to poor hygiene practices and subsequent contamination of food or water. As a biological weapon, there is a theoretical risk of aerosolization, but most likely a liquid form would be used for contamination. Infection with *S. typhi* results in a certain percentage of chronic carriers who shed bacteria through their stool or urine, potentially perpetuating or reigniting an outbreak.

OF NOTE . . .

Typhoid Mary

Mary Mallon was a cook in New York City in the early part of the 20th century. Mallon was chronically infected with *S. typhi* and spread typhoid fever through the food she prepared. She was by no means the only such person—quite likely there were over a thousand, but she, unlike many others was easily identified by public health authorities who traced an outbreak back to her. However, she later assumed a pseudonym and returned to working as a cook. Eventually she was discovered and put under permanent quarantine until the time of her death.

Pathogenesis

The ability of FBPs to inflict injury in the small intestine presumably requires the ability to survive the body's defense mechanisms: first and foremost of which is sensory input: food that smells bad. In addition, there are a number of physiologic protective measures such as gastric acidity, motility, normal flora, as well as the components of the gut-associated lymphoid tissue. These particular bacteria have two main mechanisms of gastrointestinal mucosal injury: invasion and toxin release. *Salmonella* uses invasion as its mechanism of damage, whereas *Shigella* uses both invasion and toxin release; so does *E. coli* O157:H7, producing a *Shigella*-like toxin as its primary effect with invasion occurring to a lesser degree. *Shigella* toxin and *Shigella*-like toxins possess enzymatic functions that impair protein synthesis by the host cell. Invasive or not, these bacteria injure the epithelial cells or GI tract resulting in dysregulation of absorption. The resulting state of small bowel inflammation is common among the three types of bacteria discussed.

S. typhi also acts by invasion of intestinal mucosa; however, it continues on into the submucosa where it proliferates in the Peyer's patches and is taken up into the lymphatics and spread hematogenously. Proliferation within the reticuloendothelial system is primarily responsible for the systemic symptoms caused by *S. typhi*.

Signs and Symptoms

Symptoms typically begin within 36 hours after infection and often resolve spontaneously without treatment. The primary feature of infection with these organisms is diarrhea. Invasion, and/or inflammation of the small bowel results from the bacteria's mechanism of injury and the diarrhea may be bloody. Fevers, chills, and

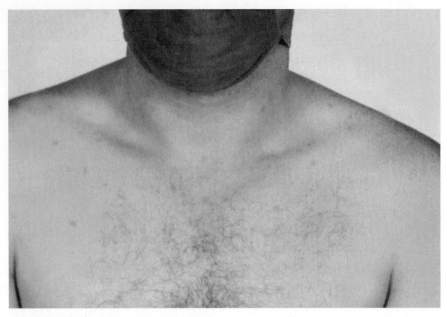

FIGURE 20-12 Patient infected with *S. typhi* show in the classic truncal "rose spots." *Courtesy of the CDC.*

myalgias occur as well. Vomiting and abdominal pain are associated most with salmonella.

If symptoms worsen or persist, complications of dehydration set in. Symptoms outside of the gastrointestinal tract can occur with bacteremia affecting any organ or tissue. When present, these systemic features tend to persist beyond the duration of the acute diarrheal symptoms. Bacteremia may also result in septicemia and septic shock.

S. typhi presents with what is termed *typhoid* or *enteric* fever, and runs typically a 3- to 4-week course. The classic presentation begins with a high "stepwise" fever concurrent with bacteremia. After about a week, the pathognomonic finding of a centrally located, macular, salmon-colored rash occurs, along with myalgias, malaise, constipation, and abdominal pain Fig. 12–12. Late in the course, intestinal ulcerations may develop with an enlargement of the spleen or liver. Symptoms usually resolve spontaneously, even in untreated cases. Approximately 15% of the time, progression occurs with systemic signs and greater risk of mortality or organ damage.

Chronic Effects

Shigella and salmonellosis have been associated with Reiter's syndrome or postreactive arthritis. *Salmonella* can lead to osteomyelitis and endocarditis. All three of these FBPs have some association with hemolytic uremic syndrome (HUS) though *E. coli* O157:H7 has the strongest association. HUS tends to be seen in children

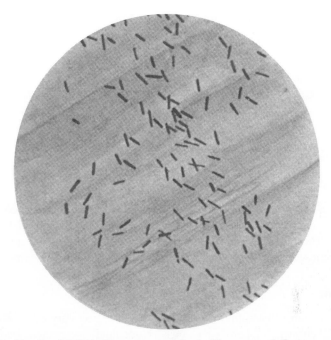

FIGURE 20-13 *E. coli* **O157:H7, a gram-negative bacillus, causes inflammatory diarrhea.** *Courtesy of the CDC.*

following infection, whereas a combined HUS–thrombotic thrombocytopenic purpura (TTP) picture is seen in adults.

Microbiology

The FBPs are all characterized as enteric gram-negative rods. *Shigella* and *Salmonella* species do not ferment lactose, whereas *E. coli* does.

Laboratory Tests

If bloody diarrhea is present, a resultant normocytic anemia is not uncommon. Should bacteremia occur, an elevated white blood cell count will likely be present. Stool samples reveal fecal leukocytes as well as RBCs. Lab findings consistent with typhi include anemia, leukopenia, and elevated liver enzymes.

Radiographic Findings

Radiologic exams are of little diagnostic value.

Differential Diagnosis

The differential diagnosis of gastroenteritis is vast (Table 20–17).

TABLE 20–17 Common Food-Borne Pathogens

Staphylococcus aureus	*Escherichia coli*
Bacillus cereus	*Listeria monocytogenes*
Clostridium perfringens	*Clostridium botulinum*
Salmonella enteritidis	Hepatitis A and enteric viruses
Shigella spp.	
Yersinia enterocolitica	

Diagnosis

Diagnosis of *Shigella* and *Salmonella* species from fecal cultures is available routinely in most commercial and hospital laboratories. Testing for *E. coli* O157:H7 can also be done in these BSL-1 labs, although some require specific notification. *S. typhi* can be diagnosed by culturing blood, stool, urine, or skin lesion specimens.

Treatment

Most experts agree that supportive care is all that is necessary as most cases will resolve on their own. Supportive measures include fluid resuscitation and electrolyte correction. The use of antibiotics is efficacious in more severe cases and serve to shorten the duration of the illness. However, there are risks to antibiotic use with one of the most common side effects being diarrhea. If treatment is warranted, ciprofloxacin 500 mg orally every 12 hours for up to 7 to 10 days is the drug of choice. Bismuth subsalicylate two to four tablespoons every thirty minutes as needed up to eight doses can be used for symptom control. *S. typhi* can be treated with the same antibiotic regimen, with intravenous use recommended for more severe presentations. Chronic carriers can be treated with ciprofloxacin (500 mg orally every 12 hours for 4 weeks) and cholecystectomy should be considered as this organ is often the reservoir of the pathogen.

Vaccine

No vaccine is currently available for these Category B agents. Most experts agree that, if an outbreak occurs, prophylactic use of antibiotics on a mass scale is risky because of the development of resistance and the possibility of superinfection.

TABLE 20–18 Food-Borne Pathogens: Clinical Features

Diarrhea, possibly bloody; fever; chills. If diarrhea is severe enough, dehydration can set in. Shock and septicemia can develop

Typhoid (*Salmonella typhi*)—high fever with bacteremia, truncal rash with salmon-colored maculopapular lesions, myalgias, abdominal pain, and constipation

Infection Control

Transmission is through the fecal–oral route so routine and diligent hand washing and fluid containment will prevent transmission. Contact precautions are indicated should health care workers and others come in contact with infected stool.

Water-Borne Pathogens

The water-borne pathogens are far more numerous than the two chosen for discussion here, including all the food-borne pathogens from the previous section. *Vibrio cholera* and *Cryptosporidium parvum* are mentioned because of their higher likelihood of use as biological weapons and also because they are reasonably representative of other water-borne pathogens.

Vibrio cholera

Background

From a historical and pathophysiologic perspective, cholera has been one of the most studied of pathogenic microbes in humans. Discussion here will be largely restricted to those aspects relevant to potential use as a biological weapon (although readers interested in the social and cultural history of cholera are referred to Charles Rosenberg's classic *The Cholera Years*). Identified first by Robert Koch in 1883, *V. cholera* inhabits both marine and freshwater environments. Cholera outbreaks have been described for thousands of years with seven (some argue eight) large-scale outbreaks identified since the 1800s, the most recent finishing in 1998. Epidemics continue to occur sporadically throughout the world, generally in less industrialized nations.

Sources

Vibrios are one of the most common water-based organisms in the world. They exist in virtually all aquatic milieus. Among industrialized nations, consumption of contaminated shellfish is primarily responsible for outbreaks, whereas in less industrialized countries, infected water sources are largely responsible.

Properties

Facultative anaerobes, *V. cholera* are motile by means of a single polar flagellum. *V. cholera* are highly sensitive to acidity, so few survive passage through the stomach. Large amounts of bacteria, somewhere on the order of 10^3–10^6 organisms, must be ingested to cause clinical infection. Once in the small intestine, *V. cholera* is resistant to protective measures such as bile salt and complement, and is able to adhere to

TABLE 20-19   Cholera: Clinical Features

Sudden onset of voluminous, nonbloody, watery diarrhea; abdominal pain may be present. Vomiting and fever are typically absent.
Signs and symptoms of dehydration are commonly seen secondary to the large volume loss including orthostasis, tachycardia, and shock in severe cases.

intestinal epithelial cells. Strictly speaking, cholera is a self-limiting disease—if patients remain adequately hydrated, they will likely recover even without antibiotic treatment.

Epidemiology

Sporadic cholera outbreaks occur worldwide involving numerous biotypes. In the United States, cholera has been virtually eliminated through water treatment and sewage-handling systems. Since the mid 1990s, there have been roughly sixty cases in the United States. About two-thirds have been secondary to consumption of contaminated seafood and one-third from infected travelers entering or reentering the country.

Means of Transmission

Transmission occurs through the fecal–oral route and infection occurs typically via contaminated food and water supplies.

Pathogenesis

V. cholera, like many bacteria that cause diarrhea, injures the gastrointestinal epithelium through the release of a toxin. The highly potent enterotoxin released by *V. cholera* blocks synthesis of glutamine triphosphate (GTP), a compound required for breakdown of cyclic AMP (cAMP). With GTP unavailable, cAMP accumulates at high levels stimulating enzymatic activity of adenylate cyclase within intestinal epithelial cells. Disruption of osmotic and electrolyte gradients results, impairing normal membrane control of fluids and ions. The luminal cells pump electrolytes hyperactively from plasma and nearby tissues into the lumen of the intestine causing massive water excretion. The bacteria and the endotoxin do not induce an inflammatory response, and bacterial invasion does not occur, so no direct epithelial injury is seen. Systemic effects due to substantial electrolyte and volume loss are possible, including organ system failure.

Signs and Symptoms

Following an incubation period of up to 5 days, cholera begins with sudden onset of massive, nonbloody, watery diarrhea described classically as having a

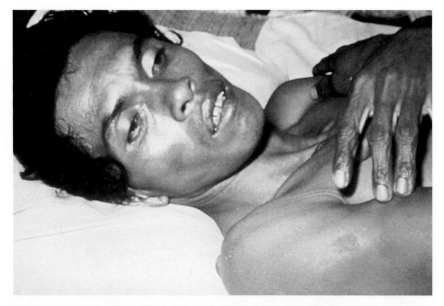

FIGURE 20-14   Patient with cholera showing "Washer's Hands"—pruning of fingers secondary to turgor loss.
Courtesy of the CDC.

"rice-water" appearance. The diarrhea typically lacks a foul odor. Vomiting or abdominal pain may be present. Within the first 48 hours patients lose fluid at a rate of approximately one liter per hour, as well as large amounts of electrolytes and nutrients.

Because no inflammatory response is elicited, fever and other systemic features are conspicuously absent in cholera. Complications are the result of volume and electrolyte loss. Other features vary according to the degree of dehydration, including loss of skin turgor—classically seen as wrinkled hands (hand-washing sign), oliguria or anuria, dry mucus membranes, tachycardia, orthostasis, and when severe enough, hemodynamic features of shock.

Chronic Effects

Chronic carriers of *V. cholera* are extremely rare, and long-term sequelae does not occur if patients recover. Cholera does not confer lifetime immunity following infection.

Laboratory Tests

An elevated hematocrit is seen secondary to a decrease in the circulating volume. Similarly, an elevated blood urea nitrogen and creatinine may be noted as a result of decreased renal perfusion consistent with prerenal azotemia. Depending on the degree of dehydration or hemoconcentration, electrolytes can appear low, normal or elevated so careful monitoring of electrolyte status is needed. Blood pH also varies according to the extent of dehydration: classically, acidosis results from a loss of bicarbonate ions and a concomitant hyperchloremia.

Microscopy

V. cholera is morphologically a gram-negative straight or curved rod (comma shaped), with a single polar flagellum. When cultured, vibrios may lose their curved appearance.

Radiographic Findings

Imaging is of no particular diagnostic or therapeutic value in cholera.

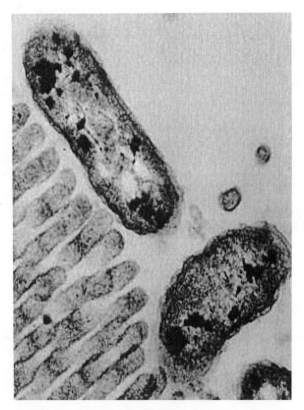

FIGURE 20–15 **Transmission electron microscopy of *Vibrio* along the intestinal brush border.**

Courtesy of Nelson ET. Infect Immun 14:527, 1976.

TABLE 20–20   Common Water-Borne Pathogens

Vibrio cholerae	Norwalk/Norwalk-like
Campylobacter	Hepatitis A
Yersinia enterocolitica	*Isospora belli*
Legionella pneumophila	*Cryptosporidium parvum*
Salmonella spp	*Cyclospora cayetanensis*
Shigella spp	Microsporidia
Escherichia coli (EHEC)	*Giardia lamblia*
Rotavirus	*Entamoeba histolytica*

Differential Diagnosis

See Table 20–20 for a list of common water-borne pathogens.

Diagnosis

The sheer volume of watery diarrhea with a rice-water appearance is often enough to make the diagnosis in the right clinical context. Conversely, the presence of blood in the stool makes cholera highly unlikely. Stool studies are the best diagnostic tool. Because no inflammation occurs, stools have little to no leukocytes in the blood. For rapid diagnosis, dark-field examination of a wet mounted, fresh, unstained stool specimen is highly sensitive and specific. Classically, diagnosis is made by recognition of vibrio's chaotic and rapid motion, sometimes termed the *shooting star* appearance, resulting from the use of its flagellum. Commercially available kits for subtyping the bacteria are available. In a wide-scale outbreak, diagnosis is often clinical and subtyping rarely alters management.

Treatment

The single most important management consideration is fluid resuscitation with electrolyte repletion. With careful attention to fluid and electrolyte balance, mortality is less than 1%; without fluid resuscitation mortality reach 40%. Fluid resuscitation is dependent on the degree of dehydration as determined by clinical findings. Including glucose in IV is important as it serves as a nutrient and as a physiologic means of enhancing intestinal water uptake. Intravenous, nasogastric, or oral rehydration may be used depending on the severity of dehydration, available resources, and the clinical setting. Oral rehydration solution is used in treating cholera until diarrhea ceases. Once the patient is rehydrated, however, fluid replacement can be reduced to amounts that match ongoing losses. Antibiotics are not curative, but do shorten the disease course. Doxycycline given as a one-time dose (300 mg orally) is the antibiotic treatment of choice, according to the WHO. Other authorities suggest that ciprofloxacin (250 mg every 12 hours for 3 days) is equally effective.

Vaccine

Historically, vaccines provided only temporary protection. New oral vaccines, that induce mucosal immunity are showing greater long-term benefit.

Infection Control

Person-to-person transmission of cholera occurs only through fecal-oral contact, therefore contact precautions are indicated. Good hygiene and thorough and frequent hand washing are effective infection control measures.

Cryptosporidium parvum

Background

Cryptosporidium is a commonly water-borne agent that causes diarrhea. It takes up residence in the liver and is shed through fecal material. Infection tends to disproportionately affect the young, the elderly, and the immunocompromised. In rare cases, infection has occurred in the respiratory tract, usually in immunocompromised hosts.

Sources

Untreated water often contains cryptosporidial cysts.

Properties

Cryptosporidia are intracellular parasites with cysts approximately 3 microns in diameter, roughly half the size of an erythrocyte. Cystic forms are quite hardy and withstand conventional water treatments used in most industrialized nations as well as in most chemical cleaning agents. They are vulnerable to ultraviolet light, boiling, and desiccation.

Epidemiology

C. parvum is found globally. Exact incidence figures are hard to determine, though some studies have shown that upwards of 90% of all untreated water is contaminated

TABLE 20–21 Cryptosporidiosis: Clinical Features (Immunocompetent)

Moderate, self-limited watery diarrhea associated with abdominal pain, nausea, vomiting
Low-grade fever
Nonproductive cough

with cysts. Based on serologic studies, estimates are that 80% of the U.S. population has been exposed to *Cryptosporidium*.

Means of Transmission

Transmission occurs through the fecal–oral route and thus follows ingestion of contaminated food or water. Infection typically requires relatively small numbers of cysts; fewer are needed if the host is immunocompromised. Once infection and replication of the parasite begins, shedding occurs in feces allowing for propagation of the disease.

Pathogenesis

The mechanism of injury from cryptosporidial infection is poorly understood. After ingestion, encystation generally occurs in the duodenum. Unlike *V. cholera*, *Cryptosporidium* are invasive—adhering to and then invading gastrointestinal epithelial cells. Invasion elicits a local inflammatory response as well as causes epithelial disruption, blunting, and even loss of microvilli.

Signs and Symptoms

In an immunocompetent person, infection may be relatively mild, even unnoticed. Following an incubation period of approximately 14 days, moderate to severe watery diarrhea sets in and is commonly associated with abdominal pain, nausea, vomiting, anorexia, malabsorption, and sometimes a low-grade fever. In immunocompetent patients, the illness will resolve without treatment within 14 days. In immunocompromised patients, however, the disease may progress to include biliary and other complications.

OF NOTE . . .

Cryptosporidium Outbreak in Wisconsin

Following an inordinate number of reports of debilitating gastrointestinal illness in the area, it was discovered in April 1993 that one of the two primary water treatment facilities in Milwaukee, Wisconsin, was malfunctioning and inadequately filtering the water supplied to residences and businesses in the area. In the weeks that followed, it is estimated that over 400,000 people developed diarrhea. This is thought to be an underestimate. Over 100 people died. The largest water-borne epidemic in the history of the United States had just occurred. The water samples revealed that although levels of the most water-borne pathogens remained stable, there was a 100-fold increase in *Cryptosporidium* in the water. Although the immunocompromised suffered the most, the vast majority of those affected were otherwise healthy people.

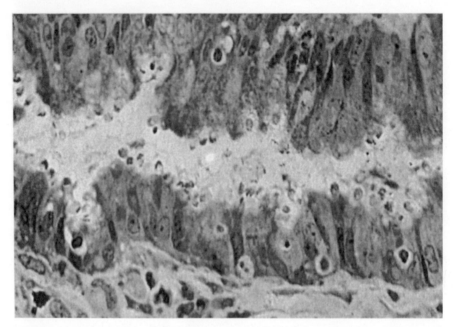

FIGURE 20–16 *Cryptosporidium* **lining and denuding the interstitial brush border.** *Courtesy of the CDC.*

Respiratory infections present with coughing and low-grade fevers, and are usually seen in conjunction with the intestinal form.

Chronic Effects

In immunocompetent adults, chronic effects are unlikely to occur. In immunocompromised patients, gastrointestinal and extraintestinal complications can occur, most notably involving the biliary tree.

Laboratory Tests

If diarrhea leads to dehydration, characteristic elevations in BUN and creatinine consistent with prerenal azotemia will be noted. Hemoconcentration of red blood cells and other components can also result from dehydration. Stool samples may be hemepositive and have WBCs since *C. parvum* is an enteroinvasive pathogen.

Microscopy

Modified acid-fast stains of fresh stool samples will reveal red oval oocysts.

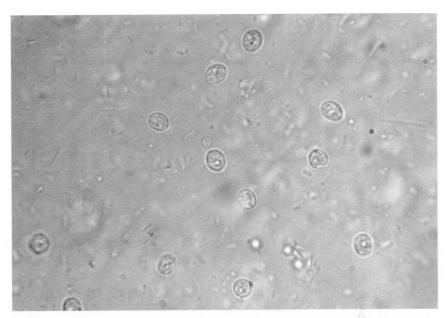

FIGURE 20–17   A stool sample positive for *C. parvum*. Note oval cysts.
Courtesy of the CDC.

Radiographic Findings

Imaging is of little diagnostic value.

Differential Diagnosis

The differential diagnosis for diarrheal symptoms or gastroenteritis is extensive. A partial list—many of which are enteroinvasive—are listed in Table 20–12.

Diagnosis

PCR or immunofluorescent antibody assays are the most sensitive means of diagnosis. Commercially available kits for antibody assays are readily available. PCR is less

TABLE 20–22   Differential Diagnosis: *C. parvum*

Amebiasis	Giardiasis
Rotovirus	Isosporia
Norwalk virus	Microsporidiosis
Campylobacter	*E. coli*
Campylobacter infections	Salmonellosis
Cholelithiasis	Shigellosis
V. cholerae	

readily available but offers epidemiological help in determining parasitic strains. Acid-fast staining of fresh stool samples are diagnostic, but this test suffers from poor sensitivity. Diagnosis of pulmonary forms require biopsy and proper staining.

Treatment

Infections in immunocompetent patients will spontaneously resolve within one to two weeks. Management is strictly supportive including fluid and electrolyte resuscitation as warranted by the clinical picture. No effective pharmaceutical therapy presently exists.

Vaccine

No vaccine currently exists.

Infection Control

Thorough hand washing and contact precautions are adequate infection control measures to limit the spread of *C. parvum*.

Alphaviruses Encephalitides

Background

Venezuelan equine encephalitis (VEE) Eastern equine encephalitis (EEE), and Western equine encephalitis (WEE) are the three types of *Alphaviruses* considered to be Category B agents. Although they are distinctive, they are discussed collectively because their similarities outweigh their differences. All three belong to the Togaviridae family of viruses and are members of the genus *Alphaviruses*—collectively referred to as the American Equine Encephalitides, or new-world *Alphaviruses*. Their similarities include characteristic molecular biology and dissemination through mosquito vectors that transmit the virus to humans and animals. There are nearly thirty viruses classified within this genus; however, only these three are commonly associated with encephalitis. Each of the three viruses contains "equine" in their name because horses are particularly susceptible to infection.

Alphaviruses are hardy, highly infectious, highly virulent, and have long been included among the agents studied by many involved in biological weapons development. Particularly concerning is that these viruses are aerosolized readily and dissemination requires little technological sophistication, money, or resources. Although any of the biological agents are altered genetically to enhance their virulence, infectivity, or any other feature, *Alphaviruses* are more likely than most biological agents to have been "weaponized" through genetic manipulation.

This belief is based on a long history of alphavirus research and development in biological weapons laboratories and their relatively simple genetic structure (single-stranded RNA). Consequently, many experts believe these viruses pose a legitimate threat as potential biological weapons.

Sources

Mosquitoes serve as the primary vectors for the viruses and natural reservoirs include equine species, such as horses.

Properties

Alphaviruses are spherical, enveloped in structure, between 60 and 70 nm in diameter and comprised of a single-stranded positive-sense RNA. *Alphaviruses* are remarkably hardy and remain stable even with long-term storage. These viruses are highly infectious, particularly VEE, as evidenced by the numerous cases of lab workers infected with aerosolized forms. In vitro studies have shown that *Alphaviruses* tend to be highly proliferative.

FIGURE 20–18   Horses serve as the primary reservoir for *Alphaviruses* and ill horses can raise the suspicion of an outbreak. Note the *paddling behavior* characteristic of Venezuelan equine encephalitis in horses.
Courtesy of Oklahoma State University, College of Veterinary Medicine.

Epidemiology

Alphaviruses are found in worldwide distribution, but the three types included here are found only in North and South America, hence their designation as "New World" *Alphaviruses*. VEE tends to occur more commonly in South America, though cases have been seen in the United States along the Gulf Coast. It has a mortality rate less than 1%. Naturally occurring cases of WEE tend to be seen, as the name suggests, west of the Mississippi River and has a mortality rate of close to 4%. EEE tends to be concentrated east of the Mississippi River from Florida to Canada, and while typically less infectious, it is more virulent than WEE or VEE with a morality rate of nearly 70%.

Cases within the United States tend to follow seasonal variation paralleling seasonal presence of mosquitoes—with cases appearing in late spring through early fall. Despite the dying out of mosquitoes each winter, the same virus spread in one season has been isolated in endemic outbreaks in subsequent seasons. How the virus survives year to year is still a mystery.

Because naturally occurring cases follow geographic and seasonal patterns, new cases that vary from the known patterns would be a critical epidemiologic clue suggestive of deliberate bioterror attack.

Means of Transmission

As mentioned, infection occurs naturally only as a result of bites from infected mosquitoes. Mosquitoes acquire the infection by feeding on the blood of infected horses. The virus then takes up residence in the cells of the mosquito's gastrointestinal tract where it replicates and remains for the life of the insect. When a carrier mosquito bites a human, the virus is regurgitated and inoculates the bite site in the dermis. With natural occurrences, environmental factors affecting mosquitoes determine the extent of an outbreak. For example, the temperate climate and seasonal variation in the United States may explain partially the relatively low number of cases and outbreaks, as these qualities minimize the life cycle and burden of mosquitoes. Additional factors such as the amount of rainfall prior to and during the mosquito life cycle are significant as well.

As a biological weapon, dissemination would in all likelihood be through aerosolized forms of the virus. Infection would occur through contact with nasal mucosal cells or the olfactory epithelial cells lining the upper portion of the airway. In a bioweapons attack, the initial distribution of cases would be a function of the nature and distribution of the exposure. However, mosquitoes would still play a role in extending or spreading infections once established.

No person-to-person transmission of *Alphaviruses* occurs.

Pathogenesis

The details of how *Alphaviruses* cause illness in humans are poorly understood. In naturally occurring cases, infection begins in the cutaneous fat, proximal to the site of the mosquito bite. *Alphaviruses* are taken into the cells by endocytosis where they replicate. Presentation of viral antigen induces uptake by tissue macrophages which then deliver the virus to regional lymph glands—the first of the two sites for which these viruses have a proclivity. In the lymphatic system the virus takes up residence in host cells and replicates further, ultimately leading to necrosis and bone marrow injury, particularly the lymphocytic cell line. Hematogenous spread may also lead to infection of the CNS. Each of the three viruses are neurotropic and infection causes neuronal apoptosis. Animal studies of aerosolized forms of the virus show evidence of direct invasion of the olfactory neurons via the nasal mucosa. Infection, necrosis, and inflammation occur throughout the brain, pathologically seen as tissue neutrophilia and gliosis.

Signs and Symptoms

Naturally occurring infection occurs following an incubation period of up to one week. Symptoms in the first few days include high fever, chills, headache—with or without photophobia, pharyngeal congestion, nausea, vomiting, diarrhea, malaise, and myalgia—and notably lower back and retroorbital pain. By day 4 or 5, impairments in mental status, fluctuating sensorium, including coma, and neurological

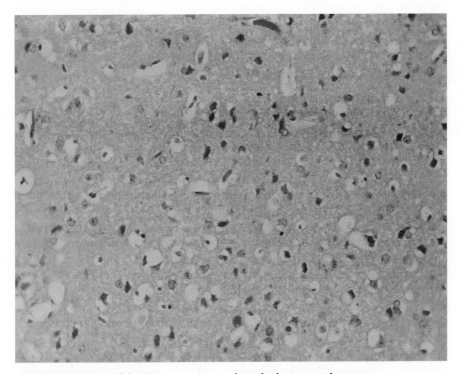

FIGURE 20–19 *Alphaviruses* **cause necrotic and edematous changes.**
Courtesy of the CDC.

dysfunction, including cranial nerve palsies, seizures, ataxias, and paralysis may be seen. Recovery occurs in one to two weeks. In general, the less severe the CNS features, the greater the likelihood of recovery.

Chronic Effects

Paralysis and other CNS deficits may result in those who survive.

☣ Deliberate Infection via Aerosolized Forms

The signs and symptoms are likely to be similar to those seen in naturally occurring cases. Animal models suggest that CNS features will appear more rapidly and greater morbidity and mortality are expected. Presentation with rapid progression of symptoms may, in fact, be an indication of a deliberate attack having occurred, as will seasonal presentations or unusual spikes in disease incidence.

Laboratory Tests

Within the first three to four days of symptoms, patient's white blood cell counts are below normal, but later they may become elevated. Other findings include mild thrombocytopenia or elevated transaminases. CSF findings are consistent with a viral infection, including elevated total protein, normal glucose, and leucocytes in the low hundreds—with PMN predominance early in the infection.

Microscopy

Standard microscopy is of little value in diagnosing *Alphaviruses*.

Radiographic Findings

Imaging plays no role diagnostically.

Differential Diagnosis

CNS findings plus fever warrants a differential diagnosis that includes causes of meningitis as well as encephalitis including all the many causes of viral encephalitis. A partial list follows:

TABLE 20–23 *Alphaviruses* **Encephalitides: Clinical Features**

High fever, chills, nausea, vomiting, diarrhea, malaise, myalgia
CNS features: headache, ataxia, seizures, paralysis, impairments in mental status, and levels of consciousness, including coma. Cranial nerves may also be affected.
Signs and symptoms are likely to occur more rapidly and more severely with bioterrorism than with naturally occurring forms.

TABLE 20-24   Differential Diagnosis: *Alphaviruses*

St. Louis encephalitis	Herpes simplex
Dengue fever	Japanese encephalitis
Encephalitis	Rickettsial infection
Malaria	Brain abscess
Meningitis	Lymphomatous or carcinomatosis
Yellow fever	meningitis

Diagnosis

A clustering of febrile CNS illnesses, coupled with recent animal illnesses (horses and birds in particular) are useful epidemiologic clues. Viral IgM may be identified using ELISA, preferably from CSF samples. A fourfold increase between acute and convalescent titers are diagnostic. PCR can also be used from CSF samples, although it is not widely available.

Treatment

Treatment is supportive. Antivirals medications have not been shown to be of clinical benefit.

Prophylaxis

Vaccines do exist, but these are not commercially available. They are recommended only for high risk circumstances under IND protocols.

Infection Control

No person-to-person transmission occurs, so standard precautions are employed. Decontamination can be done with standard disinfectants and autoclaving.

Category C Agents

Hanta and Nipah viruses are presented here as examples of infectious diseases that fall within the Category C agents, but this is not intended to imply that there are no other Category C agents that are capable of being used as bioweapons. Any new, poorly understood agents characterized as emerging diseases can be included among the Category C agents. It is thought that the changes in the public health infrastructure as a result of September 11 will serve to address not only Category C agents more effectively, but also naturally occurring emerging diseases, because the approach to an outbreak is rooted in the same epidemiological principles regardless of the source of the pathogen. The CDC examples of Nipah and Hanta viruses are reasonable ones, representing Category C criteria appropriately.

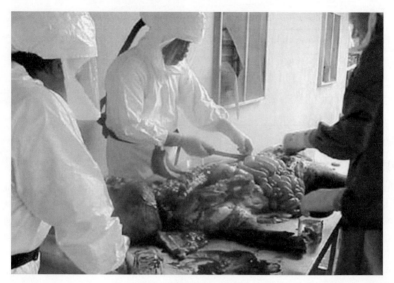

FIGURE 20-20 Scientists performing animal dissection during a Nipah outbreak. *Courtsesy of Food and Agriculture Organization of the UN.*

☣Nipah Virus

Background

A member of the paramyxovirus family, Nipah virus was first identified in 1999 following an outbreak in Malaysia and Singapore. A viral encephalitis was being spread by fruit bats in pigs at local farms. Eventually the virus jumped to humans as well as to other animals causing a severe encephalitic illness. Over the course of nine months 265 cases occurred with 105 fatalities. Animal and human CSF samples revealed a viral agent that was subsequently named Nipah, after Sungei Nipah, the Malaysian village where it was identified. In addition to high morbidity and mortality rates it caused tremendous economic damage to pig farmers and considerable fear and panic within the country.

Although listed by the CDC as a Category C agent, Nipah virus is one of several encephalitic bioterrorism agents, along with the *Alphaviruses* discussed earlier. Nipah's presentation is similar to the *Alphaviruses*. Any encephalitic clinical presentation requires that both viral families be included on the differential diagnosis.

Sources

Fruit bats belonging to the genus *Pteropu,* indigenous to Southeast Asia, Australia, and India, are likely the primary host for Nipah. These bats do not seem to fall ill despite carrying the virus. Though bats played a role in spreading Nipah infection in pigs during the Malaysian outbreak, the infected pigs seemed to be the primary means of spread to humans. There is still some uncertainty about the natural reservoir for Nipah virus.

Properties

Nipah is encapsulated, ranging in size from 100 to 500 nm, and comprises single-stranded, negative-sense RNA.

Epidemiology

Because the virus was identified only in 1999, precise data on incidence are difficult to determine. The Malaysian outbreak involved 265 infections with 105 deaths (about a 40% mortality rate). Since that identifying outbreak, there have been no further outbreaks in Malaysia, but there have been reports of three separate outbreaks in Bangladesh, India, the most recent report occurring in early 2004. This epidemic infected fifty-three people, killing thirty-five of them (66% mortality rate). Unlike the Malaysian outbreak, infections did not seem related or even secondary to pig infections. Instead, bats were implicated in the spread of the virus.

Means of Transmission

With such little data to draw from, defining routes of transmission is difficult. Based on the known outbreaks of Nipah, it is believed that both pigs and bats play a role. The Malaysian outbreak occurred from contact with infected body fluids (e.g., pig urine) and respiratory secretions are a viable means of infection as well. The mechanism of spread in the Indian outbreaks is less well understood but probably was caused by bats. Person-to-person transmission is not suspected for Nipah virus or at least has not been documented within these two communities, or nosocomially. As a biological weapon, *Nipah* virus would likely be distributed via aerosolized forms. It is unclear at this time how that might alter the typical features of the illness, if at all.

Pathogenesis

Because of its relative newness, the mechanisms of virulence for Nipah have not been fully worked out. What is known is that it has a proclivity for neuronal cells where it induces inclusion body formation, syncytial cell formation, and an inflammatory response. Commonly affected are endothelial cells, with resultant vasculitis, thrombosis, and even necrosis. Evidence of ischemia has been seen proximal to areas of infection.

Signs and Symptoms

Following an incubation of up to three weeks, signs and symptoms set in. These include fluctuating temperature, headache, nausea, vomiting, dizziness, alterations in consciousness, tachycardia, and elevated blood pressure. When present, neurological features tend to follow other symptoms. Significant neurological deficits include

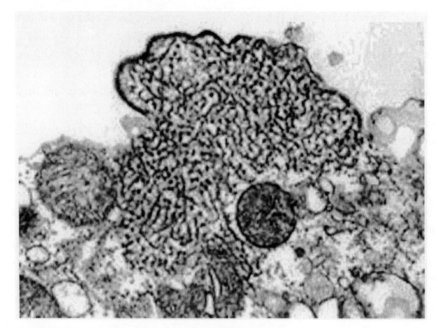

FIGURE 20–21   Electron micrograph of Nipah virus.
Courtesy of the CDC.

abnormal pupils diminished deep tendon reflexes (DTR); meningeal signs, drowsiness, coma, seizures, focal myoclonus, and nystagmus. Clinical severity of these neurologic signs varies from extremely mild to severe.

Chronic Effects

The original identification of the Nipah virus occurred only five years ago; consequently long-term sequelae remain uncertain. Follow-up studies reveal that approximately 8% of patients with encephalitic presentations suffered relapses as far out as two years. Approximately 4% of patients who never developed neurological signs and symptoms later developed encephalitis, again as long as two years following infection. In neither of these groups was Nipah isolated, possibly because of mutations in the virus. Signs and symptoms in recrudescent cases tended to be milder than initial

TABLE 20–25   Nipah Virus: Clinical Features

Fever, headache, nausea, vomiting

CNS signs and symptoms include dizziness, alterations in consciousness from drowsiness to coma; depressed DTRs, meningeal signs, pupil abnormalities, nystagus, seizures, and myoclonus.

Autonomic findings include hypertension, tachycardia.

infections and resolved spontaneously. Fifteen percent of Malaysian epidemic patients had residual neurological deficits. One-third of this group remained comatose, while the rest had varying degrees of cognitive and cerebellar impairment.

Laboratory Tests

Patients with acute Nipah infections are likely to have evidence of thrombocytopenia as well as elevated liver enzymes. CSF shows elevated total protein and leukocytes, with lymphocytic predominance consistent with viral meningitis.

Microscopy

Microscopy is of little value with Nipah virus infections.

Studies

MRI of the brain reveals multiple and widespread subcortical and white matter changes. CT of the head is often normal. EEG may show diffuse slow waves with focal sharp waves or irregular slow waves. Focal irregularities in the temporal area are common as well in EEG recordings.

Differential Diagnosis

CNS involvement, in addition to fever, warrants a differential diagnosis that includes causes of meningitis and encephalitis, including viral encephalitides.

Diagnosis

ELISA identification of both IgG and IGM can be done for Nipah antigen. PCR is useful, but should only be done at the CDC.

Treatment

Treatment is largely supportive. Preliminary trials with ribavirin show some reduction in the duration and severity of disease, but it is not yet FDA-approved for treatment of Nipah, even under IND protocol.

TABLE 20–26   Differential Diagnosis: *Nipah* Virus

St. Louis encephalitis	Herpes simplex
Dengue fever	Japanese encephalitis
Encephalitis	Rickettsial infection
Malaria	Brain abscess
Meningitis	Lymphomatous or carcinomatosis
Yellow fever	meningitis

Prophylaxis

Currently, no vaccine exists. Because of scant and unproven data, there is no role for ribavirin as a prophylaxis at this time.

Infection Control

Person-to-person transmission has not been documented, but there is a theoretical risk of infection through aerosolization of respiratory secretions. Most experts agree that droplet precautions are warranted; some suggest respiratory isolation in the event of intentional attack with Nipah.

Hantavirus

Background

Hantavirus is a term for a relatively new genus within the family *Bunyaviridae*. Hanta is the original member of the genus, but several other species have been identified. All cause similar clinical presentations. Unlike other members of the Bunyaviridae family, Hantaviruses are not carried by arthropods, but by rodents of the family Muridae. Two clinically distinct clinical syndromes have been noted with infection from these viruses: hemorrhagic fever with renal syndrome (HFRS) and the more deadly form, Hantavirus pulmonary syndrome (HPS), also known as Hantavirus cardiopulmonary syndrome. Both clinical syndromes are associated with particular species of Hantaviruses (see the following table). HPS is sometimes referred to as "new world," whereas HFRS is sometimes referred to as "old world," geographical terms based on where the virus was first identified and rooted in the history of European exploration and domination (Table 20–27).

Hanta was identified in 1993 in the southwest United States following an outbreak of an unknown infectious disease first detected by doctors from the Indian Health Service. Though this was the first identification of Hanta, retrospective analyses based on serology identified cases as far back as 1959.

Hantaviruses have been among those biological agents investigated for use as biological weapons and numerous occurrences of laboratory outbreaks have oc-

TABLE 20–27   Species of Hantavirus

HFRS Causative Species	HPS Causative Species
Dobrava	Sin Nombre
Seoul	New York
Puumala	Black Creek Canal
Hantaan	Andes

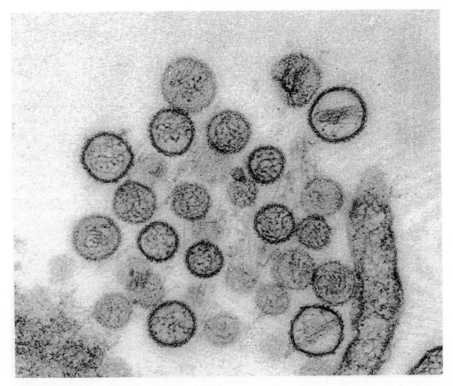

FIGURE 20-22   **Electron micrograph of Hantavirus.**
Courtesy of the CDC.

curred as a result of this research. A deliberate attack with Hantavirus would employ dissemination via aerosolized forms of the virus.

Sources

Natural forms of Hantaviruses are carried by murine hosts that suffer only a transient illness, but are believed to maintain lifelong infection. Each viral species is associated with a different mice and rat species around the world. These rodents shed the virus through bodily fluids and feces. Most human cases occur in locales where rodents are sharing or frequenting the human dwellings.

Properties

Hantaviruses are spherical with a lipid envelope, 80 to 120 nm in diameter, and possess a single-stranded, negative-sense RNA. There are greater than twenty known Hantavirus species at this time.

Epidemiology

Hantavirus infections occur in worldwide distribution with each region tending to have its local species. As many as 150,000 people are hospitalized each year for Hantavirus infection. For naturally occurring infections, incidence tends to parallel fluctuations in rodent populations. Incidence indirectly parallels those environmental factors that affect rodent populations, such as weather, food and water supplies, predation, and pest control. The viruses causing HPS tend to have a high mortality rate, upwards of 50%, though most of the causative species have not been studied adequately. The viruses causing HFRS are less virulent with a mortality rate ranging from 1% to 15% depending on the species. Although geography tends to be important with locally identified viral species causing most local infections, there is documentation of Hantavirus infections occurring globally. Sporadic cases regularly occur; the most recent Hantavirus outbreak occurred in March 2000 in Panama with twelve infections and three deaths, all presenting with the HPS form.

Means of Transmission

The precise means of transmission from rodent to human is not known. Transmission probably occurs via aerosolization of contaminated urine and feces or direct inoculation from contaminated rodent saliva through biting of humans. Person-to-person transmission has not been documented within the United States, but it has been seen in human epidemics elsewhere in the world.

In the event of biological attack, aerosolized dissemination of Hantavirus is the anticipated route of exposure. However, rodents may still play a role in terms of perpetuation or recurrence of the outbreak.

Pathogenesis

The precise pathogenic mechanisms of Hantaviruses are poorly understood. It is known that they have a predilection for endothelial cells into which they gain entry. Endothelial infection of the kidney, heart, and lungs tends to be the most common sites. Animal studies indicate that the species involved in HPS are found commonly within pulmonary capillary beds. Presumably, inhalation of contaminated particles brings the virus into the lungs where it is taken up by phagocytic cells and carried to regional lymph nodes. From there, hematogenous spread occurs, seeding the endothelium systemically, but preferentially in the lung, heart, and kidneys. Inside the endothelial cells, membrane permeability is disrupted. There is also evidence that the virus directly impacts both the Golgi apparatus and the endoplasmic reticulum. Replication occurs within the endothelial cells leading to further hematogenous spread. Additionally, experimental studies in mice suggest that the virus suppresses CD8+ lymphocytes, a feature that no doubt contributes to virulence of the agent and delays in host clearance of the infection.

Not surprisingly, the two clinical forms reflect their pathophysiology. In HFRS, for example, glomeruli and the renal endothelium are particularly sensitive to Hantavirus. Renal disease (e.g., decrease in glomerular filtration) is secondary to hemodynamic effects rather than to direct glomerular injury. Cytokine dysregulation plays an important role as well, causing capillary permeability, vascular leak, and subsequent hemodynamic changes.

Signs and Symptoms

The incubation period for *Hantaviruses* is extremely broad, extending from 1 to 7 weeks, though 1 to 4 weeks is more typical. In the event of an intentional attack with aerosolized Hantavirus, it is not clear how the initial presentation or clinical course will differ, but presumably incubation time will be shorter. Factors such as the type of species involved and genetic modifications could affect the disease as well. What follows is based on information derived from naturally occurring outbreaks.

֎HPS

Following a mean incubation period of roughly 10 days, infection with HPS-causing viruses begins with a viral prodrome of fever, chills, myalgias, malaise, headache, abdominal pain, and anorexia. A nonproductive cough, with accompanying rales, may worsen as the disease progresses. Within a week, the illness marked by hypotension and tachycardia with fluid shifts into the interstitium of the lung may ensue. The presence of hypoxia and cyanosis is dependent on the severity of the pulmonary edema. Bleeding is seen in roughly one-third of patients, most commonly in the form of hemoptysis, hematuria, petechiae, or bleeding from venopuncture sites. For this reason, Hantavirus are on the differential list of hemorrhagic fevers.

֎HFRS

The clinical severity of HFRS varies greatly according to the species of virus. The mean incubation is approximately 2 weeks, but can extend weeks longer. Viruses causing HFRS present with distinct phases that are marked by fever, hemodynamic instability, oliguria or polyuria. An essential feature is clinical prostration. Patients may take months to fully recover. The febrile phase of HFRS is further marked by headache, myalgias, visual and vestibular problems, central truncal hyperemia, and petechiae.

The extent of renal insufficiency varies from oliguria to renal failure with secondary electrolyte imbalances. This phase is commonly marked by a disseminated intravascular coagulation (DIC)-like picture as well as by frank bleeding.

Hemodynamic instability may occur secondary to cytokine dysregulation of membrane permeability and subsequent fluid shifts, leading in severe cases to respiratory failure or ARDS.

Mortality is usually secondary to worsening pulmonary edema, septic shock, and DIC, often with hemorrhaging.

Chronic Effects

There is insufficient evidence regarding the long-term sequelae of Hantaviruses. Long-term renal impairment resulting from renal insufficiency associated with HFRS is not uncommon.

Laboratory Tests

Blood work reveals thrombocytopenia of less than 100,000/μl. An elevated white blood cell count with increased bands is common. Peripheral blood smears are often helpful in diagnosing Hantavirus from other causes in the differential diagnosis, particularly HPS where peripheral blood smears show immunoblasts, myelocytes, metamyelocytes, or promyelocytes. Hemoconcentration may accompany fluid shifts secondary to capillary leakage. Elevations in liver transaminases are common. HPS is likely to have an elevated partial thromboplastin time (PTT) and/or prothrombin time (PT) whereas abnormalities in blood urea nitrogen (BUN) and creatinine are common with HRFS, as is proteinuria.

Microscopy

Standard microscopy is not useful in identifying Hantaviruses.

Radiographic Findings

HPS and HFRS may cause pulmonary congestion with bilateral interstitial edema on clear radiographs.

Differential Diagnosis

The differential diagnosis for *Hantavirus* is extensive and making the diagnosis is difficult on clinical grounds alone. Any differential should include atypical pneumonias, rickettsial infections, relapsing fevers, plague, and tularemia. Additionally, causes of pulmonary hemorrhage—such as Goodpasture's syndrome—should be considered.

Diagnosis

Because the virus is rarely detectable in the blood, diagnosis must be based on immunological markers. PCR is ideal for diagnosis but not yet widely available. ELISA may be used as well for both IgG and IgM during acute illness.

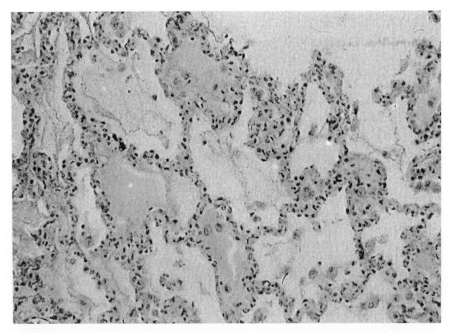

FIGURE 20-23   **Interstitial pneumonitis seen in biopsy of Hantavirus pulmonary syndrome.**
Courtesy of the CDC.

HPS

Hantavirus-specific IgM is usually positive at presentation. A fourfold rise in Hantavirus-specific IgG is also considered diagnostic.

HFRS

Any increase in antigen-specific IgG is diagnostic.

Treatment

Treatment is primarily supportive and may involve intensive care and hemodialysis. Trials with ribavirin have showed varying degrees of benefit depending on the species of Hantavirus. Mortality improved significantly in a double-blind study of ribavarin in the treatment of HRFS. Therapeutic trials with HPS are underway. At present, there is no FDA-approved treatment for Hantavirus, and IND protocols are not in place. Nonetheless, ribavarin treatment would likely confer some degree of clinical benefit if it were available.

Prophylaxis

No vaccine is available in the United States, although research is ongoing. There is clinical evidence of the benefit of ribavirin for post-exposure treatment of HFRS, and possibly for HPS. At present, there is little evidence regarding the use of ribavirin prophylactically.

Infection Control

Although person-to-person transmission has not been documented in the United States, it is thought to have occurred elsewhere. Consequently, respiratory isolation is recommended, especially in the event of a deliberate attack when the species involved or whether the virus has been genetically modified is unknown.

Should a biological attack occur, rodent control will play an important role in preventing further spread of the virus during the outbreak, as well as helping to decrease the likelihood of future recurrence.

Decontamination with standard commercial disinfectants is adequate.

☣Additional Biotoxin Agents

Trichothecene (T-2) mycotoxins, aflatoxin, and saxitoxin are included among the "additional" agents and were chosen from the myriad of potential biological weapons for inclusion in Category C agents, based on WHO, the U.S. military, and other sources. These three agents were deemed worthy of inclusion in the source book because they possess features making them more "useful," or more likely to be selected, as biological weapons.

☣Mycotoxins

Background

Mycotoxins are products of fungal metabolism that function as poisons when they come into contact with host cells. Generally, these toxins provide survival advantage to the fungus by impeding competitors, such as bacteria, or predators, such as herbivores.

"Mycotoxicosis," the poisoning of metabolism secondary to fungal proteins should not be confused with "mycosis," which is the growth of fungus on or in the host, such as tinea cruris or aspergillosis. Penicillin historically is the most famous mycotoxin, the antibiotic properties of which have had a profound impact on health and disease. Of course, not all mycotoxins benefit humanity. Many mycotoxins are quite harmful or even deadly. In the context of bioterrorism, our chief concern is with those toxins capable of use as biological weapons. The major mycotoxins include: trichothecenes, aflatoxins, satratoxin H, fumonisins, vomitoxin, and ergot alkaloids.

Trichothecene (T-2) Mycotoxin

The virulence of trichothecene mycotoxins is an historical fact, most dramatically illustrated following consumption of contaminated grains in the former Soviet Union in the years following WWII. More than 100,000 people in Siberia (greater than 10% of the regional population) died from its poisonous effects. The only cases of trichothecene-affected individuals in the United States are those of home-owners, whose homes have sustained water damage allowing growth of the fungus developing signs and symptoms of mycotoxicosis.

Mycotoxins, trichothecenes in particular, have long been researched as potential biological agents. Despite some controversy, most military experts believe they were used in Laos during the Vietnam War (Fig. 20–24; see also "of Note . . ."). There are also unofficial reports of use during the first Gulf War. This history, combined with their ionate hardiness and toxicity, make trichothecenes viable threats as biological weapons. T-2 mycotoxin could be used in aerosolized forms or detonated with conventional explosives. According to military experts, individuals or groups

OF NOTE . . .

Yellow Rain in Laos

Contradictory accounts obfuscate the claim that in the mid-1970s, the Hmong people of Laos were victims of trichothecene mycotoxin attacks. Southeast Asia was rife with political struggle at the time; the United States was beginning to pull out from Vietnam, though the CIA continued to support and train local anti-Communist factions throughout the region, including the Hmongs in Laos. The Soviets, meanwhile, supported and trained Communist insurgents in the area. Out of Laos came reports of "poisonous, yellow rain" falling on secluded Hmong strongholds. The Hmongs described the attacks as occurring following missile attacks or following planes flying overhead. The yellow rain was said to leave an oily residue where it fell. The Hmongs, a poor people in a poor country, had no protective gear of any kind. Those exposed were said to have experienced seizures, bleeding, blindness, and other injuries. Over 6,000 deaths were attributed to yellow rain. Similar reports were heard when the Vietnamese Communists invaded Cambodia and from Soviet-occupied Afghanistan with thousands of deaths said to have resulted. When trichothecene were eventually found in samples tested by American scientists, the United States formally accused the Soviets of violating the treaty banning the use of chemical weapons. The Soviets denied the charge and Russia denies it to this day. Another group of American scientists criticized the original findings explaining that the samples were positive because the mycotoxin is indigenous to the area and that, furthermore, indigenous bee species swarms were known to produce "yellow rain"—a mixture of bee feces and pollen. The controversy remains unresolved.

using available pharmaceutical methodology could feasibly produce tons of myco-toxin and use it either directly on populations or on food sources.

Sources

There are forty types of trichothecenes, including *Stachybotrys* and *Fusarium*, a common grain mold species. Trichothecene (T-2) mycotoxins is distinctive in that it is the only one that has dermal effects (see Signs and Symptoms section).

Properties

T-2 mycotoxins are heat and ultraviolet (sunlight)-resistant, and resistant to common disinfectants as well. In cases of ingestion, no foul odor or taste is reported. Based on animal studies, the LD_{50} for ingestion is 4 mg/kg; for inhalation it is 1.2 mg/kg; for dermal absorption, it is around 7 mg/kg. T-2 is highly soluble in commonly available organic solvents such as methanol. As a toxin, T-2 is metabolically active in its absorbed form and even its breakdown products possess some toxicity. T-2 mycotoxin is the only agent among the potential biological weapons that can be absorbed through the skin.

FIGURE 20-24 An example of the "yellow rain" produced by bees on snow; given by some as the explanation for the yellow rain seen in Laos and Cambodia in the 1970s.

Courtesy of Bee Works.

Epidemiology

Found in worldwide distribution, T-2 mycotoxins occurr naturally on grains and cereals. Data regarding incidence are strikingly absent in the medical literature, but exposure is generally said to have a mortality rate of around 35%.

Means of Transmission

From a biological weapons standpoint, there are three modes of transmission: cutaneous, ingestion, and inhalation. Naturally occurring cases result most commonly from ingestion of contaminated grains; however, inhalation and dermal absorption are possible. In a deliberate attack on a population, the majority of cases would have respiratory features, possibly with secondary ingestion. Skin exposure is possible with variable degrees of systemic features. If food or water sources are attacked, the majority of cases would demonstrate gastrointestinal features.

Pathogenesis

T-2 mycotoxin has multiple effects on cell functions, but tends to have the greatest impact on cell lines with high turnover such as epithelial cells, where they act similar to chemotherapeutic agents or radiation. Toxicity causes breaks in DNA, irreparable mutations in DNA sequences, changes in the phospholipid bilayer, impaired mitochondrial respiration, inhibition of RNA function, and polypeptide elongation leading to a loss of normal protein synthesis.

Signs and Symptoms

What is known about the clinical effects of T-2 mycotoxin is based on case reports from accidental research laboratory exposures, as well as descriptions of mass exposures from Russia and Laos.

Cutaneous exposures are associated with burning, pain, and erythema at the site of contact, often progressing to blistering and necrosis. With high enough cutaneous absorption, systemic features occur, including nausea, vomiting, diarrhea, and bleeding. Ocular exposures cause lacrimation and conjuctival irritation.

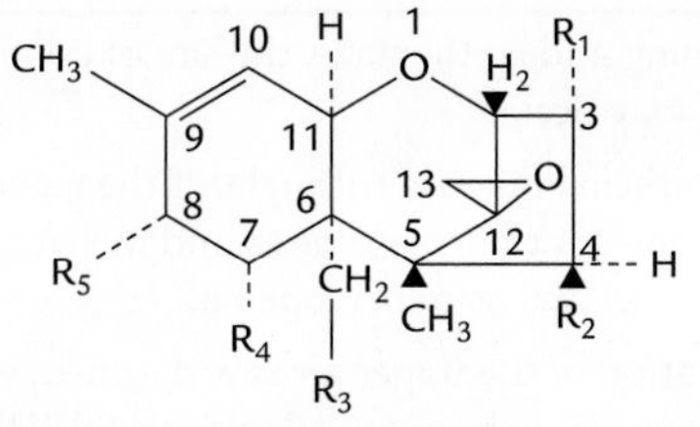

FIGURE 20–25   Chemical structure of T-2 mycotoxin.
Courtesy of RAND Organization.

Ingestion is associated with ulcerations of the oropharynx and gastrointestinal tract, nausea, and vomiting, with hematemesis and melena. Systemic absorption results in fever, chills, seizures, CNS impairment, hypotension, epithelial necrosis, and myelosuppression.

Inhalational exposure results in oropharyngeal itching and pain, rhinorrhea, sneezing, epistaxis, dyspnea, wheezing, and cough. Hemoptysis may be present. Severe exposures may result in pulmonary edema and even respiratory failure.

Chronic ingestion results in a clinical syndrome termed alimentary toxic aleukia (ATA). This entity follows a four-stage chronology:

Stage 1. Onset occurs within hours and may last for over a week. Signs and symptoms include mucosal irritation of the gastrointestinal tract with fever, nausea, vomiting, diarrhea, abdominal pain, and malaise.

Stage 2. Onset occurs as early as two weeks after exposure and results from the myelosuppressive effects of the mycotoxin, and is reflected in a pancytopenia.

Stage 3. This stage is characterized by hemorrhagic features, with progressively worsening petechial lesions seen. Necrotizing lesions may also develop in this stage.

Stage 4. The convalescence stage often takes months to complete.

Systemic Effects

Regardless of the route, sufficiently absorbed doses of T-2 mycotoxin will result in systemic symptoms and signs of malaise, dizziness, cerebellar impairments, myelosuppression, bleeding, and hemodynamic instability.

Chronic Effects

Though it may take several months for full recovery, no long-term sequelae have been documented to date.

TABLE 20–28 T-2 Mycotoxicosis: Clinical Features (Acute)

Cutaneous exposure—burning and erythema at the site of contact; progression to blisters and/or necrosis may occur.

Ingestion—ulceration of epithelial mucosa throughout the gastrointestinal tract; nausea, vomiting, sometimes with hematemesis and melena. Fever, chills; hypotension may be seen, as well as pancytopenia.

Inhalation—burning, ulceration of the upper airway, dyspnea, wheezing, cough. In more severe cases, hemoptysis, pulmonary edema, and ARDS may result.

Systemic effects—dizziness, prostration, coagulopathies, and/or hemodynamic instability

Laboratory Tests

Lab findings are nonspecific and vary according to absorbed dose. Suppression of the bone marrow results in any or all of the cell lines being depressed. The CBC may show pancytopenia, or isolated thrombocytopenia, leukopenia, or anemia.

Microscopy

Microscopy is not of benefit in diagnosing mycotoxin exposares.

Radiographic Findings

Nonspecific changes may be noted on chest radiograph if inhalation occurs and is severe enough. Otherwise, routine imaging is not helpful.

Differential Diagnosis

Depending on the nature and degree of exposure, the differential diagnosis will vary substantially. Cutaneous exposures warrant consideration of chemical agents, particularly the vesicants such as lewisite or mustard gas, though these typically have an odor. Otherwise, the gastrointestinal symptoms warrant a very large differential. In terms of a deliberate attack, both ricin and SEB could present similarly, though these are likely to be painless and lack skin findings.

Diagnosis

Making a diagnosis of T-2 mycotoxin is difficult, although cutaneous findings narrow the differential considerably in the right context. A mycotoxin syndrome in the setting of multiple concurrent cases, evidence of yellow rain, or a terrorist attack followed by a cluster would be highly suspicious. There are no commercially available diagnostic tests, although laboratory resources through the CDC can confirm the presence of T_2-mycotoxin.

Treatment

Treatment should not be undertaken until the patient is fully decontaminated (see Decontamination). Ongoing exposure to contaminated clothing or skin could worsen the victim's symptoms, and cause secondary exposures to health care personnel.

Medical management consists of supportive care. There are no specific interventions or antidotes at this time. Measures such as activated charcoal and other poison control measures have a role in treating ingestions. Respiratory support may also be warranted depending on the clinical picture.

Prophylaxis

No vaccine currently exists.

Decontamination

If there is any question about possible contamination, patient care should be deferred until decontamination is complete. While no person-to-person transmission occurs, contaminated clothing serves as a reservoir for further exposure of the patient and a source of exposure to those in contact with victims, such as health care workers or first responders. Clothing should be removed and the patient should be washed thoroughly with soap and water. Contact precautions should be used until decontamination is done, after which time standard precautions are adequate. Removed clothing should be secured in rubber basins or double bagged before being incinevated.

Aflatoxin

Background

Though not considered an imminent biological weapon threat by the U.S. military or by the CDC, the WHO considers aflatoxin a potential biological agent in the biotoxins family. Furthermore, a 1999 report by the United Nations Special Commission on Iraq reported finding a government document implicating the Iraqi

FIGURE 20-26 **Peanuts are one of the most common sources of naturally occurring** *Aspergillus flavus,* **the fungus that produces aflatoxin.**
Courtesy of the Department of Primary Industries Queensland, Australia.

military in developing weapons with aflatoxin. Indeed, a host of bombs and missiles were filled with the toxin, however, there is no unequivocal evidence that aflatoxins were ever used against the U.S. or Allied Forces during the first golf war.

Aflatoxin is a mycotoxin produced by the fungus *Aspergillus flavus* and other species. Aflatoxin is a known carcinogen. Dissemination could occur as a form of agroterrorism through contamination of crops or food supplies, or directly against populations as a biotoxin.

Sources

Aspergillosis is found worldwide, and so aflatoxin is available globally as well. Aspergillosis tends to grow on or in peanuts and peanut products, grains, corn products, milk, and other foods where it produces aflatoxin and contaminates the food products.

Properties

Although the precise mechanism of toxicity is not known, the LD_{50} is estimated in the range of 0.5 to 10 mg/kg of body weight. There are four aflatoxin forms of concern; in descending order of toxicity, they are B1, G1, B2, and G2. The only one of these of interest is B1; it is the primary naturally occurring threat and the likeliest to be used as a biological weapon.

Epidemiology

Because of the ubiquity of the toxin and the often subclinical, mild, unreported, or undiagnosed cases, precise incidence data are lacking. In the United States, no outbreaks have been reported among humans.

Means of Transmission

Typically, ingestion is the primary means of exposure secondary to consumption of contaminated foods. As a biological weapon, inhalation forms are also a possibility.

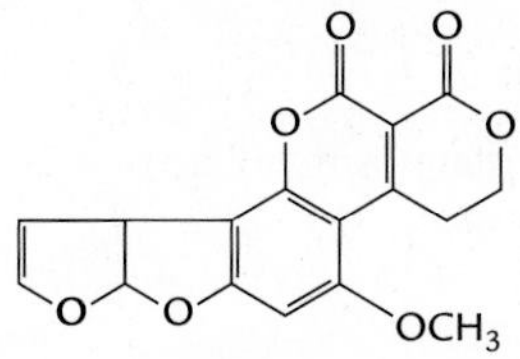

FIGURE 20–27   Chemical structure of aflatoxin.
Courtesy of the Center for Food Safety and Applied Nutrition/FDA.

TABLE 20-29 Aflatoxicosis: Clinical Features

Ingestion—nausea, vomiting, hepatic injury, hemorrhage. With severe exposures, hepatic necrosis and coma may occur.

Inhalation—unknown; though, no doubt respiratory signs and symptoms would be noted accompanied by hepatic injury.

Pathogenesis

Aflatoxin is metabolized by the cytochrome P-450 system generating toxic metabolites that bind to proteins and impair cell function. Aflatoxins are also mutagenic, causing DNA changes, mutations of hepatocellular p53 tumor suppressor genes, and consequently hepatocellular carcinoma.

Signs and Symptoms

Aflatoxicosis occurs following ingestion of food contaminated with aflatoxin. The typical presentation includes nausea, vomiting, fever, hepatitis, jaundice, peripheral edema, hemorrhage, and acute hepatic injury. With more severe exposures, hepatic necrosis, seizures, coma, cerebral edema, and death occurs. High morbidity and mortality are associated with high-dose exposures. It is not clear how exposure through aerosolized toxin would alter the clinical presentation, although respiratory injury is probable.

Chronic Effects

Long-term affects are primarily seen in regards to the liver where cirrhosis can develop and with which there is a well-known association with hepatocellular carcinoma. When large-scale exposures occur, many of those exposed will remain asymptomatic, but may still be at high risk for chronic effects.

Laboratory Tests

Possible elevations in liver function tests may be seen. Prolonged prothrombin times have been noted, as has mild hypoglycemia.

Microscopy

Microscopy is not relevant in diagnosing aflatoxicosis.

Radiographic Findings

Imaging may be useful in assessing the extent of the disease, but there are no pathognomonic radiographic findings for aflatoxin.

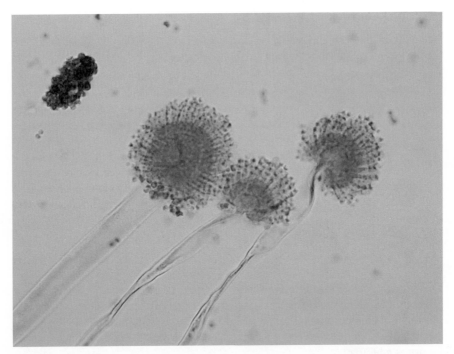

FIGURE 20-28 Light microscopy of *Aspergillus flavus*, a species responsible for producing aflatoxin.
Courtesy of Department of Laboratory Medicine at the University of California at San Francisco.

Differential Diagnosis

The differential diagnoses for aflatoxin includes the various causes of hepatitis, including viral, bacterial, alcohol, and other toxins.

Diagnosis

Diagnosing aflatoxicosis is difficult unless an overt exposure has occurred. A suggestive clinical picture can be confirmed by serum assays, but these are not yet readily available to clinicians.

Treatment

Treatment is supportive. Acute ingestions should be managed with standard poisoning protocols.

Prophylaxis

No vaccine or antidote exist currently.

Infection Control

Aflatoxin is not transmissiable person-to-person. Assessing food supplies for possible contamination may limit further exposure.

Saxitoxin

Background

Saxitoxin is a naturally occurring and potent neurotoxin produced by blue-green algae species. Most authorities consider its use as a biological weapon unlikely, but the WHO and some U.S. military experts believe it to be a viable threat and so a brief discussion is warranted.

Sources

Saxitoxin is produced by dinoflagellates of the genus *Gonyaulax,* a type of algae that serves as food for shellfish, such as mussels. The shellfish accumulate the toxin and predators, including humans, are affected following ingestion of the shellfish. The dinoflagellate that produces the toxin is easily grown and its toxin readily harvested for use as a biotoxin.

Properties

Consuming as little as 1 mg of saxitoxin is potentially lethal for humans. Higher doses may prove fatal within minutes. It is a highly stable compound, and because it is soluble in water, the toxin is cleared fairly rapidly through the urine. Mortality is low in naturally occurring cases of saxitoxicosis, but it may be considerably higher in a biological attack if a sufficient dose is delivered.

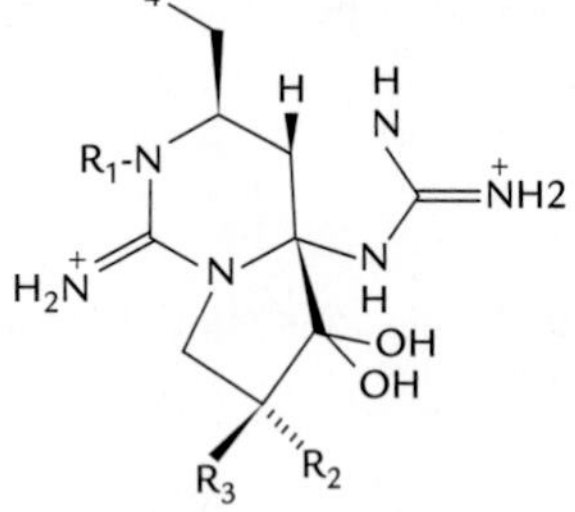

FIGURE 20–29   The chemical structure of saxitoxin.
Courtesy of the Medical Research Council/Research and Bioinformatics Divisions.

Epidemiology

Outbreaks occur periodically, with cases typically clustered in and around seafood restaurants serving contaminated foods. Such outbreaks usually occur following alterations in local ocean conditions favoring growth of the dinoflagellate. Precise incidence data are unavailable.

Means of Transmission

Natural transmission occurs via consumption of shellfish, such as clams, mussels, and scallops that subsist on the dinoflagellates. As a biological weapon, saxitoxin would most likely be disseminated in aerosolized forms. Deliberate contamination of food and water is also possible, but would likely cause only a very small scale epidemic or sporadic illness.

Pathogenesis

Saxitoxin is a potent neurotoxin that binds reversibly adjacent to sodium channels on neural and skeletal tissues. Binding causes impairment of Na-channel

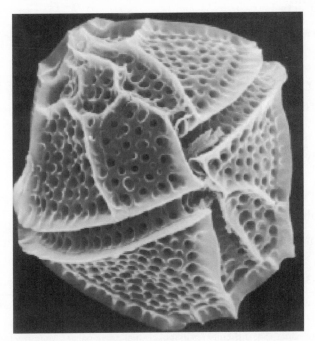

FIGURE 20-30 **Electron micrograph of *Gonyaulax*, the genus that produces saxitoxin.**
Courtesy of the Department of Biology/Whitman College.

function, limiting sodium permeability, and thereby preventing action potential conduction.

Signs and Symptoms

Following ingestion, signs and symptoms begin within minutes or as long as an hour. Numbness, burning, and tingling are noted, particularly of the mouth and lips, with ocular irritation. Shortly thereafter, similar sensations are noted peripherally along with a loss of muscular coordination. Cranial nerve palsies can occur, including diplopia, incoherent speech, dysphagia, dysarthria, or loss of speech. Nausea and vomiting also occur. Other neurological features—such as dizziness, paralysis, memory loss, disorientation, and headache—are also noted. Loss of consciousness is not seen with saxitoxin. Paralysis of respiratory muscles leading to respiratory failure may result. Signs and symptoms progress over a 12 hour period, and take up to a week resolve completely.

The extent and severity of signs and symptoms depends on the dose of toxin ingested or inhaled. It is not clear if aerosolized saxitoxin would present differently. However, animal studies indicate that signs and symptoms, including respiratory failure, progress more rapidly in aerosol exposures.

Chronic Effects

Full convalescence may take up to a few months. No long-term sequelae have been noted.

Laboratory Tests

Based on case reports, no consistent or pathognomonic lab abnormalities have been identified.

Microscopy

Microscopic studies are unrevealing with saxitoxin exposure.

Radiographic Findings

Imaging shows no consistent or typical features.

TABLE 20–30 Saxitoxicosis: Clinical Features

Progressively worsening neurological signs and symptoms, beginning with paresthesias of the extremities and lips
Worsening coordination, difficulty speaking, cranial nerve palsies, memory loss, headache
Signs and symptoms typically resolve within a week.

Differential Diagnosis

Acute neurological findings warrant consideration of other poisonings, such as those caused by neurotoxic shellfish. It is also important to consider nerve gas and botulinum. Unlike botulinum, however, which only affects motor function, sensory findings occur with saxitoxin. For syndrome surveillance purposes, saxitoxin is best grouped with the nerve agents.

Diagnosis

In naturally occurring cases, dietary history and clinical features provide the most help. Making a bioterror diagnosis of exposure to saxitoxin is difficult without a sentinel event or clustering of cases. Diagnostic clinical and environmental assays exist, but these are available only at specialized laboratories.

Treatment

No antidote exists presently. Medical management is supportive, including mechanical ventilation. Use of activated charcoal, gastric lavage, and other standard poison interventions are recommended if ingestion is suspected. With naturally occurring cases, patients who survive the first 24 hours have an excellent prognosis. In contrast, animal studies with aerosolized exposure to saxitoxin demonstrate a poor prognosis.

Prophylaxis

No vaccine or antidote exists.

Infection Control

Saxitoxin is not transmitted by person-to-person contact. Patients may be treated with standard precautions. If the source is thought to be contaminated food, obviously it should be destroyed.

TABLE 20–31   Differential Diagnosis: Saxitoxin

Botulism	Nerve gases
Poliomyelitis	Polymyositis
Encephalitis	Diabetic neuropathy
B-12, folate deficiencies	West Nile virus
Hyper/hypokalemia	Spinal cord infections
Hypophosphatemia	Iatrogenic causes
Meningitis	Systemic lupus erythematosus
Lyme disease	Alcohol abuse
Myasthenia gravis	Heavy metals, toxicity

Abrin

Background

Like ricin, abrin is a plant phytotoxin known by a variety names such as jequirity bean, Indian licorice seed, rosary pea, Indian bead, Seminole bead, and Buddhist rosary bead, to name several. It is produced as a powder and is extremely toxic. As little as .007 mg/kg of abrin is potentially fatal. Pathophysiologically, abrin acts as a ribosome inhibitor interfering with normal cellular protein metabolism akin to ricin. Routes of exposure include dermal absorption, inhalation, and ingestion.

The clinical presentation of abrin may be delayed one to three days, but then profound effects on multiple organ systems ensue. Ocular effects include redness, lacrimation, edema of the eyelids and pain, as well as dilated pupils and retinal hemorrhage. Abrin causes severe allergic reactions, ranging from local urticaria and rhinitis to anaphylaxis. If ingested, abrin causes profound hemodynamic instability, including hypotension, severe dehydration, and tachyarrhythmias. Central nervous system effects such as sleepiness, disorientation, hallucinosis, seizures, and coma occur as well. Gastrointestinal complications range from abdominal pain to nausea, vomiting, and diarrhea. Luminal edema and necrosis of the liver are also seen. Gross hematuria, anuria, and renal necrosis occur, as does myopathy, tremors, and tetany.

Treatment is again largely supportive. Clinicians should bear in mind the need for aggressive volume repletion with abrin exposure. Workers exposed to abrin must be scrupulous about avoiding contamination with the allergic and toxic dust. First responders should wear chemical protective clothing and pressure-demand self-contained breathing apparatus (SCBA) gear or HEPA filter purified air-powered respirators (PAPRs). Decontamination is achieved by removing all clothing and thoroughly washing the exposed worker or victim with soap and water.

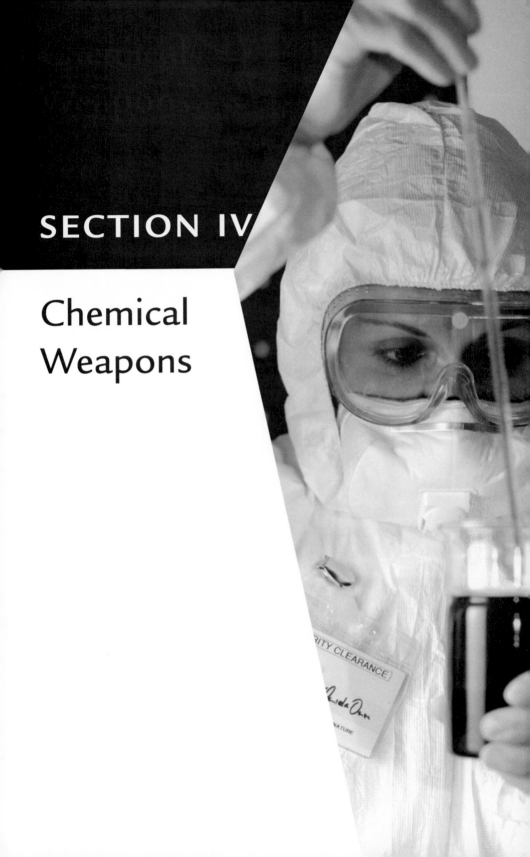

SECTION IV

Chemical
Weapons

CHAPTER 21

Introduction to Chemical Weapons

Chemical agents differ from biological agents in a number of significant ways. To begin with, the onset of symptoms may occur instantly or within a few hours of exposure, whereas biological agents can take up to weeks to present. Chemical agents are synthesized through some type of industrial process, whereas biological agents are either self-replicating microbes or compounds formed by these microbes such as botulinum toxin (produced by *Clostridia*; see Chapter 17) or ricin, a natural metabolic product of the castor bean plant, *Ricinus communis*. Chemical agents act by disrupting normal metabolic function at varying sites throughout the body according to the type of agent used and are categorized into roughly eleven categories (Table 21–1). From a weapons standpoint, not all chemical agents pose a threat as a likely choice for terrorism or as a WMD. In large part this is because of logistical or technical considerations—some require too much volume for easy concealment, others pose difficulties of dispersal, and others require such quantity as to be impractical for small, poor groups with limited technical or financial resources.

One of the considerations that makes chemical weapons a legitimate threat is the sheer number of chemicals that, if used in the "right" way, could function as a weapon. These agents could also be used in combination, potentially increasing their potency, or altering physical properties such as to volatility in order to augment abortion. The most likely means of disseminating chemical agents are sprays, or through detonation of explosives. Spraying distributes the chemicals more efficiently. They are often classified according to how well they dissipate once released, that is, persistent or nonpersistent agents.

TABLE 21-1   Comprehensive List of Designated Chemical Weapons

Blister Agents/Vesicants
Mustards
  Distilled mustard (HD)
  Mustard gas (H) (sulfur mustard)
  Mustard/Lewisite (HL)
Mustard/T
  Nitrogen mustard (HN-1, HN-2, HN-3)
  Sesqui mustard
  Sulfur mustard (H) (mustard gas)
Lewisites/chloroarsine agents
  Methyldichloroarsine (MD)
    Ethyldichloroarsine (ED)
  Lewisite (L, L-1, L-2, L-3)
  Phenodichloroarsine (PD)
  Mustard/Lewisite (HL)
Phosgene oxime (CX)

Blood Agents
Arsine (SA)
Cyanides
  Cyanogen chloride (CK)
  Hydrogen cyanide (AC)
  Potassium cyanide (KCN)
  Sodium cyanide (NaCN)
  Carbon monoxide

Caustics (Acids)
Hydrofluoric acid

Choking/Lung/Pulmonary Agents
Ammonia
Bromine (CA)
Chlorine (CL)
Methyl bromide
Methyl isocyanate
Osmium tetoxide
Hydrogen chloride
Nitrogen oxide (NO)
Perfluororisobutylene (PHIB)
Phosgene
  Diphosgene (DP)
  Phosgene (CG)
  Phosphine
  Red phosphorous (RP)
  Sulfur trioxide-chlorosulfonic acid
    (FS)

Sulfuryl flouride
Teflon and PHIB
Titanium tetrachloride (FM)
White phosphorus
Zinc oxide (HC)

Incapacitating Agents
Agent 15
BZ
Cannabinoids
Opioids (e.g., fentanyl)
LSD
Phenothiazines

Metals
Arsenic
Thallium
Mercury
Barium

Nerve Agents
G agents
  Cyclohexyl sarin (GF)
  Sarin (GB)
  Soman (GD)
  Tabun (GA)
V agents
VX

Organic Solvents
Benzene

Riot Control Agents/Tear Gas
Bromobenzylcyanide (CA)
Chloroacetophenone (CN)
Chloropicrin (PS)
CS

Toxic Alcohols
Ethylene glycol

Vomiting Agents
Adamsite (DM)
Diphenylchloroarsine (DA)
Diphenylcyanoarsine (DC)

Modified from CDC Chemical Agents. (http://bt.cdc.gov/agent/agentslistchem—category.asp)

FIGURE 21-1   **Chemical stores present a risk both for theft and for onsite sabotage.**
Courtesy of the CDC.

The signs and symptoms for many of the agents are nonspecific and may be difficult to identify, especially with an index case. We have tried to cover the most likely of these agents, and when possible, to use one agent from a class as a means of providing information regarding the other members of that class. By virtue of the number, discretion is needed to determine which agents, from the comprehensive list in Table 21–1 are worth familiarizing oneself with out of necessity. We have selected for indepth discussions those chemical classes and agents that are generally considered the most likely to be used in a terrorist attack, and thus designed to be weapons, or those that are sufficiently prototypical for a given class to help clinicians understand class effects. Industrial and commercial agents could also be used, as they can be quite harmful and may be easier to acquire or to target through sabotage (Fig. 21–1). Many bioterrorism experts believe that the most viable threat from chemical weapons is from commercial sources, either through theft of existing stores of chemicals (waste or storage warehouses) or sabotage. It is incumbent on community physicians to educate themselves about the types of chemicals used in local industries in their communities, as well as about the type of chemicals that pass through the town, be it by truck, rail, or boat (Fig. 21–2).

FIGURE 21–2 Chemical hazard symbol.
Courtesy of Howard University College of Medicine.

CHAPTER 22

A Brief History of Chemical Weapons

The crude use of chemical weapons is documented in written and visual form as early as 600 BC. That year, the Athenian general Solon contaminated the water supply of the besieged Greek city of Cirrha with black hellebore root. Crippled by severe diarrhea, the Cirrhaeans were defeated. This may have been the first known use of incapacitating agents as chemical weapons. During the Peloponnesian War (420 BC), Spartan forces overran an Athenian stronghold by using irritating fumes created by the burning of sulfur, coals, and tars. Nearly a millennium later, this practice evolved into a common military tactic referred to as "Greek fire," the formula that has forever been a secret, but is thought to be a precursor to napalm involving burning a mix of sulfur, resin, pitch, naphtha, lime, and saltpeter. Greek fire was used in large scale first by the Byzantines in naval wartime. They would place the materials in bronze tubes, ignite the materials, and direct the flame through the tube at enemy fleets.

The great Carthaginian general Hannibal (184 BC) used belladonna plants to induce disorientation in opposing troops during the Punic Wars. In 1672 AD, the Bishop of Muenster used crude grenades filled with belladonna to attack the city of Groningen. In 1881, men conducting a railway survey in North Africa became ill after eating dried dates offered to them by local tribesmen. It turns out that the fruit had been tainted with *Hyoscyamus falezlez*, a plant containing a scopolamine-like chemical. In 1908, a troop of French Colonial soldiers stationed in Hanoi were poisoned with a phytotoxin that induced hallucinations and acute mental status changes.

FIGURE 22–1 Depiction of the use of Greek fire as documented in the Madrid Skylitzes, an illustrated historical text of the Byzantine empire at the end of the first millenium.

Courtesy of University of North Florida/Paul Halsall.

Modern Chemical Warfare

The Industrial Revolution included in its momentum a revolution in the development of chemical weapons. Although there had been threats of use during the Civil and Napoleonic Wars, it wasn't until WWI that chemical weapons were actually utilized. Ironically, a Nobel Prize–winning scientist, Fritz Haber, is largely responsibility for their development and implementation (Fig. 22–2). The irony lays within Haber's having won a Nobel Prize for discovering a process for extracting nitrogen out of air. This discovery led to the industrial production of fertilizer, which allowed Western agricultural output to increase immensely, resulting in an abundance of food for a hungry world. Many view this as one of the most valuable discoveries in human history.

Haber, who considered himself a great German patriot, knew that the war was at a stalemate. He knew, also, that the German military had already failed twice in attempts to use bromide against the French. Haber proposed the use of chlorine gas, and development was immediately begun. On April 22, 1915, Fritz Haber stood on the Western front, at Ypres, Belgium, facing British and Canadian troops. When he felt wind conditions were at their best, he ordered the gas to be released (Fig. 22–3). Based on the "success," the German army then asked him to lead a similar assault on the Eastern front against the Russian army. The night he left, his wife, also a respected chemist, committed suicide; an act many historians believe was out of disgust at his inhumanity. For his actions, Haber would later earn the distinction

FIGURE 22-2 Nobel Prize–winning chemist Fritz Haber, later charged with crimes against humanity.
Courtesy of Saskatchewan Department of Learning.

of being the only Nobel Prize winner to be accused of war crimes. He was never prosecuted, however within 6 months, the Germans had developed phosgene and by 1917, they had developed mustard gas—a misnomer as it is, more accurately, an aerosol (see Glossary). Though the Germans led the way, the British and later, the United States, were never far behind. By the conclusion of WWI, an estimated 100,000 tons of chemical agents had been used, including phosgene, mustard gas,

FIGURE 22-3 Release of chlorine gas at Ypres, 1915.
Courtesy of The World War I Document Archive.

FIGURE 22–4   WWI soldier and horse donning gas masks.
Courtesy of The World War I Document Archive.

arsine, and chlorine. Altogether chemical gas attacks were responsible for 90,000 deaths and 1.3 million injuries (Fig. 22–4).

The inhumanity of such weapons united the world into signing the 1925 Geneva Protocol for the Prohibition of the Use of Asphyxiating, Poisonous or Other Gases, and Bacteriological Methods of Warfare. This international treaty prohibited the use of chemical agents during wartime, but put no restrictions on researching, manufacturing, or stockpiling of these weapons. In the context of a war, the treaty did not proscribe the use of such weapons used for retaliation against enemies who employed them. As a means of responding to such an attack or as a means of deterring attacks, many of the signatories felt that it was necessary to research, manufacture, and stockpile such weapons.

Just prior to the start of WWII, a German scientist discovered what would become a whole of new class of chemical agents: nerve gas. Interestingly, during

WWII, no chemical agents were used in the Western theater. Horrifingly, they were against civilians. The Nazis used Zyklon B, a commercial form of hydrogen cyanide, to exterminate millions of Jews and others in the concentration camps (Fig. 22–5). At the end of the war, allied troops recovered approximately 20,000 tons of tabun, a type of nerve gas, as well as significant amounts of sarin, another nerve agent, contained in German munitions. WWII's Eastern theater played out very differently. The Japanese government acknowledges that during the war, it oversaw the production of roughly 7,000 tons of lewisite and mustard gas, but the evidence indicates that it has refused comment as to if and how these were used. Despite this silence, these weapons were used by the Japanese (Fig. 22–6). The Chinese government claims that during the war, Japan used chemical weapons on civilians and soldiers upward of 2,000 times, causing approximately 100,000 Chinese casualties. Both sides in the global conflict conducted intense research into chemical defenses as well. In the United States, over 60,000 servicemen were used as experimental subjects in the army's chemical defense research program. Recent studies of the long-term effects of chemical defense tests from WWII have been conducted by the U.S. Veterans Adminstration (see chronic effects sections in chemical agents chapters).

After WWII and into the 1950s, British scientists discovered one of the most potent nerve agents known: VX. The United States undertook large-scale manufacturing

FIGURE 22-5   A Nazi warehouse filled with Zyklon B canisters at the Majdanek death camp.
Courtesy of the Main Commission for the Investigation of Nazi War Crimes/USHMM Photo Archives #50575.

FIGURE 22-6 A canister of mustard gas left from WWII by the Japanese military in Jilin province, China.
Courtesy of Kanagawa University/Professor Keiichi Tsuneishi.

of VX starting in 1961. Out of growing concern for the widescale distribution and availability of chemical and biological agents—including nations and groups considered extreme or violent—came the 1972 Biological and Toxin Weapons Convention (BTWC). This act expanded the restrictions begun in the 1925 treaty. The BTWC prohibits the production, development, and stockpiling of chemical weapons. Unfortunately, no provisions were made for enforcement. Furthermore, as with biological weapons, much of the problem lies in the unclear distinction between military and legitimate public and industrial development.

Following WWII, the U.S. military initiated a substantial research and development effort aimed at creating an arsenal of nonlethal, psychoactively incapacitating chemicals. The well-known psychedelic agent lysergic acid diethylamide (LSD), marijuana derivatives, and anticholinergic agents such as 3-quinuclidinyl benzilate (QNB, also known under the NATO classification as BZ) emerged from this collaborative effort between the U.S. military and the nation's pharmaceutical industry. The latter chemical was weaponized successfully, and there are conflicting reports as to whether it saw use during the Vietnam War. *Jacob's Ladder,* a 1990 movie starring Tim Robbins, implicates BZ as the cause of deaths and psychotrauma in an American battalion. Although most experts believe BZ never saw operational use, the film portrays accurately the uncertainty inherent in the battlefield use of these agents. This fact is one of the primary reasons that the Department of Defense destroyed its stockpiles of BZ and other incapacitating agents beginning in 1988.

During its Afghan occupation, the Soviet military stood accused of using incapacitating agents. Although these claims were never fully substantiated, the Russian military did use the incapacitating BZ (or a BZ-like chemical agent) in an aborted attempt to free over 100 hostages held by Chechnyan terrorists in a Moscow theater. The rescue attempt backfired horribly, killing 115 hostages and incapacitating 50 rebels. During the late 1990's war in Bosnia and Herzegovina, the army of the Federal (Serb) Republic of Yugoslavia may have used incapacitating agents on 15,000 Bosnian refugees attempting to escape the war. Bosnian survivors describe the acute onset of hallucinosis, disorientation, and bizarre behavior, and international observers suspect that a BZ-like agent may well have been loosed on the fleeing civilians.

FIGURE 22–7   After the Sarin gas attack, victims receive assistance on a Tokyo subway Platform.
Courtesy of National Academics of Press/Source unknown.

In the past 20 years, the significant use of chemical agent weapons worldwide has come from two distinct areas: one, military, the other, terrorist-based. Throughout the 1980s, Iraq used both mustard and nerve agents against Iranian forces during the Iran-Iraq War, upwards of 10,000 Iranian soldiers were killed or injured. The Iraqui government also used chemical agents against Kurdish citizens residing in northern Iraq, killing thousands of civilians and injuring tens of thousands. A 1998 British intelligence report accused the Iraqi government of stockpiling large quantities of a glycolate anticholinergic incapacitating agent known as Agent 15, considered to be closely related to BZ. Finally, it should be noted that during the first Gulf War, coalition forces uncovered 50,000 shells and bombs containing mustard gas, sarin, and cyclohexyl sarin. No such similar caches were found during the second Iraqi war.

The other notable use of chemical agents was by Aum Shinrikyo, a Japanese cult that manufactured sarin nerve gas. In 1994, the cult released sarin gas in the Japanese city of Matsumoto, killing 7 and injuring about 600. Again in 1995, it released sarin gas into the Tokyo subway system, killing 12 and injuring 3,800 (Fig. 22–7). At present, it is thought that some twelve countries either probably or certainly possess an arsenal of chemical agents. Some of the countries with known offensive chemical warfare programs—such as North Korea, Libya, Egypt, Israel, Myanmar, and Taiwan—have either not signed or not ratified the BTWC.

The availability of chemical weapons from military and industrial sources creates a distinct concern. What's more, the number of chemicals that could be used to inflict injury is extensive, and many have distinct means of causing injury to a variety of organ systems, with both acute and chronic effects. For the most part, treatment regimens are supportive in nature. Chemical weapons, then, create a particularly troubling threat.

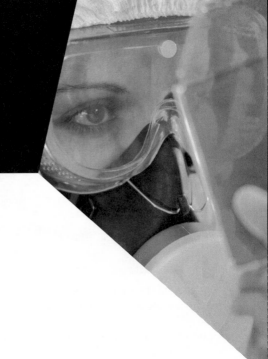

CHAPTER 23

Pulmonary Agents

Clinical Vignette

A 65-year-old patient not known to you arrives in your office complaining of dyspnea, chest tightness, and burning of his nose and eyes. He appears rather dyspneic and anxious and is immediately put into an exam room. The patient reports that he had just left a major industrial fire where he had been watching firefighters attempt to control the blaze. He noticed an acrid odor in the air, and when he began coughing, he decided to leave. Within a few minutes he noticed increasing difficulty breathing, and seeing your office, he pulled in to request urgent evaluation. His past medical history is significant only for childhood asthma. He specifically denies any foreign travel and has never been a smoker. On exam, he is tachypneic and tachycardic; you hear wheezes but no crackles. His pulse oximetry finds an SaO$_2$ of 88%. You immediately call for an ambulance. What should you tell the emergency room?

Background

On April 22, 1915, the German army released over 150 tons of chlorine gas downwind toward British and Canadian troops at the Battle of the Marne in Ypres, France (Fig. 23–1). Of the 7,000 soldiers present at the front that day, 5,000 suffered injury or death (see Of Note . . . box below). The modern era of chemical weapons had begun. Later that same year on the very same battlefront, the German army released

FIGURE 23–1 United States Marines wearing gas masks on the frontline during WWI.
Courtesy of The World War I Document Archive.

phosgene gas, with far more deadly effects. Further development of chemical weapons proceeded throughout the war, prompting continuous refinement and development of field protective gear in order to limit casualties (Fig. 23–2). The devastating effect of chemical weapons during WWI mobilized the international community and prompted the League of Nations in 1918 to institute the first international treaty to ban chemical weapons. Relative to the most potent nerve agents, pulmonary agents are easy to make and inexpensive to buy. Chlorine and phosgene, two agents first introduced as chemical weapons in WWI, are still available commercially and are used widely in plastics, rubber, and dye industries worldwide. Industrial or transportation accidents with these agents are much more likely to occur than deliberate releases, but their potential as chemical weapons must be considered.

OF NOTE . . .

April 22, 1915, Yypres, France

We knew there was something wrong. We started to march towards Ypres but we couldn't get past on the road with refugees coming down the road. We went along the railway line to Ypres and there were people, civilians and soldiers, lying along the roadside in a terrible state. We heard them say it was gas. We didn't know what the hell gas was. When we got to Ypres we found a lot of Canadians lying there dead from gas the day before, poor devils, and it was quite a horrible sight for us young men. I was only twenty so it was quite traumatic and I've never forgotten nor ever will forget it.

Private W. Haye
Royal Scots

FIGURE 23-2   British soldier in gas mask.
Courtesy of the World War I Document Archive.

Properties

A wide variety of chemical agents damage lung tissues and have the potential to be used as pulmonary agents. Tissue damage is driven primarily by two physicochemical properties: reactivity and solubility. Two organohalides, phosgene ($COCl_2$) and chlorine (CL_2), are considered prototypical pulmonary agents and will be detailed for illustrative purposes.

Phosgene (Fig. 23–3) is often described as having the sweet odor of freshly cut hay; however, the odor may be faint and in general is an unreliable means of identification. Phosgene is generally stored as a liquid. At temperatures above 47°F

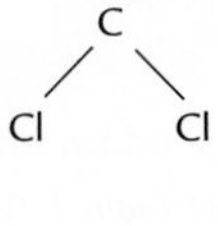

$$\underset{\underset{\displaystyle Cl \qquad Cl}{\diagup \quad \diagdown}}{\overset{\overset{\displaystyle O}{\|}}{C}}$$

FIGURE 23-3   Phosgene gas.

$$Cl = Cl$$

FIGURE 23-4 Chlorine gas.

(8.2°C), phosgene is a colorless gas that is nearly four times denser than air, and so it accumulates close to the ground. When combined with water, phosgene breaks down into hydrochloric acid and carbon dioxide. If the gas is released and there is enough ambient humidity, phosgene would have a whitish color instead of being colorless.

Chlorine (Fig. 23–4) is a greenish-yellow gas that is widely used in water purification, disinfection, and the synthesis of other chemicals (plastics, rubber, and dyes). Nearly 12 million tons are produced annually for industrial, commercial, and household use. Chlorine is 2.5 times heavier than air, a property that has relevance to its clinical effects. It is also highly reactive and has the familiar pungent, irritating odor of bleach.

Sources

A wide range of industries produce or use pulmonary agents as chemical intermediates. For chlorine gas, these include petroleum refining, photographic processing, pulp and paper mills, sewer and waste-water treatment plants, and disinfectants (Fig. 23–5). Phosgene is a byproduct of industrial fires in which

FIGURE 23-5 Chlorine tank railcar used for emergency response drills.
Courtesy of The Disaster Assistance and Rescue Team/NASA.

synthetic chlorinated polymers are consumed, posing a risk to firefighters and bystanders. Pesticide and pharmaceutical manufacturing are common examples in which these exposures may occur. Welders and petroleum refinery workers are also at risk. Chloropicrin is a fumigant used in agriculture, and organofluoride polymers (PFIBs such as Teflon) are used in literally dozens of industrial processes. Workers at risk include materials scientists and engineers, as well as individuals involved in plastic, polymer, resins, and elastomer manufacturing. Combustion of any of these widely used compounds may result in airborne exposures to a wide variety of pulmonary irritants, all of which may cause the characteristic pulmonary toxidrome.

Pathogenesis

For nearly all biological and chemical agents considered in this book, host and tissue factors and properties of the agents themselves provide the relevant clinical parameters that define toxicity (Table 23–1). This is certainly the case with pulmonary agents. Pulmonary agents' toxicity is influenced by the site of action and the properties of the specific agent, in particular, water solubility and chemical reactivity. The large surface area of the lung (50–150 m^2), compared to the surface area of the skin (2 m^2) and eye (.0002 m^2), makes the respiratory system an efficient route for chemical agent exposure. Unlike other chemical agents, however, the effects are primarily local; that is, they occur at the site of contact within the respiratory tree.

The second element of toxicity relates to properties of the chemical itself. Highly water-soluble agents, such as ammonia or HCl, have their greatest impact on the mucus membranes and upper airways. Intermediate soluble agents, such as chlorine gas, impact the respiratory tree both proximally and distally. Low water-soluble agents, for example, phosgene or oxides of nitrogen, bypass the water-rich upper respiratory epithelium and exhibit their most damaging effect at the lung periphery. At sufficiently high concentrations, however, all of these agents may cause asphyxiation and acute respiratory distress syndrome.

TABLE 23–1   Clinical Presentation: Pulmonary Agents

	Site	Symptoms	Signs
	Nasopharynx	Sneezing, pain	Erythema
Central	Oropharynx	Odynophagia	Inflammation
	Larynx	Choking	Hoarseness, stridor
	Trache/bronchi	Pain, cough	Wheezing, rhonchi
Peripheral	Bronchioles	Dyspnea	Crackles
	Alveoli	Tightness	

Chemical reactivity defines the other key element to understanding pulmonary agents' toxicity. In the tissues, phosgene undergoes hydrolysis to hydrochloric acid (HCl) that overwhelms the cell's normal buffering mechanisms and thereby damages the alveoli-capillary membrane. Hydrolyzed phosgene also forms free radicals that react with cellular macromolecules containing amine, hydroxyl, and other groups, setting up an inflammatory and chemotactic response. Cellular injury and chemokines increase the permeability of the capillaries, allowing fluid to accumulate in the interstitium resulting in one of the cardinal features of pulmonary agents: pulmonary edema. Distally acting chemicals, of which phosgene is an example, deplete surfactant and thereby contribute to alveolar collapse. Free radicals also consume antioxidants, decrease ATP, and disrupt red cell membranes leading to iron leakage.

Although the most damaging injury from phosgene occurs far down the broncheoalveolar tree, at higher inhaled concentrations it has more proximal effects, including bronchospasm or laryngospasm. Phosgene is nearly two times as potent as chlorine gas and has been known to cause death with just a few breaths (Fig. 23–6).

Chlorine is of intermediate water solubility and affects the central and peripheral areas of the lung. Like phosgene, chlorine undergoes hydrolysis, forming HCl and free radicals. These contribute directly to cytotoxicity and the inflammatory responses already described.

Other lung irritants—HCl, ammonia, chloramine gas—are more water soluble and highly reactive. They react with the respiratory epithelium proximally causing airway irritation, akin to a chemical tracheobronchitis. The highly irritating nature of the exposure usually leads to rapid avoidance and lower exposure to the individual. If exposure is of sufficient concentration, however, peripheral effects such as pulmonary edema or pneumonitis may ensue. The severity and speed of the inflammatory changes are directly proportional to the intensity of the exposure. Progression to overt

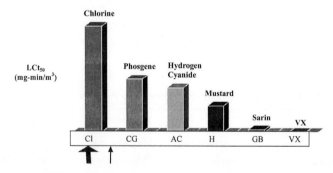

FIGURE 23–6 This chart compares the toxicity of chlorine and phosgene to other chemical weapons. The LCt_{50} refers to the vapor concentration required to kill 50% of exposed arsenals. The lower the LCt_{50}, the more toxic the substance. Chlorine and phosgene are the least toxic of the chemical weapons. Even the least toxic of the nerve agents is about 60 times more toxic than chlorine vapor.

clinical symptoms and signs is dependent on lymphatic clearance of the interstitial fluid generated from the inflammatory process. Once these regulatory mechanisms are overwhelmed, fluid accumulates in the interstitium, and the typical respiratory symptoms of shortness of breath or dyspnea occur. With these chemicals, the time frame until symptoms appear runs from as little as thirty minutes to twenty-four hours. This latency is one of the reasons why firefighters and others exposed to fires are often observed for a day before being discharged since chemical pneumonitis may progress rapidly to ARDS.

The presence of pulmonary edema presents the same potential consequences from these agents as it does with any other etiology. The primary problem is impaired oxygen diffusion across the alveoli with subsequent hypoxia. A second concern especially with severe exposures is there can be excessive accumulation of fluid in the lung, as much as a liter per hour, which can decrease the effective circulating volume leading to hypotension and multiorgan system failure. Pulmonary agents act within the respiratory system and are not absorbed systemically.

Signs and Symptoms

The presentation of clinical features following exposure occurs after a variable lag time, dependent largely on the dose and duration of exposure. With low-level exposure, an asymptomatic period of up to 24 to 72 hours occurs when dry cough and dyspnea can appear. With increased exposure, the time until symptoms appear decreases proportionately. Higher intensity exposures may show early symptoms, within a few minutes, such as pronounced shortness of breath, frothy cough, or signs, such as rales indicative of alveolar fluid or loss of surfactant. Severe exposures may begin with symptoms of upper airway irritation followed by an asymptomatic period and then the onset of lower respiratory symptoms.

Symptoms may also begin with conjunctival irritation with accompanying lacrimation, mild coughing, chest pressure or tightness, and shortness of breath. Skin irritation may be noted if the skin has sufficient moisture on it to initiate the hydrolytic process. These early symptoms are a result of superficial irritation and are not lethal. However, if severe enough, irritation of the airway epithelium can result in laryngospasm, which is potentially fatal. Findings on auscultation of the lungs may be unremarkable early on, but crackles may be noted that can increase in intensity and distribution in concordance with the severity of exposure. The presence of pulmonary edema from exposure to these agents poses the same risks as from any other etiology: poor oxygenation, which may be noted by respiratory distress and even cyanosis. As the edema worsens, the associated cough may produce white-yellow frothy sputum.

Hypotension, tachycardia, decreased urine output, and other signs of a low-volume state are seen when the amount of sequestered fluid becomes large and indicate a poor prognosis. Hypovolemic shock can develop leading to organ hypoperfusion and multiorgan system failure. Remember that the fluid accumulation may not cause radiographic or clinical findings for some time.

Laboratory Tests

Decreased alveolar gas exchange causes hypoxia that is evident in arterial blood gases and pulse oximetry. Normal oxygenation 4 to 6 hours following exposure is a good prognostic sign.

Radiology

There are two primary chest radiograph findings associated with phosgene exposure. Phosgene causes air trapping through its effect on the distal airways, radiographically manifestated as hyperinflation. The other chest radiograph findings, which tend to be delayed in onset or following heavy exposure, are infiltrates and ARDS due to the alveolar collapse and capillary leakage. ARDS indicates the individual's exposure was more severe. More central effects of chlorine gas are harder to see radiographically. Platelike atelectasis is the norm. Subtle and more extensive radiographic changes may be delayed in onset, depending on intensity and duration of exposure. This radiographic lag may limit chest radiograph utility in diagnosis and management of pulmonary agent exposures.

Differential Diagnosis

The types of respiratory symptoms seen in pulmonary agent exposure are quite broad, a fact that makes for a broad differential diagnosis. Even in known chemical exposures, distinguishing the causes of cough and lacrimation are rather nonspecific. Included in the differential are acute asthma exacerbations or other allergic-type reactions; upper respiratory infections; and cardiac, respiratory, and other noncardiac causes of pulmonary edema. In the setting of an unknown but patently deliberate attack, clinicians should be aware that nerve agents may present with significant respiratory symptoms and signs. Rapid onset of symptoms should cause consideration of tear gas, chlorine, ammonia, or chloroamines. Delayed symptoms are most consistent with phosgene gas exposure.

Diagnosis

There is no definitive test or pathognomonic clinical syndrome that clinches a diagnosis of poisoning with pulmonary agents. An exposure history coupled with a constellation of respiratory signs and symptoms will aid in diagnosis (Table 23–2).

Chronic Effects

Long-term pulmonary damage may follow any acute injury or inflammatory or infectious process. These include asthma, reactive airways disease, chronic bronchitis, or progressive lung function loss as documented by serial spirometry. Desquamation

TABLE 23-2	Making the Diagnosis: Pulmonary Agents
Pattern of Presentation	Multiple, concurrent cases of respiratory symptoms
Features	Variable according to dose and exposure route
	Early: rhinorrhea, lacrimation, cough—nonproductive.
	Late: dyspnea, cough productive white/yellow sputum
Findings	Early: lung exam normal
	Late: rales, decreased breath sounds, dullness to percussion
	CXR Early: normal
	Late: infiltrate bilaterally
	Hypoxia

and chronic skin pigmentation changes were seen in survivors of the Iraqi gas attacks during the early 1980s.

Vulnerable Populations

Individuals with preexisting respiratory conditions are likely to experience more severe adverse clinical effects from pulmonary agents at lower concentrations. This includes individuals with asthma, chronic obstructive pulmonary disease (COPD), or interstitial lung disease (ILD). Both children and adults may have asthma triggered by exposures as well. Significant hypoxia represents a risk to pregnant women, fetuses, and individuals with heart disease. There is no evidence that pulmonary agents are carcinogenic, mutagenic, or teratogenic.

Decontamination

As quickly as possible, ensure that ongoing exposure to the agent ends (Table 23–3). Presumably patient evaluation is taking place in a safe or secured medical facility. Consulting medical toxicologists, the local poison control centers, or knowledgeable experts in hazmat or occupational/environmental medicine is important. If grossly contaminated, use appropriate PPE and remove patient's clothing to avoid trapped pockets of air that may be inside clothing. Clothes should be double

TABLE 23-3	Critical First Steps: Pulmonary Agents
Medical Treatment	Decontamination with soap and water, supportive care
Infection Control	None if properly decontaminated
Public Health	Contact SHD

bagged. Should any concerns exist about the presence of residual liquids, flush the area with copious amounts of water.

Management

The two overriding concerns for management of pulmonary agent exposures are eliminating further exposure of the patient and making sure the environment does not place first responders or treating clinicians at risk. Most pressing is the patient's capacity to oxygenate—a patent airway must be present and maintained. The need for intervention is based on the clinical scenario. If laryngospasm or stridor is present, intubation may be needed. High-flow oxygen is critical and must be delivered either by mask or by intubation. Providing positive airway pressure, either via a mask or endotracheal intubation may be needed, if ARDS or pulmonary edema is present. Diuretics are of little value in improving the pulmonary edema and may worsen the volume status as is discussed later.

Hypovolemia caused by extensive extravascular fluid losses is a major clinical issue in severely exposed individuals. It is essential to assess the patient's volume status with fluid repletion as dictated by the overall clinical picture. Suctioning may be needed to maintain a clear airway if secretions are copious as may occur when edema is severe. The use of pressors may be needed.

Medical means of managing bronchospasm may include beta agonists or theophylline. Intravenous (IV) steroids may be of value as well by reducing the inflammatory response, but this recommendation is not evidence based. There is some evidence in animal studies that N-acetylcysteine and other medicines that modify leukotriene production, (e.g., montelukast) may be of value. Nonsteroidal anti-inflammatory drugs (NSAIDs) may reduce fluid volume in the lung. Supplemental oxygen is a mainstay of therapy with exposure to each of these agents.

Strict bed rest is vital after any exposure as physical exertion can shorten the time until symptom onset and increase the severity of the symptoms. Sedation as a means of inducing rest is not recommended, however, unless the patient is intubated. Atropine, barbiturates, stimulants, and antihistamine are contraindicated in the management of toxicity from pulmonary agents.

The onset of symptoms for the peripherally acting chemical agents may be delayed as much as twenty-four to seventy-two hours. Identifying the window of exposure for the patient or emergency personnel is important. As is the case with asymptomatic firefighters presenting with smoke exposure, it is medically appropriate to observe patients with definite exposure for twenty-four hours, even if asymptomatic. If there is any question about exposure, patients without signs or symptoms twelve hours after the possible exposure are unlikely to need medical care. Conversely, patients who have respiratory symptoms within six hours of exposure have a poor prognosis and are candidates for intensive care observation. If respiratory symptoms present within twelve hours of exposure, the patient should be monitored carefully.

TABLE 23–4   Quick Reference Guide: Pulmonary Agents

Diagnostic Keys	Lacrimation, cough, dyspnea, wheeze, crackles
Risk Factors	Patients with asthma COPD; and children
Transmittable	No, if decontaminated
Transmission	Contact with liquid droplets, if not decontaminated
Management	Bronchodilators, steroids, O_2 supplementation, supportive care as needed
Prevention	
Preexposure	None
Postexposure	Medical observation for up to 72 hr
Containment	Patient: none: just decontamination
	Secondary contacts: unnecessary if decontamination has occurred
Populations at Heightened Risk	Pulmonary co-morbidities, CAD
Long-term Issues	Reactive airways disease, impaired lung function
Contacts	Local, state DPH, CDC, FBI

Summary

Pulmonary agents act primarily as direct irritants to the upper and lower respiratory tract. The location of their action is contingent largely on solubility and reactivity. Acute effects range from mild irritation to fulminant ARDS and asphyxiation. Clinical clues include cough, laryngitis, dyspnea, and wheezing, and if sufficient hypoxia is present, cyanosis and changes in sensorium. Treatment is largely supportive, and delayed ARDS is common. Steroids may prevent delayed fibrosis associated with specific agents, such as oxides of nitrogen. Children, individuals with chronic lung disease, and heavy physical exertion may all increase the risk of injury due to increased respiratory rates. Chronic respiratory sequelae, such as ARDS, asthma, and accelerated ventilatory loss, are common.

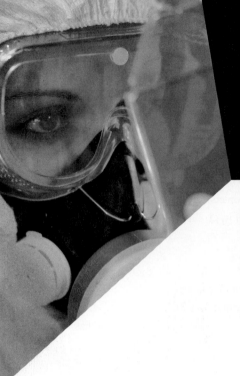

CHAPTER 24

Vesicants or Blistering Agents

Clinical Vignette

A 40-year-old Kurdish émigré presents to your office accompanied by his daughter who translates for him. As a teenager, his village was gassed by the Iraqi military. He developed diffuse skin erythema over most of his body, profound eye pain and photophobia, and blepharospasm. No blisters developed, but within several hours he noticed mild upper airway irritation and a dry cough. He did not have any shortness of breath. Within a few days, he noticed his skin became dark and some areas desquamated. In the following six months he had recurrent respiratory and skin infections, but he has been otherwise well for the past twenty-four years. His exam is significant for patchy areas of hyperpigmentation.

There are two major classes of vesicants: the mustards and the arsenicals. Each class is discussed using a prototypical agent. A third agent, often classified with the vesicants, is phosgene oxime. This compound is poorly understood, but a brief discussion is warranted and can be found at the end of this chapter. Vesicants are more "strategic" in so far as they tend to injure rather than kill (compared to other classes of chemical weapons.) They cause injury by blistering whatever part of the body is exposed—internal or external. They may be released via bombs or aerosolized through sprays.

☣Sulfur Mustards

Background

Mustards were identified in the early to mid-1800s, while arsenicals were developed in the United States during WWI. Although first used in 1917, three years into WWI and three years after the military use of chlorine gas began, mustards resulted in more WWI casualties than all the other chemical weapons combined—accounting for upward of 80% of all chemical agent causalities. Vesicants have been used since WWI, most notably by Iraq during the 1980s. The Iran–Iraq war inflicted an estimated 50,000 casualties, mostly against Kurdish Iraqis in northern Iraq. At present, no antidote exists for mustard gas. This is unfortunate as nearly a dozen countries maintain mustard as part of their cadre of chemical agents. In contrast to their use as chemical weapons, members of this class have found use as chemotherapy for certain types of cancers.

Properties

Mustards, including the nitrogen and sulfur types, are thought to be named for the pungent odor or from the yellow-brown color. They are among the most persistent of the chemical agents. Historically and militarily, sulfur mustards have played a bigger role and are felt to be more of a threat as a weapon of terror. Nitrogen mustards are considered battlefield agents and so are less likely to be encountered by the public (Fig. 24–1), but are discussed briefly as well.

The chemical structure of mustard results in a highly reactive sulfur moiety, which may account for its toxicity (Fig. 24–2). Despite having a characteristic smell, odor is not felt to be a reliable means of detecting mustards. Mustards are oily liquids at room temperature because of their high boiling point, but as ambient temperatures rise, vapors can form. Mustard vapor can linger for days and, because of its density, it accumulates in low-lying areas. The LD_{50} for liquid mustard is

FIGURE 24–1　Gas shells exploding on WWI frontlines.
Courtesy of the World War I Document Archive.

TABLE 24–1 Summary of Vesicant Properties[a]

Properties	Impure Sulfur Mustard (H)	Distilled Sulfur Mustard (HD)	Phosgene Oxime (CX)	Lewisite (L)
Chemical and Physical				
Boiling Point	Varies	227°C	128°C	190°C
Vapor Pressure	Depends on purity	0.072 mmHg at 20°C	11.2 mmHg at 25°C (solid) 13 mmHg at 40°C (liquid)	0.39 mmHg at 20°C
Density:				
Vapor	Approx 5.5	5.4	<3.9?	7.1
Liquid	Approx 1.24 g/mL at 25°C	1.27 g/mL at 20°C	ND	1.89 g/mL at 20°C
Solid	NA	Crystal: 1.37 g/mL at 20°C	NA	NA
Volatility	Approx 920 mg/m³ at 25°C	610 mg/m³ at 20°C	1,800 mg/m³ at 20°C	4,480 mg/m³ at 20°C
Appearance	Pale yellow to dark brown liquid	Pale yellow to dark brown liquid	Colorless, crystalline solid or a liquid	Pure: colorless, oily liquid As agent: amber to dark brown liquid
Odor	Garlic or mustard	Garlic or mustard	Intense, irritating	Geranium
Solubility:				
In Water	0.092 g/100 g at 22°C	0.092 g/100 g at 22°C	70%	Slight
In Other Solvents	Complete in CCl₄, acetone, other organic solvents	Complete in CCl₄, acetone, other organic solvents	Very soluble in most organic solvents	Soluble in all common organic solvents
Environmental and Biological				
Detection	Liquid: M8 paper Vapor: CAM	Liquid: M8 paper Vapor: CAM, M256A1 kit, ICAD	M256A1 ticket or card	Vapor, M256A1 ticket or card, ICAD
Persistence:				
In Soil	Persistent	2 wk? 3 yr	2 h	Days
On Material	Temperature-dependent; hours to days	Temperature-dependent; hours to days	Nonpersistent	Temperature-dependent; hours to days
Skin Decontamination	M2581 kit Dilute hypochlorite Water M291 kit	M258A1 kit Dilute hypochlorite Soap and water M291 kit	Water	Dilute hypochlorite M258A1 kit Water M291 kit

TABLE 24–1   Summary of Vesicant Properties[a] (Continued)

Properties	Impure Sulfur Mustard (H)	Distilled Sulfur Mustard (HD)	Phosgene Oxime (CX)	Lewisite (L)
Biologically Effective Amount:				
Vapor (mg? min/m^3)	LCt_{50}: 1,500	LCt_{50}: 1,500 (inhaled) 10,000 (masked)	Minimum effective Ct: approx 300; LCt_{50}: 3,200 (estimate)	Eye: <30 Skin: approx 200 LCt_{50}: 1,200– 1,500 (inhaled) 100,000 (masked)
Liquid	LD_{50}: approx 100 mg/kg	LD_{50}: 100 mg/kg	No estimate	40–50 mg/kg
Onset of Pain:	Hours later	Hours later	Immediate	Immediate
Onset of Tissue Damage	Immediate; onset of clinical effects is hours later; fluid-filled blister	Immediate; onset of clinical effects is hours later; fluid-filled blister	Seconds; solid wheal	Seconds to minutes; fluid-filled blister

CAM: chemical agent monitor
ICAD: individual chemical agent detector
LD_{50}: dose that is lethal to 50% of the exposed population (liquid, solid)
LCt_{50}: (concentration? time of exposure) that is lethal to 50% of the exposed population (vapor, aerosol)
NA: not applicable
ND: not determined

[a] Modified from Textbook of Military Medicine: Medical Aspects of Chemical and Biological Warfare, Chapter 7.

approximately 1.5 teaspoons, enough liquid to cover up to 20% of the body. Mustard vapor's LCt_{50} is considerably lower and represents a more potent risk of injury than many other chemical agents (Table 24–1). Mustards are vulnerable to water, which dissolves this class of compounds rendering them harmless; the warmer the water, the faster the rate of destruction.

Sources

Commercial production of mustards does not occur in the United States. Individuals involved in securing, transporting, or destroying existing military

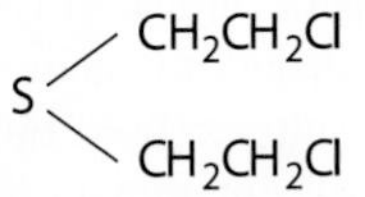

FIGURE 24–2   Chemical structure of sulfur mustards (H, HD): CH$_3$NO$_3$.

stockpiles may be exposed accidentally. Exposed individuals may inadvertently carry the oily compound home on contaminated clothing, although this is unlikely. Individuals involved in laboratory and toxicologic testing of mustards represent another potential at-risk group. The general public may be exposed to sulfur mustard at hazardous waste sites that contain or are involved in the destruction of sulfur mustard. Mustards do not occur in nature. Water contamination is unlikely as mustard is rapidly hydrolyzed and rendered harmless.

Pathophysiology

It is not known with certainty how mustard causes cellular damage. This partly explains why no effective therapy exists. Two theories presently exist regarding the mechanism of injury. The most commonly accepted theory is that mustards act by creating reactive intermediary compounds that alkylate DNA, RNA, proteins, and other macromolecules in the cell, thereby inhibiting normal glycolysis, breaking DNA strands, and activating apoptosis. A second hypothesis is that mustard consumes intracellular glutathione, one of the prime cellular defense mechanisms against free radicals. Such injury leads to a loss of homeostatic mechanisms—calcium regulation and membrane integrity, in particular. Soon thereafter, cell death occurs. These agents are referred to as radiomimetic because they mimic the clinical effects of ionizing radiation. Though both theories have some basis in animal studies, neither sufficiently explains the histopathologic changes seen in mustard victims. What is certain is that mustard has its greatest insult on rapidly dividing cells (hence, the therapeutic role as chemotherapy). The effects occur within a few minutes of tissue contact, and its presence in the body quickly becomes undetectable. Such rapid reactivity and elimination is important to keep in mind because it means that blood, skin, blisters, or blister fluid are not contamination risks to those aiding the victims.

The skin blistering caused by mustards is thought to result from two effects: damage to the dermal proteins that anchor the epidermal layer and liquefaction necrosis of the deeper cell layers of the skin. The fluid within the blister is a consequence of an inflammatory response to the cellular damage and does not contain mustard itself.

Signs and Symptoms

The organ systems affected by mustards are those that come into direct contact with the liquid or vapor: such as the eyes, skin, or respiratory system (Table 24–2). More severe exposures may result in hematologic, nervous, and gastrointestinal toxicity. A distinctive clinical aspect of mustard gas exposure is a latent period before the onset of injury. Tissue damage can occur without immediate symptoms of burning, pain, or itching. As a result, exposed but asymptomatic individuals may not take necessary avoidance or decontamination steps. Only the mustard agents are initially pain free—arsenicals and phosgene oxime (both discussed later) are painful immediately.

TABLE 24–2	Making the Diagnosis: Blistering Agents
Pattern Identification	Multiple, concurrent cases of erythematous, blistering skin
Features	Skin—spectrum from erythema to bullae Eyes—conjunctivitis, ophthalmitis, lacrimation, irritation Respiratory—cough, hoarseness, SOB, ARDS GI—nausea, vomiting
Findings	Blistered skin, pulmonic involvement, rhinitis, lacrimation Mild to moderate exposure—leukocytosis Severe exposure—pancytopenia. ARDS, death

Skin

Sites most prone to injury are those that tend to be warm and moist: neck, antecubital fosse, and perineum. With mild exposure, skin erythema, similar to that seen with sunburn occurs and is associated with burning or itching. The area may be mildly edematous as well. More commonly, skin exposure results in the formation of blisters that begins with the formation of multiple, pinpoint vesicles on the periphery of the site that coalesce into bullae, ranging in size from 0.5 to 5.0 cm and usually sitting on an erythematous base. The wall of the blisters are typically thin and translucent and contain fluid that initially is clear but later turns yellow and coagulates. Mustard agents are not present in the blister fluid. The extent and severity of the skin injury is a consequence of the intensity of exposure and of the skin sensitivity of the individual. Severe exposure may cause centrally necrotic lesions (Fig. 24–3). Mustard vapor is less toxic to the skin and more likely to result in first- or second-degree burns, whereas liquid mustard is more likely to cause severe skin injury, that is, third-degree burns, and often develops quickly (Fig. 24–4).

As noted previously, mustard's clinical impact occurs after a latency period lasting from a few hours to a day. A distinguishing feature is that during this latency period despite cellular insult, the patient feels little pain and indeed may be altogether unaware of exposure. This lack of awareness may prevent the exposed individual from leaving a contaminated area or taking appropriate decontamination procedures. The latency period is followed by the emerging skin erythema, and, if blisters follow they often tear, or become encrusted (Fig. 24–5). Reepithelization occurs as it would after any kind of burn, with erythema resolving within a few days and uncomplicated blisters (i.e., without secondary infections) resolving within a month. As with other burns, the extent of skin involvement determines mortality rates. Burns and blisters covering more than 25% of the body are associated with higher mortality rates.

Eyes

Mustards affect the eyes at concentrations 100-fold lower than that needed to generate skin blistering. Ocular signs, therefore, are a useful early warning sign for

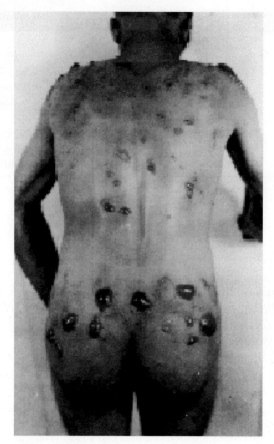

FIGURE 24–3　Mustard induced blisters in truncal distribution.
Courtesy of Treatment of Chemical Agent Casualties and Conventional Military Chemical Injuries;
Departments of the Army, the Navy, and the Air Force, and Commandant, Marine Corps.

mustard exposure. The latency period for ocular injury is also shorter, ranging from as little as a few minutes to an hour. Eye injury includes conjunctivitis, corneal damage, pain, lacrimation, photophobia, and blepharospasm. Mustard appears to have some cholinergic properties as miosis may be noted. Blistering of the eye does not occur. The intensity and severity of ocular signs and symptoms are dose dependent. Visual impairment due to corneal edema and clouding caused by the in-migration of neutrophils may also be seen. Recovery from ocular injury is variable, but symptoms may persist for more than several weeks. Direct eye contact with mustard liquid is the most potentially serious eye exposure, causing generalized inflammation of the eye with possible perforation of the cornea. These patients are at risk for loss of the exposed eye.

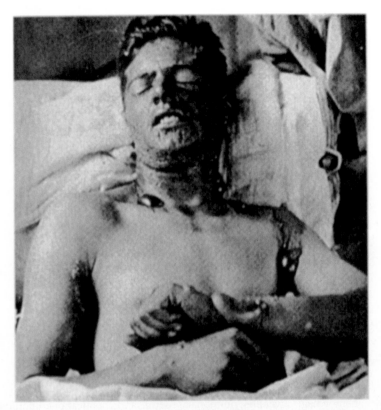

FIGURE 24–4 Mustard gas victim from WWI with diffuse blistering over his upper body.
Courtesy of Malaspina University/Department of History.

Respiratory

Injury to the respiratory system typically occurs following exposure to mustard vapor and is invariably accompanied by ocular findings. Milder exposures preferentially affect the upper respiratory tract, with sinusitis, pharyngitis, or epistaxis. As exposure intensifies, more distal structures such as the larynx, trachea, and bronchi are affected. Hoarseness that progresses to aphonia is a classic feature of sulfur mustard exposure, and a dry, barking cough, chest tightness, and airway hyperreactivity may develop. Irritation of the respiratory epithelium causes a sterile toxic bronchitis, with clinical signs such as productive sputum, fever, and leukocytosis. Later, the affected epithelium sloughs off with pseudomembrane formation, fibrinous exudate, and, possibly, necrosis. The latter puts the victim at risk for airway obstruction. Sulfur mustard has somewhat of a dose-dependent latency period, ranging from four to six hours. More severe exposures intensify the milder clinical findings and shorten the latency period. Severe exposure may result in laryngospasm,

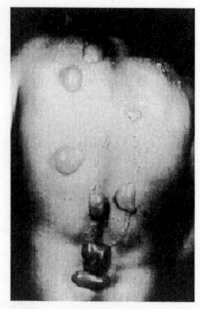

FIGURE 24–5 Mustard gas-induced blisters.
Courtesy of Treatment of Chemical Agent Casualties and Conventional Military Chemical Injuries;
Departments of the Army, the Navy, and the Air Force, and Commandant, Marine Corps.

dyspnea, hemorrhagic alveolitis, pulmonary edema, and chemical pneumonitis, the latter usually occurring within twenty-four hours if at all. Secondary infection may occur at any point, aided in part by the immunosuppressive effects of the chemical. Early mortality with mustard is due to laryngospasm and airway obstruction. Delayed mortality is usually caused by secondary respiratory infections. As a general rule, early onset of respiratory symptoms confers negative prognosis.

Gastrointestinal

Transient and mild nausea and vomiting are common with mustard exposure, possibly due to stress or to the cholinergic properties of mustard agents rather than to direct toxicity. Direct toxicity to the gastrointestinal tract is uncommon, but can occur with more intense exposures, causing nausea, vomiting, diarrhea, or constipation beginning several days after exposure. The prognosis in such a case is poor.

Central Nervous System

The signs and symptoms of CNS involvement are highly nonspecific. Lab animal studies indicate that severe exposures may cause seizures.

Hematologic

Severe mustard gas exposure causes bone marrow suppression, and the resulting pancytopenia puts victims at risk for infection. When necrosis and involution in the bone marrow and other reticuloendothelial structures (such as the spleen, thymus, and lymph nodes) occurs, it augurs a poor prognosis.

Chronic Effects

During the Iran–Iraq War, Iranian mustard gas victims noted transient hyperpigmentation of the affected skin that often resolved after a few months. Chronic hyperpigmentation, scarring, and wound carcinomas were seen in WWI gas attack survivors. Severe ocular exposures may result in corneal opacities or chronic conjunctivitis, and scarring of the iris or lens may impair vision and place victims at increased risk of glaucoma. Delayed keratitis has been described. Chronic respiratory effects, including chronic bronchitis and reactive airways, may result from severe acute exposures. Further, munitions workers involved in the production of mustards during WWI showed evidence of increased lifetime risk of cancer of the airways. Nitrogen mustards are used as chemotherapeutic agents and are considered mutagenic. Sulfur mustards have not been associated with reproductive cancers or other cancers other than those arising from the site of direct toxicity. Localized chromosomal injury may be the reason for the increased incidence of airway and skin cancer seen with exposure to mustards. Lingering psychological affects, such as depression and anxiety, are considered posttraumatic stress responses and have been documented. Both human epidemiologic and animal data support mustard agents as genotoxic, mutagenic, and teratogenic.

Vulnerable Groups

The immunosuppressive effects of vesicants poses particular risk to individuals with preexisting immunosuppression or chronic medical conditions prone to infections, such as those with chronic respiratory conditions. Children and adults with chronic respiratory conditions (e.g., asthma) or immunodeficiency disorders and geriatric patients with age-related immunosuppression may also be at greater risk. Individuals with preexisting dermatologic conditions, especially those conditions that are immunologic mediated, may experience exacerbation of their underlying disorders. Toxicologic studies have demonstrated that nitrogen mustards are fetotoxic, making pregnant women and unborn children particularly vulnerable. Children may have unique susceptibilities because of their large surface-area-to-volume ratio and lower height.

Differential Tests

The prominence of skin findings with vesicants reduces greatly the differential diagnosis, particularly in instances of a known chemical attack. Diagnosing index or early cases where multiple bullae are present must include consideration of T-2 myco-toxin, as well as infectious causes such as staphylococcal scalded skin syndrome. Toxic epidermal necrolysis, other allergic drug responses (e.g., erythema multiforme or Stevens-Johnson syndrome), and severe contact dermatitides, such as these from poison ivy or poison sumac, are also in the differential. Bullous autoimmune disorders—bullus pemphigus vulgaris or bullous pemphigoid—and chemical burns from strong irritants, such as acids and caustics, are also considerations.

Laboratory Tests

There are no laboratory tests specific enough to diagnose mustard gas. Confirmatory testing is possible through gas chromatographic monitoring, but such specialized testing is available only to those involved in civilian hazmat or military biodefense teams and not possible at most local health departments, clinics, or hospitals. In the first hours after exposure, a leukocytosis proportional to the severity of exposure begins. This leukocytosis wanes starting on the third day. A significant systemic exposure may cause pancytopenia, and an absolute neutrophil count below 500 cells per cubic millimeter (mm^3) indicates a poor prognosis. With severe vesicant exposure, renal function and serum chemistries will need to be followed to address fluid and electrolyte abnormalities. Monitoring of oxygenation and serial CXR are indicated if significant respiratory symptoms and signs are present.

Biological monitoring for thiodyglycol, a urinary metabolite of mustard, is feasible, but this test is not widely available, and its role in diagnosis and management is uncertain. Portable field environmental testing with chemical detection devices is possible for mustards, as with other chemical agents.

Decontamination

Decontamination begins with making sure that patients and health care workers and emergency personnel are away from the source of exposure and qualified hazmat personnel and other officials are contacted. Decontamination must be achieved swiftly in order to have any benefit (Table 24–3). Contaminated objects and clothes can be made safe with boiling or can be stored safely in doubled plastic bags or a rubber container for later destruction. Affected areas should be cleaned as soon after exposure as possible, but even delayed washing is warranted to limit ongoing exposure. Skin sites should be flushed generously with soapy water or

 TABLE 24–3   Critical First Steps: Vesicants

Decontamination—removal of clothes followed by thorough cleaning with soap and water

Medical measures—supportive

Exposure control—decontamination

with a 0.5% sodium hypochlorite solution and water. Affected eyes should be rinsed generously with saline.

Treatment

No clinical trials or standards of care are established for vesicants. No antidotes exist and care is supportive based on organ system involvement and severity of exposure. Severe exposures should prompt the implementation of standard burn care and supportive care such as is available in burn units or ICUs. Total burn surface areas exceeding 20% are associated with greater than 50% mortality. Early airway or respiratory tract symptoms are bad prognostic markers. Airway symptoms within 6 hours of exposure strongly predict mortality. Early lymphopenia indicates substantial bone marrow toxicity, and the risk of superinfection and death. Bone marrow transplantation or the use of marrow proliferative drugs such as granulocyte colony-stimulating factor (GCSF) or erythropoietin are untested treatment modalities but may have utility.

Skin

For simple exposures, calamine or other soothing agents (i.e., 0.25% camphor, menthol) may be used. Blisters/bulla less than 1 cm in size should be left alone, whereas those larger than 1 cm should be deroofed and treated with topical antibiotics every six to twelve hours. Sterile petrolatum may be used if no antibiotic ointment is available. Adequate pain control should be maintained. Whereas fluids and electrolytes should be monitored, this is less of a concern here than for thermal burns; in fact, overhydration should be avoided.

Eyes

By the time a patient is seen for eye exposure to mustard agents, it is unlikely to be present on the conjunctival surface. Irrigation with saline is still warranted to flush away any residual mustard, as well as foreign bodies, such as eyelashes or dirt. Pain should be controlled with systemic analgesia. Lubrication and soft lighting will help with pain and photophobia. Topical steroids are of uncertain value after the first forty-eight hours, and topical anesthetics should be avoided. Topical antibiotics are used to prevent secondary bacterial infection. Topical atropine will reduce postinflammatory scarring and blepharospasm. Normal vision returns anywhere from days to months, depending on the severity of the ocular injury. Due to mustard's potential for serious acute and long-term consequences, early consultation by an ophthalmologist is suggested.

Respiratory

Mild upper airway symptoms (pharyngitis, sinusitis) can be treated with moist air and cough suppressants. As discussed earlier, a sterile pneumonitis may develop that can seem infective because of a transient leukocytosis and fever. However, antibiotic therapy should be avoided since infection is unlikely in the first seventy-two hours. Infection may be likely if the sputum becomes purulent, fevers persist or worsen, or if radiographs show increasing infiltrates. Antibiotics are appropriate provided sputum cultures and gram staining are done.

Continuous positive airway pressure (CPAP) or intubation should be decided according to the clinical picture, but may be necessary if laryngospasm, edema, or airway obstruction (during the sloughing of epithelium) occurs.

Steroid therapy has not been shown to improve bronchospasm, but bronchodilators may be effective. The brisk inflammatory response and cytoxicity initiated by vesicant's tissue damage may be favorably influenced by early use of NSAIDs.

Gastrointestinal

Nausea and vomiting may be treated with antiemetics. Atropine subcutaneously (0.4–0.6 mg) may be useful for GI symptoms secondary to systemic effects of mustard agents.

Bone Marrow

Bone marrow transplantation/transfusion may be of benefit because of the short half-life of mustard (minutes) after its absorption and thus should not harm the new cells. Sodium thiosulfate has been shown to elevate the LD_{50} as well as decrease systemic effects in animal models when given either before exposure or within twenty minutes of exposure. Dialysis has been shown to be of no benefit and may in fact be detrimental. Activated charcoal has been shown to be ineffective.

Background

Developed in the 1930s as potential chemical weapons, nitrogen mustards derive from the same class of compounds as sulfur mustards (Fig. 24–6). Nitrogen mustards have never been used as weapons, in large part because they would not be effective. Their only use, in fact, has been as a chemotherapeutic agent primarily with the agent mechlorethamine (the M in MOPP therapy for treating Hodgkin's lymphoma) (Fig. 24–7). Nevertheless, the possibility exists that a group or individual might use nitrogen mustard if it were accessible, so a brief discussion is provided.

Properties

Nitrogen mustards are oily, clear yellow liquids with a musty, fishy, or fruity smell, depending on the specific agent. Unlike other chemical agents, including sulfur mustards, nitrogen mustards are relatively water insoluble and hydrolyzed intermediates maintain toxicity even in water.

Nitrogen Mustard (HN-1)

$(C_6H_{13}C_{12}N)$

Nitrogen Mustard (HN-2)

$(C_5H_{11}C_{12}N)$

Nitrogen Mustard (HN-3)

$(C_6H_{12}C_{13}N)$

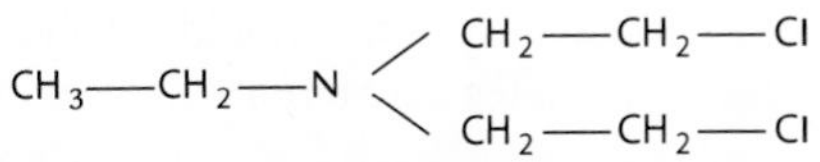

$$CH_3 - CH_2 - N \left\langle \begin{array}{l} CH_2 - CH_2 - Cl \\ CH_2 - CH_2 - Cl \end{array} \right.$$

FIGURE 24-6   Nitrogen mustard (HNs): N = CRC₁.

Sources

Nitrogen mustards have found limited medical use. They are not part of the U.S. chemical weapons stockpile and to date have not been used as battlefield chemical weapons. Laboratory workers doing toxicologic and treatment research may be exposed accidentally.

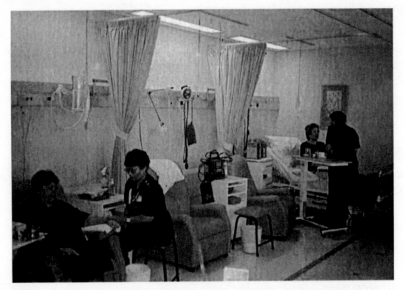

FIGURE 24-7   Nitrogen mustard is used as part of the MOPP chemotherapy regimen.
Courtesy of Southern Cross Wynberg Hospital.

Pathophysiology

The mechanisms of injury from nitrogen mustards are poorly understood but are probably analogous to those summarized for sulfur mustards.

Signs and Symptoms: Eyes

Injury to the eye occurs with significantly lower doses than would be required to affect other organ systems. Although sulfur injury can take up to an hour to present, nitrogen mustard exposures within minutes cause pain and erythema but may not peak for upward of ten hours. Injury is generally worse than that caused by sulfur mustard. With more severe exposures, ocular hemorrhaging and necrosis is seen with resultant blindness. Milder exposure should resolve fully in ten to fourteen days, whereas more severe exposures may require upward of four months to heal.

Signs and Symptoms: Skin

Skin pathology requires moderate to high levels of exposure and for the agent to be in liquid form. Signs and symptoms are similar to those discussed for sulfur mustards.

Signs and Symptoms: Respiratory

Signs and symptoms are similar to those discussed for sulfur mustards.

Signs and Symptoms: Hematopoietic

Once absorbed, nitrogen mustards, like their sister agent, cause pancytopenia, including leukopenia, neutropenia, anemia, and thrombocytopenia. This is due to damaged stem cells in the bone marrow and direct injury to the reticuloendothelial and lymphatic systems (hence the use of mechlorethamine in Hodgkin's lymphoma treatment). Such manifestations are indicative of high levels of exposure and are associated with poor outcomes.

Signs and Symptoms: Central Nervous System

Like the sulfur mustards, nitrogen mustards' effects on the CNS are not well described but may include seizures, ataxia, or mental status changes.

Treatment: Eyes

See management of sulfur mustard-induced eye injury.

Treatment: Skin

Treatment is as with sulfur agents, but decontamination is especially useful with nitrogen mustards since absorption occurs more slowly than with sulfur mustards.

Treatment: Respiratory

Treatment is similar to that as discussed for sulfur mustards.

Treatment: Hematopoietic

Transfusions should be given as clinically warranted. Use of colony-stimulating factors such as erythropoieten, pegfilgrastim, and filgrastim has not been studied in this contest but may be helpful. Antibiotics should be considered based on lymphocyte and neutrophil counts, or documented infection.

Treatment: Central Nervous System

Management is primarily supportive in nature.

Chronic Effects

Nitrogen mustards are potential teratogens and are linked to decreased fertility in humans. Bone marrow toxicity and immunosuppression also occur, as is true with sulfur mustards.

Decontamination

Decontamination measures are similar to those described for sulfur mustards.

☣ Vesicants Part II: Arsenicals

Background

The American military developed the arsenicals during WWI, although they were not used at the time and clinical experience is scant. It is considered by many to be of secondary importance as a terrorist or military threat; however, arsenicals are readily available and many nations stockpile them. For this reason, a brief discussion of the distinctive features of the prototypical arsenical, lewisite, is warranted. Though very similar in properties, presentation, and management to the mustards, arsenicals do have some distinctive features.

Properties

Named after the American chemist who discovered it, Winford Lee Lewis, lewisite (Fig. 24–8) is an oily, clear, or brown liquid with a geranium-like odor (although, again, odor is an unreliable means of detection). Arsenical vapors are more volatile than mustards and are much less toxic when volatilized than when in liquid form.

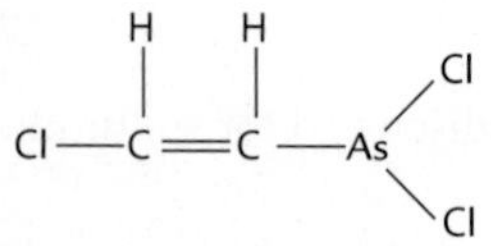

FIGURE 24–8 Lewisite (β-chlorovinyldichloroarsine (CH)$_2$AsCl$_3$).

Like the mustards, blisters caused by lewisite exposure do not contain the agent. Lewisite is effectively neutralized when mixed with water.

Sources

Lewisite is not produced in the United States, but it is stored at least in one military facility. The Chemical Weapons Convention requires the United States to destroy existing stockpiles of lewisite by 2007. Arsenic is used widely in industry as chemical intermediates in laboratories and hospitals. Pesticide manufacturing and metal mining and smelting are common industrial sources. These industrial processes are widely used throughout the world and can be directed with relative ease into the production of arsenical chemical agents. Outside the United States, arsenicals have been found as environmental contaminants.

Vulnerable Populations

Military or civilian contractors involved in securing, transporting, or destroying stockpiles of arsenicals may be exposed through accidents or deliberate sabotage.

Mechanism of Injury

The mechanism of lewisite toxicity is unknown, although glutathione depletion and reactivity with intracellular enzymes containing sulfhydryl groups are postulated.

Signs and Symptoms

The crucial distinguishing feature between lewisite and mustard agents is that lewisite causes immediate pain. With the more potent liquid forms, arsenols result in blisters and skin burning, whereas vapors cause immediate respiratory irritation. The immediacy and severity of lewisite's effect has a benefit insofar as victims will be aware of having been exposed and can minimize further exposure by leaving a contaminated area. Lewisite increases capillary permeability and consequently substantial fluid extravasation occurs. Third spacing of extravasated fluid decreases effective circulating volume causing hypovolemia, shock ("lewisite shock"), and organ hypoperfusion. In contrast to mustard, lewisite does not cause bone marrow suppression. These important distinctions aside, the clinical presentation management of mustards and arsenicals are quite similar (Fig. 24–9).

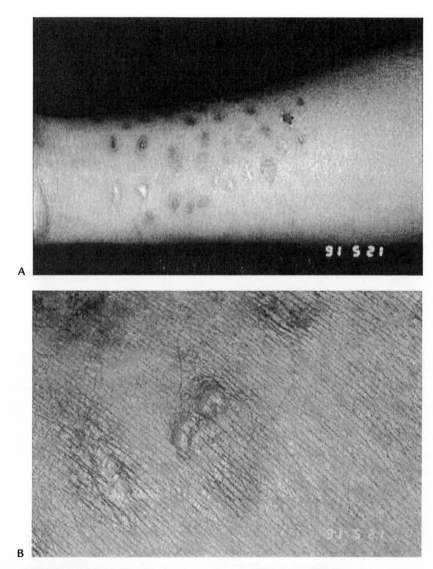

FIGURE 24-9 Both photos are of a Gulf War soldier from 1991 thought to have been exposed to lewisite.
Courtesy of NIOSH/CDC.

Laboratory Tests

In the proper clinical or epidemiologic context, urinary excretion of arsenic is a useful marker of exposure, but its value to estimate dose or to prognosticate is unknown. Several field detection devices are available for identification and environmental monitoring for various chemical weapon agents, including arsenicals. Colorimetric devices are also available (see Appendix).

Differential Diagnosis

Arsenical burns can be distinguished from mustard burns by the fact that pain is noticed instantly. The differential diagnosis is similar to that of mustard agents.

Chronic Effects

Knowledge of the long-term effects of lewisite are limited by the scantiness of direct human experience. Toxicologic studies suggest that lewisite has mutagenic, carcinogenic, and teratogenic effects. Lewisite is a sensitizer and can cause asthma.

Based on lewisite's acute clinical effects, chronic ocular and respiratory sequelae are biologically plausible, but once again firm observations cannot be made due to the absence of human experience.

Decontamination

See mustard agents.

Treatment

Treatment is similar to mustards with one important difference: a specific treatment for lewisite exists. British-Anti-Lewisite (BAL or dimercaptrol) was invented by British scientists during WWI as a means of protecting allied soldiers. BAL chelates arsenic (and other heavy metals), but must be administered either topically or intramuscularly to be effective. BAL has little impact on the ocular, dermal, or respiratory symptoms of lewisite but may forestall some systemic effects such as diarrhea. Side effects of intramuscular BAL, such as hypertension and tachycardia, can be severe. The role of BAL in treating lewisite exposure is disputed. Some experts recommend that BAL be used only in patients with severe respiratory symptoms or shock. Others recommend it for even moderate exposures. There is general agreement that BAL is beneficial only if used early on. Topical BAL ointment has been shown to be effective for treating ophthalmologic and skin manifestations, but it is no longer available commercially.

☣Phosgene Oxime (CX, Dichloroformoxime, $CHCl_2NO$)

Background

Phosgene oxime is often grouped with the vesicants, though it does not cause blisters. More properly, it is referred to as a corrosive, urticant, or nettle agent. Phosgene oxime's primary value as a chemical weapon is its ability to breach protective gear,

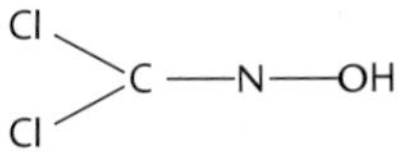

FIGURE 24-10 Chemical structure of phosgene oxime
(CX, dichloroformoxime, $CHCl_2NO$).

rapid absorption, and near instantaneous clinical manifestations. Phosgene oxime should not to be confused with the pulmonary irritant gas phosgene ($COCl$ or CG; see Chapter 23).

Properties

Phosgene oxime (Fig. 23–10) may exist in solid, liquid, and vapor form. At ambient temperatures, volatilization occurs and the resultant vapor is denser than air. As is true with the other agents discussed in this chapter, phosgene remains close to the ground. Phosgene oxime is not very biopersistent, degrading quickly in typical soil conditions and within a few days in water.

Sources

Phosgene oxime has never been known to be used on a battlefield and has virtually no role in commercial processes.

Mechanism of Injury

Although not known with any certainty, the clinical effects of phogene oxime are thought to be due to direct toxic effects from the chlorine or oxime groups, or indirectly through inflammatory mediator-induced tissue injury. Corrosive findings similar to those seen with acid burns are the hallmark of phosgene. Compounding phosgene oxime with other chemical agents (e.g., VX gas) potentiates the clinical effects of the second agent because of the dermal compromise caused by phosgene oxime.

Signs and Symptoms

Phosgene oxime causes immediate and intense clinical manifestations on surfaces with which it comes into contact: the upper and lower respiratory tract, the skin, and the eyes. The rapid onset of pain makes phosgene more like lewisite. Pain, edema, and the rapid onset of localized tissue necrosis (likened to a corrosive acid) is the dermal sin qua non of phosgene oxime exposure. Ocular exposure results in immediate pain, chemical conjunctivitis, and keratitis. Respiratory distress and acute noncardiac pulmonary edema results from either systemic absorbtion or inhalation of phosgene oxime. Parenchymal necrosis and pulmonary thrombosis may accompany acute pulmonary edema. Based on lab animal studies, ingestion of phosgene oxime may result in hemorrhagic gastroenteritis.

Chronic Effects

Phosgene oxime has yet to be used as a battlefield weapon, and long-term clinical effects are unknown. Toxicologic studies of phosgene oxime are limited to acute studies. Chronic effects, carcinogenicity, mutagenicity, or teratogenicity are unknown. No data from the manufacturer of phosgene oxime are available.

Treatment

Phosgene oxime's rapid local and systemic absorption negatively impacts clinical management and decontamination. Decontamination is nearly worthless unless performed immediately and scrupulously. Treatment of necrotic skin lesions and eye injury are analogous to burns and corrosive injuries, respectively. The pulmonary syndrome is managed supportively.

Laboratory Tests

No specific tests for phosgene oxime are available. Diagnostic testing is driven by clinical manifestations. For example, early chest radiographs and arterial blood gas testing are warranted with vapor inhalation.

Differential Diagnosis

Corrosives or acid burns are present.

TABLE 24–4   Quick Reference Guide: Blistening Agents

Diagnostic Keys	Skin blisters, eye irritation, respiratory symptoms; onset, immediate or delayed determines specific agents
Transmittable	No, if decontaminated
Transmission	None, if decontaminated
Management	Supportive
Prevention	
Preexposure	None
Postexposure	Decontamination
Containment	Patient: none Secondary contacts: none
Populations at Heightened Risk	Immunocompromised, elderly, young
Long-term Issues	Chronic ocular, dermal, respiratory effects, possible carcinogen
Contacts	Local, State DPH, CDC, FBI

Decontamination

Basic decontamination of clothing and skin and other contaminated objects should be undertaken immediately, although given the agent's rapid absorption, the practical impact of this recommendation on other than a war zone is debatable.

☣Summary

Blistering agents are strategic chemical weapons that cause a sublethal debilitating skin syndrome in exposed individuals. Acute effects also include upper and lower respiratory tract irritation and immunosuppression. Although human epidemiologic data are limited, blistering agents may be considered carcinogenic, mutagenic, and teratogenic. Chronic damage to tissues directly contacted by blistering agents results is common. Permanent respiratory effects, such as asthma or chemical bronchitis, have been described. Treatment requires rapid decontamination.

CHAPTER 25

Nerve and Incapacitating Agents

Clinical Vignette

A 47-year-old Japanese man new to your office arrives complaining of fatigability, lassitude, poor memory, irritability, and hypervigilance whenever he has to go to public places. As part of your standard social history, you learn that he was formerly a maintenance worker in the Tokyo subway system before emigrating to the United States in 1996. He tells you he was present during the Aum Shinrikyo attack in 1995. At that time, he experienced eye irritation, rhinorrhea, wheezing, cramps, nausea, and diarrhea. He was treated overnight at a local hospital. He has never felt the same and wonders if his current problems are related to the sarin gas attack. What do you tell him?

Background

In 1936, German scientist Gerhard Schrader discovered tabun, the first in a class of chemical weapons known as nerve agents. The discovery of a compound with similar effects, sarin, was made in 1938 (Fig. 25–1). Sarin would come into world-wide notoriety in the terrorist attack on the Tokyo underground system in March 1995 (Fig. 25–2). Nerve agents are compounds classified as organophosphates (OPs) which were originally developed for use as pesticides. The code names GA, GB, and GD, representing tabun, sarin, and soman, respectively, were standardized by the Tripartite pact after WWII. The "G" in these code names simply stands for German. Rapid analysis and trials of antidotes for these agents were begun as the

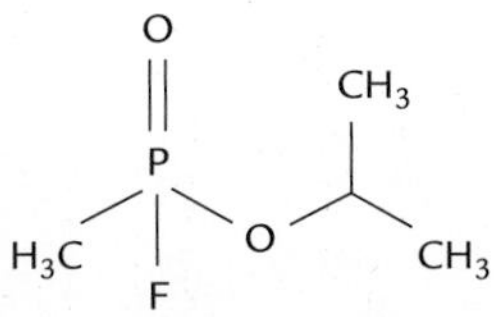

FIGURE 25-1   Chemical structure of sarin gas (GB).

Cold War loomed, though Soviet production of nerve agents began as early as 1946. As a practical matter, clinicians are far more likely to encounter occupational or environmental organophosphate poisoning than nerve agent exposure from terrorist incidents (Table 25–1). However, the knowledge gleaned is generalizable to their use as weapons.

Nerve agents are the most toxic of the chemical agents. Companies involved in the manufacturing of these or similar agents are at high risk for sabotage or theft by those wishing to exploit chemical weapons.

Properties

Nerve agents exist as liquids at room temperature and are usually diluted in an aqueous solution. The liquids vary in color from colorless to brown. Although several of these agents have odors described as "fruity," odor is an unreliable means of identifying most chemicals in either an occupational or a bioterrorist context. The hydrophilicity and lipophilicity of nerve agents are responsible for their rapid absorption through the skin, mucous membranes, and clothing.

TABLE 25-1   Quick Facts: Nerve Agents as Bioweapons

June 1994. Aum Shinrikyo releases sarin gas in Matsumoto, exposing over 300 persons, hospitalizing 56, and killing 7. First responders and health care workers also develop mild symptoms of nerve gas exposure.

March 1995. The same terrorist group releases sarin gas in the Tokyo subway, affecting 5,000 to 6,000 riders, hospitalizing nearly 500, killing 12, and causing permanent neurologic deficits in many.

Sarin gas is part of both the U.S. and Russian chemical weapons stockpiles, with 5,000 tons and 11.7 tons, respectively.

Countries suspected of stockpiling nerve agents: India, South Korea, Iraq, Syria, Egypt, Iran, Libya, and North Korea.

Saddam Hussein launched 2 major and 280 smaller chemical weapons attacks on the Kurds of northern Iraq during the 1980s using mustard gas and sarin gas.

Russian military used "knockout gas" (fentanyl) to end a hostage situation in a Moscow theater, killing 115 hostages and incapacitating 50 Chechnyan terrorists.

FIGURE 25-2   The headquarters of Aum Shinrikyo.
*Courtesy of the Department of Health and Human Services/Office of Public Health Emergency
Preparedness.*

Volatility and biopersistence varies among the different categories of nerve
agents. G agents exhibit less biopersistence compared to V agents (Table 25–2).
Chemical modification or alterations in carrier agents may affect volatility and biopersistence as well. Thickening agents, such as acrylates, increase viscosity and thus
prolong biopersistence. They also add to the risk of secondary aerosolization, given
the right environmental conditions. All nerve agents are rapidly inactivated by
strong alkalis and chlorines, a fact that has implications for clinical management
and decontamination (see Decontamination and Treatment). Nerve agent vapors
are designed to be denser than air, and if undisturbed by wind, vapors remain close

TABLE 25-2   Properties of Nerve Agents

	Odor	Biopersistence	Aging Time
G Agents			
Cyclohexyl sarin (GF)	Odorless	Persistent	>4 hr
Sarin (GB)	Odorless	Persistent	3–4 hr
Soman (GD)	Fruity odor	Persistent	2 min
Tabun (GA)	Fruity odor	Persistent	>4 hr
V Agents			
VX	Odorless	Nonpersistent	>40 hr

OF NOTE . . .

Aum Shinrikyo

After several failed attempts to release botulinum toxin in a Japanese government building, the worldwide religious cult based in Japan, Aum Shinrikyo, which preaches an apocalyptic message and whose name means Supreme Truth, levied a successful attack.

On June 27, 1994, members of the sect drove a truck into a suburban neighborhood of Matsumoto, Japan, where three of the judges presiding over a lawsuit against the sect were being housed. They released sarin gas, which reached a housing complex, killing 7, hospitalizing 200, and sending 500 people in seek of medical attention.

On March 20, 1995, Aum Shinrikyo members placed packages containing sarin gas in five different trains of the Tokyo subway system, one of the five trains service many key governmental agencies. Riders first noticed the odor of solvents and acute eye irritation, but for some this was soon followed by dyspnea, muscle weakness, and unconsciousness. Most affected individuals managed to escape the train before succumbing. Nicotinic symptoms of pallor, tachycardia, hypertension, muscle fasciculation, and muscular weakness predominated, with less prominent muscarinic effects (e.g., sweating, secretions, and bradycardia). More severely affected individuals experienced seizures. Within hours it was over: somewhere between 5,000 and 6,000 commuters were affected. Over 550 individuals were transported by ambulance, and nearly 3,000 more managed to find their own way to one of Tokyo's 41 hospitals, area urgent care facilities, or private physician offices. Four hundred ninety-three were admitted to hospitals. Seventeen required intensive care, and 12 died. The majority of those presenting for evaluation had mild symptoms and were sent home. The majority of those admitted were observed and discharged within 48 hours. All told, 12 Japanese commuters died, most within the first 24 hours. Two individuals died 2 weeks later of hypoxic brain injury, and others experienced permanent residual neurologic deficits. One hundred thirty-five ambulance drivers, police, and health care workers also developed symptoms, and 33 were hospitalized. Calls flooded Tokyo's telephone system, making it nonfunctional for hours following the attack.

When investigators located and raided the site of sarin production, they found a crude production facility capable of producing very large quantities of chemical weapons using readily available equipment and under the supervision of cult members who were themselves scientists. Evidence showed that Aum Shinrikyo was attempting to develop biological weapons, including botulinum toxin, anthrax, cholera, and Q fever. They even made efforts to acquire Ebola virus. Although arrests and greater surveillance has damaged the group, they continue to recruit members and spread its apocalyptic message.

to the ground. This property imposes greater risk of exposure to children or to adults in low-lying spots. Vaporization of nerve agents occurs when heat from an explosive device disperses its vaporization of droplets aerosolized by a sprayer. Secondary vaporization can occur from the ground or from other surfaces that have been coated with the liquid or droplets.

Sources

Short of industrial or military sabotage, or a chemical terror event, only those individuals involved in the production, transportation, or storage of nerve agents are likely to be at risk for inadvertent exposure to nerve agents. Production, storage, and transportation practices are similar to those involved with organophosphate pesticides, and individuals require full hazmat gear and work in highly controlled settings with engineering equipment and PPE to limit inadvertent exposure and industrial accidents. Storage facilities must be vigilant regarding security measures.

Pathophysiology

Nerve agents' physiologic and clinical effects mirror those of the organophosphate family of pesticides. They act by irreversibly inactivating esterase enzymes, the most clinically relevant one being acetylcholinesterase (AChE). AChE acts in the synaptic cleft throughout the peripheral nervous system, as well as in the autonomic nervous system and central nervous system (CNS). Its function is to degrade the neurotransmitter acetylcholine (ACh). Nerve agents bind covalently to AChE within seconds or minutes, causing structural changes that render the enzyme useless. After the formation of the covalent bond, an isopropyl (alkyl) group detaches from the enzyme as part of a process referred to as "aging." Once aging occurs, the enzyme is irreversibly impaired. The loss of enzymatic breakdown results in the accumulation of ACh in the synapse and hyperstimulation of muscarinic and nicotinic receptors throughout the body. This hyperactivity, as well as the direct disruption of neurons in the CNS, define the clinical presentation of nerve agents. Although inhibition of AChE is responsible for the clinical syndrome, inhibition of other esterase enzymes is of clinical value in diagnosing exposure to either organophosphates or nerve agents. Specifically, both serum and red blood cell (RBC) cholinesterase levels are useful as markers of exposure to organophosphates and nerve agents.

Toxicity

The extent and severity of the consequences from an exposure are dependent on the particular nerve agent used. VX is by far the most potent of the nerve agents currently known to exist (Table 25–3).

There are two primary means of exposure. The G agents are vapors under most conditions and so are inhaled or absorbed through the conjunctiva, whereas the V agents are liquids and are absorbed through the skin. Ingestion is a possible but

FIGURE 25-3 Pesticide use in the agricultural setting.
Courtesy of USDA Natural Resources Conservation Service/Jeff Vanuga.

unlikely route of exposure. Inhalational exposure is most likely to occur when nerve agents are released in an aerosolized form or if sprayed (Fig. 25–3), such as from a helicopter or crop duster. Systemic absorption occurs within minutes. The rapidity with which clinical effects occur is dose and time dependent. In general, vapor exposure elicits symptoms immediately or within minutes, with peaks occurring within a few minutes after exposure. Vapor can also be absorbed through the eyes concurrent to respiratory absorption. Percutaneous exposure occurs when liquid forms of the nerve agents come into contact with the skin, eyes, mouth, or membranes of the nose (Fig. 25–4).

Aside from being the most potent of the nerve agents (note the LD$_{50}$), VX is also the least volatile of the agents, thereby requiring smaller amounts of absorption for effects to occur. Conversely, sarin is far more volatile and is more likely to evaporate before absorption through the skin can take place.

Clothing or other barriers to the skin reduce exposure, depending on the nature (material, layering, etc.) of the barrier. An important exception is wet cloth, which can increase exposure and minimize evaporation. Temperature plays a role in the rate at which nerve agents are absorbed: absorption increases as temperatures increase. Absorption through the skin yields signs and symptoms within minutes, but may take as long as thirty minutes. With lesser exposure, symptom onset may be many hours and will typically involve GI features. A rule of thumb is that the longer the delay in symptom onset, the less severe the clinical course.

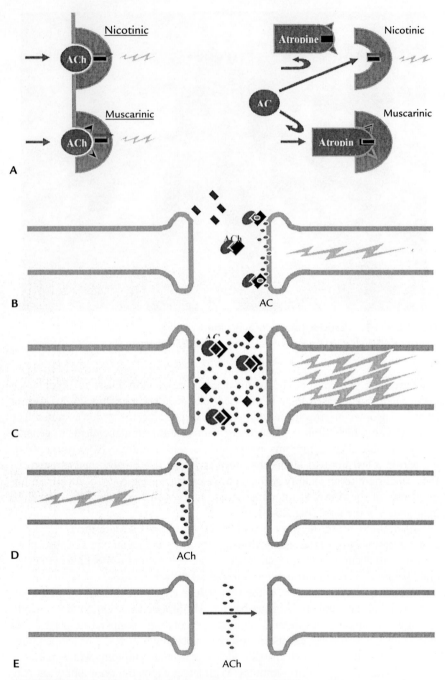

FIGURE 25–4 (A) ACh and atropine at receptors. **(B and C)** Exposure to nerve agent.
(D–F) Nerve transmission: nerve to nerve. **(G)** Impulse termination: the role of ACh.
Courtesy of US Army Medical Research Institute of Chemical Defense.

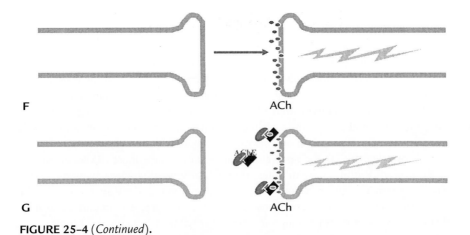

F   ACh

G   ACh

FIGURE 25–4 (*Continued*).

Ingestion results from consumption of contaminated food or water. Ingestion may also occur secondarily from ingestion of droplets from aerosolized forms.

Signs and Symptoms

Broadly speaking, the presentation of signs and symptoms for nerve agents may be divided into local and systemic effects. If the exposure level is small, then local symptoms may be all that are present. For example, a very small dermal exposure might present with localized muscle fasciculations and sweating. Far more likely is that there will be sufficient exposure to result in systemic symptoms. The route of exposure dictates whether systemic or local effects are seen.

Local Symptoms and Signs: Vapor Exposure

Nerve agents in vapor form result in exposure and absorption through the respiratory tract and eyes. Symptom onset occurs almost immediately and may last up to

TABLE 25–3 Relative Lethality of Nerve Agents by Route of Exposure

	Dermal (LDt$_{50}$)	Vapor/Aerosol (MCt$_{50}$)
Sarin	1,700 mg	2–3
Tabun	1,000 mg	3
Soman	50 mg	<0.1
Cyclohexyl sarin	30 mg	<0.1
VX	6–10 mg	.04

Modified from Medical Management of Chemical Casualties Handbook; United States Army Medical Research Institute of Chemical Defense. Chemical Casualty Care Division.

forty-eight hours. By the time a patient is seen in an office or hospital setting after exposure to vapor, it is quite likely the peak effect has occurred and no new symptoms should be expected.

It is important to note that ophthalmic findings can occur strictly as a local response and are not, of themselves, proof of systemic effects. When the eyes are exposed to nerve agent vapor, miosis, conjunctival hyperemia, eye pain, and a frontal lobe headache will be noted within a few minutes of exposure. With a minimal exposure, the signs and symptoms will likely resolve within twenty-four hours. With more intense exposures, signs and symptoms may last up to seventy-two hours.

Absorption through the respiratory tract represents the most efficacious means of absorption. Within a few minutes, rhinorrhea, wheezing, chest tightening, and nasal hyperemia occur. If the exposure is relatively limited, these symptoms should end within several hours after exposure. As with the eyes, it is important to note that upper respiratory tract symptoms can occur without systemic symptoms.

Liquid Exposure

Large liquid exposures result in the onset of symptoms within half an hour, although it is possible for symptoms to be delayed for as long as a day after exposure. The shorter the time until onset, the more severe the clinical features. Should exposure to nerve agents occur by contact with liquid forms, signs would be noted focally with profuse sweating and fasciculations at the site of contact within a couple of minutes to hours and last upward of five days. If liquid nerve agents come in contact with the eyes, the signs and symptoms and their respective onset and duration is the same as those listed for vapor exposure to the eye. The only significant difference is that the onset occurs instantaneously. If liquid nerve agents are ingested, the signs and symptoms are similar to the systemic affects as discussed later, though, not surprisingly, the first organ system to be affected will be the GI tract.

Systemic Symptoms and Signs

The ubiquity of muscarinic, nicotinic, and cholinergic receptors throughout the body explains the wide range of symptoms and signs experienced by individuals exposed to nerve agents. Substantial variability in clinical presentation is common, however, and driven by host factors (e.g., genetic receptor heterogeneity or comorbidities) and environmental factors (e.g., route of exposure and dose). Due to the clinical variety adopting a syndromic diagnostic approach is useful; a comprehensive list is provided (Table 25–4), although selected clinical syndromes are discussed in more detail. Children may not show the characteristic miosis and hypersecretion of adults and may present with more of the neurologic symptoms and signs.

Muscarinic receptor blockade (Fig. 25–5) results in ocular, mucus membrane, cardiac, respiratory, and GI effects. Miosis occurs from direct exposure of the eye to vapor or liquid, or may result if systemic absorption is significant. The constriction is bilateral and occurs immediately following direct contact. Associated ocular symptoms and signs include pain, diminished vision, and injected conjunctivae.

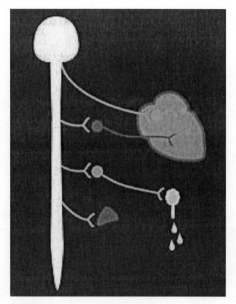

FIGURE 25-5 **Distribution of muscarinic receptors.**
Courtesy of Washington University.

The upper respiratory tract, including the mucus membranes of the nasal passages and pharynx, is often affected, producing copious rhinorrhea. Involvement of the lower respiratory tract induces bronchospasm, mucus production, and chest tightness, which can present clinically similar to asthma or asthmatic bronchitis. If

TABLE 25-4 Nerve Agents: Symptoms and Signs by Receptor Type

Target Organ	Symptoms and Signs	Receptor Type
Eyes	Miosis, pain, decreased vision, conjunctival hyperemia	Muscarinic
Mucus membrane	Rhinorrhea	Muscarinic
Lungs	Bronchospasm, mucus production Tightness, apnea	Muscarinic
Skin	Sweating (local or general)	Muscarinic
GI	Nausea, vomiting, diarrhea, cramping	Muscarinic
Muscles	Fasciculations (local or general)	Nicotinic
Cardiac	Bradycardia	Muscarinic
	Tachyarrhythmias	Nicotinic
CNS	Irritability, mood changes, insomnia, seizures, presyncope, syncope, central apnea	Cholinergic and nicotinic

FIGURE 25–6　Distribution of nicotinic receptors.
Courtesy of Washington University.

the dose is sufficient, respiration may slow or cease altogether from a combination of central and peripheral nervous system effects. If respiratory symptoms occur first, inhalational exposure is probable. GI symptoms and signs, such as nausea, vomiting, diarrhea, or cramping, may occur following nerve agent ingestion, or with significant systemic exposure.

Nicotinic receptor blockade (Fig. 25–6) may present with dermal, muscular, or cardiac symptoms and signs. Dermal exposure produces localized sweating, or with more substantial exposures, generalized diaphoresis. Localized muscle fasciculations suggests focal skin contact, but at higher doses fasciculations may generalize and are followed by muscle fatigue and even flaccidity. Fasciculations tend to persist even after other symptoms have resolved. These features are the sequelae of nicotinic poisoning at motor end-plate receptors. Heart rate is a variable clinical sign. Autonomic nervous system innervation to the myocardium includes muscarinic and nicotinic input. Bradycardia results from unopposed muscarinic receptor blockade, but this may be offset by the adrenergic stimulation caused by nicotinic poisoning of autonomic ganglia. Bradycardia, as well as bradyrythmias and heart block, are seen with nerve agent exposure.

Cholinergic receptors in the CNS are responsible for the CNS findings in nerve agent poisoning, including seizures, apnea, mental status changes, and diminished or loss of consciousness. CNS signs may begin within the first half hour after severe exposures. Cholinergic receptors are found throughout the CNS, and cause diverse neurologic signs and symptoms. Exposures to moderate or small amounts result in nonspecific mental status changes such as irritability, mood changes, insomnia, and headaches. Sometimes these occur without any focal signs or symptoms, or they may develop after a latency period and last after the resolution of other symptoms and signs. Neurologic symptoms often persist for months.

Laboratory Studies

At present there is no test to directly measure nerve agents or their degradation products. As is true with organophosphate and carbamate pesticides, RBC and plasma cholinesterase (ChE) activity can be used as markers for exposure to nerve agents. Unfortunately, significant intraindividual and interindividual variabilities exist. Correlation between RBC or plasma ChE and clinical findings is poor, particularly after vapor exposure. Plasma ChE activity is reduced by liver disease, infection, and oral contraceptives. However, if exposure to nerve agents is suspected, treatment should not be delayed even if cholinesterase levels are normal. If significantly depressed, RBC or plasma ChE activity may confirm a diagnosis and may be used to monitor the clinical course. For example, rising enzyme activity suggests the nerve agent has peaked and the enzymes are regenerating. Blood gases and pulse oximetry help determinine the severity of respiratory impairment. Elevation in creatine kinase (CK), WBC, and ketonuria were seen in the Tokyo sarin attack and may help identify a nerve agent toxidrome, especially if associated with low ChE levels and case clustering. After the Tokyo sarin attack, ChE levels took up to three months to normalize.

Differential Diagnosis

The clinical signs and symptoms of nerve agents are diverse and affect multiple organ systems (Table 25–5). Poisoning from organophosphate and carbamate pesticides is indistinguishable from nerve agent exposure. In the setting of mass casualties or an obvious chemical attack, but where the exposure is unknown, neurologic or respiratory syndromes suggest pesticide poisoning, nerve agents, or even hydrogen cyanide (see Chapter 25). Cyanide and nerve agents are distinguishable more by the time of onset of the neurologic features: Nerve agents produce an immediate neurologic syndrome, whereas the neurologic symptoms with either pulmonary or blood agents occur somewhat more gradually as the disruption of oxidative metabolism generates histotoxic anoxia (see Table 25–6).

Diagnosis

Making the diagnosis of nerve agent exposure (Table 25–5) requires recognition of the array of symptoms and signs. This is easier to do if ChE clustering occurs following a sentinel event. An index case or unidentified small-scale exposures require that clinicians maintain a differential that includes such chemical weapons.

TABLE 25–5   Making the Diagnosis: Nerve Agents

Pattern of Presentation	Multiple, concurrent cases of multiple organ system symptoms, especially respiratory, GI, and CNS
Features	Variable according to dose and exposure route. Key triad: papillary constriction, hypersecretion, and fasciculations
Dermal	Burning, itching
Ocular	Blurring, miosis, lacrimation
Pulmonary	Mucus membrane irritation, SOB, wheezing
Neurologic	Syncope, apnea, convulsions, prostration
Hypersecretion	Drooling, lacrimation, diarrhea
Findings	Miosis, drooling, rhinorrhea, fasciculations—focal or generalized, tachycardia, or bradycardia Decreased RBC or serum ChE activity

As always, a thorough occupational and environmental history can be of enormous help (see Chapter 2). Serum or RBC ChE levels may be helpful if suspicion is high, but even a positive result will only confirm an exposure to a cholinesterase inhibitor and not necessarily to a terrorist attack.

Chronic Effects of Nerve Gas

Relatively little is known about the long-term effects on survivors of nerve gas exposure. What little is known is extrapolated from what was seen following occupational exposures and the Tokyo sarin attacks in 1994 and 1995. These include delayed neuropathy, EEG changes, cancer, and keratitis. Long-term neuropsychological and neurophysiologic changes include decreased memory and concentration, confusion, and irritability. Delayed but ultimately reversible muscle weakness also occurs.

Significant respiratory exposure to nerve agents may precipitate asthma or reactive airways dysfunction syndrome (RADS). Fasciculations and tremors may persist for months after exposure. Individuals who survive apnea, significant hypoxia,

TABLE 25–6   Differentiating Cyanide versus Nerve Agents

Findings	Cyanide	Nerve Agents
Pupil size	Normal, mydriasis	Miosis
Acid–base balance	Anion gap	Respiratory alkalosis
Neurologic symptoms	Twitching	Fasciculations

or coma may have persistent neurocognitive injury, including mood or personality changes, anxiety or depression, loss of intellectual function and judgment, or Parkinsonism. Sleep disorders, chronic headaches, difficulties with concentration and memory, slurred speech, and gait ataxia may also result. Survivors should be evaluated serially with neurologic evaluations, formal neurocognitive testing, and magnetic resonance imaging (MRI) similar to the protocols used in carbon monoxide poisoning.

Clinical experience following the sarin attack in Tokyo provides useful information on what might be expected following nerve gas exposure. Treatment with atropine and pralidoxime was instituted. One comatose patient regained consciousness within eight hours and was ambulatory after two days. Initially, disorientation, impaired short-term memory, and abnormal EEGs were found, as well as depression of plasma and RBC ChE levels. Many of these effects persisted. Memory and learning difficulties, as well as retrograde amnesia and personality changes, were documented. Follow-up studies as far out as two years indicate an unusually high prevalence of a neurasthenic syndrome with fatigability, visual problems, headaches, anxiety, and neurocognitive and psychological symptoms. Posttraumatic stress syndrome or poorly understood chronic toxicities are possible explanations for these persistent effects. Sister chromatid exchanges in sarin-exposed individuals were seen, as were MRI changes in the anterior cingulated cortex, a locus for attention, emotional regulation, and fear response in animals. These findings may explain the long-term neuropsychological sequelae seen in sarin-exposed individuals. Other than the sister chromatid exchanges, nerve agents are not known to be teratogenic, carcinogenic, or mutagenic.

Vulnerable Populations

Individuals with preexisting neurologic, psychological, and neurocognitive impairment are likely to suffer more lasting and extensive sequelae from nerve agent exposures. Children, because of the relative immaturity of their nervous system, and asthmatics because of the potential for more severe respiratory distress, may also represent uniquely vulnerable populations.

Decontamination

By the time a clinician is seeing a patient, he or she has already been through full decontamination (Table 25–7). If not, the initial management should include rapid disrobing of the patient and thorough washing with soap and water. If there is any question of contamination, measures should be immediately undertaken to decontaminate. Time is of the essence—the sooner patients are decontaminated, the less their exposure.

TABLE 25-7   Critical First Steps: Nerve Agents

Medical Treatment	Atropine, PAM titrated to signs, diazepam, intubation
Infection Control	Decontamination if liquid exposure
PEP	None
Public Health	Contact SHD, FBI

Secondary exposure to health care providers is unlikely so long as decontamination has occurred. Vapor exposure would not be an issue assuming the examination is occurring in a safe and clear site (however, clothes should still be removed to eliminate the possibility of "trapped" vapor). Liquid forms present the greatest risk to health care workers or emergency responders. The volatility of nerve agents promotes rapid evaporation and offers a measure of protection by the time most patients are evaluated. The U.S. military recommends removal and double bagging of clothes and washing of the skin in alkaline solution (1:9 dilution of chlorine bleach will suffice) to neutralize and inactivate residual nerve agent. If bleach is unavailable, water with household detergents is an acceptable substitute. Focal sites of exposure may be washed with pads soaked in the washing solution for twenty to thirty minutes and then rinsed copiously with water. If bleach or detergents are unavailable, flushing with large quantities of water is advised. When in doubt as to whether dermal exposure occurred, rinsing the patients' entire body with water is prudent.

Treatment

Nerve agent mortality is primarily a function of respiratory failure. Consequently, initial assessment of exposed patients must begin with the ABCs: airway patency, breathing, and circulation. Management is approached no differently from poisonings or toxicities of any kind: after removing the source or moving away from the source, implement pharmacologic and medical management. Medical care should only begin if the care provider(s) is protected from both primary and secondary contamination.

Pharmacologic treatment is the primary means of intervention. All the drugs used target the effects of the cholinergic excess at one of the three affected neural sites: muscarinic receptors, nicotinic receptors, or the CNS. The three drugs that are used in the treatment of nerve agent exposures are atropine, pralidoxime chloride, and diazepam. Because no one drug addresses all the sites of poisoning, management of nerve agent exposures will almost always be a combination of the drugs discussed later (Table 25–8). Men and women should receive standard adult dosing schedules, whereas children should receive standard pediatric dosing where applicable.

Muscarinic symptoms are managed largely by atropine (Fig. 25–7). With its high affinity for both central and peripheral muscarinic receptors, atropine is a competitive inhibitor of acetylcholine. By binding to muscarinic receptors, atropine

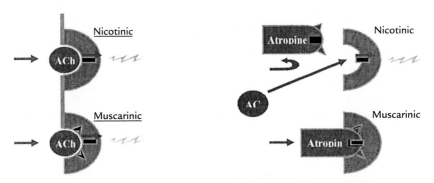

FIGURE 25-7 **Mechanism of action: atropine. ACh and atropine at receptors.**
Courtesy of US Army Medical Research Institute of Chemical Defense.

reduces the cholinergic hyperstimulation caused by nerve agents. It follows that atropine has no effect on nicotinic symptoms such as muscle fatigue, fasciculations, apnea, or paralysis. The half-life of atropine is approximately four hours so that repeated dosing is sometimes necessary. Unlike organophosphates, whose lipid solubility enhances their persistence within the body and often requires substantial repeated doses of atropine, nerve agents are more readily treatable because they are water soluble and cleared more rapidly from the body. For this reason, much less atropine is usually needed to reverse the cholinergic syndrome. Nowadays, most rapid response units or hazmat teams are taught to carry atropine.

Atropine administration is generally reserved for mild to moderate exposures. Mild symptoms such as rhinorrhea need not be treated with atropine. If mild

TABLE 25-8   Nerve Gas Treatment

Atropine
   Mild to moderate symptoms
   Adults: 2 mg IM
   Children: 0.02 mg/kg q 2-5 min until resolution dyspnea.
   In the severely poisoned patient, dosages may need to be increased and given more
     rapidly (5 mg in adult or 0.05 mg/kg in children every 2 to 5 min).

PAM
   Adults: 1-2 g IV in 250 ml of normal saline or D_5W for adults over 30 min
   Children: 15 to 25 mg/kg for children over 30 min
     If IV access is not possible, IM administration is an option.
   In moderate or severe nerve agent poisoning, initial PAM administration may be
     followed by an infusion of 200-500 mg/hr.

Diazepam
   Adults: 5-10 mg IV q 10-20 min, until seizures resolve (not to exceed 30 mg/8 hr);
     may repeat in 2-4 hr, as occasion requires.
   Pediatric 0.2 mg/kg

symptoms consist primarily of ophthalmic features, then antimuscarininc eye drops may be used. Patients with mild respiratory distress need not be given atropine; instead, they can be observed closely for improvement. Moderate respiratory distress is treated with atropine, titrating to eliminate symptoms (see Appendix for dosing schedule and Table 25–8).

Excessive atropine may cause anticholinergic symptoms, but these are self-limiting and largely innocuous. Physostigmine, which can reverse anticholinergic symptoms, is contraindicated with nerve agents.

☣Pralidoxime (PAM)

A synthetic pyridinium compound and an oxime, PAM effectively regenerates ChE after poisoning by nerve agents. It acts by breaking the nerve agent–ChE bond, thereby displacing the nerve agent. Displacement of nerve agents by PAM occurs only if the enzyme has not yet aged—that is, the alkyl group is still in place (see Pathophysiology). Consequently, PAM's therapeutic utility is time dependent: the sooner it is given, the greater the benefit. Regardless, all patients should receive PAM, as aging is slower with nerve agents than with pesticides.

PAM tends to exert its greatest effect at nicotinic receptors at the motor endplates, and therefore improvement can be tracked clinically. Because it acts selectively, PAM should not be used as monotherapy as it will not address the muscarinic or CNS features. If symptoms continue or worsen, PAM may be given at hourly intervals. Hydration should accompany PAM administration as the drug is cleared rapidly through the kidneys. Side effects of PAM include GI distress, nausea, vomiting, headache, high blood pressure, and blurry vision.

☣Benzodiazepines

Benzodiazepines are used to control and to prevent seizures caused by all the nerve agents. Diazepam readily crosses the blood–brain barrier to block the effects of acetylcholine on the CNS. Diazepam should be given to individuals with severe exposure to prevent seizures. The U.S. military recommends administering diazepam in nerve agent exposures prior to onset of seizures. Diphenylhydantoin (dilantin) and phenobarbital are contraindicated in nerve agent exposure.

☣Ipratroprium

Nebulized ipratropium bromide can be used as supplemental therapy for treating bronchospasm caused by nerve agent exposures.

Nonpharmaceutical interventions, such as endotracheal intubation or frequent suctioning, are often necessary, depending on the patient's clinical status and response to pharmacologic interventions.

Summary

Nerve agents are potent, rapidly acting toxins whose clinical effects are due to acetylcholine excess. Nerve agent toxidromes are treatable with specific therapies, but treatment must be prompt. Chronic effects include somatic, neurocognitive, and psychological symptoms. Treatment should focus on removal from exposure, field decontamination, and supportive measures (basic and advanced life support). Specific antidotes exist for many of these agents, including pralidoxime, atropine, and diazepam.

Incapacitating Agents BZ or QNB (3-Quinuclidinyl Benzillate)

Background

As mentioned in the history of chemical warfare, shifting political winds and humanitarian mores in the late 1950s and 1960s led the U.S. military to collaborate with the pharmaceutical and chemical industries in the research and development of nonlethal chemical agents. The concept of a "humane war" in which enemies could be incapacitated temporarily without the permanent disfigurement or death typical of the first generations of chemical agents was viewed with great promise by military planners. Most of the military records of these efforts, including medical data on individuals exposed experimentally to these pharmacoweapons, still remain classified. In the intervening years, incapacitating agents have been developed for use as chemical weapons, riot control strategies, or personal protection.

The purpose of incapacitating agents is to render the exposed individual temporarily incapable of functioning. Incapacitating agents have a wide variety of disabling effects—ranging from mucus membrane irritation to nausea, vomiting, and diarrhea. From a strategic military perspective, ideal incapacitating agents impair performance by disrupting higher-order CNS activities. The psychological, behavioral, or mental status changes in those exposed do not usually cause death or permanent harm, such as occurs with nerve agents. Among the classes of chemicals investigated for use as incapacitating agents are anticholinergic agents (e.g., physostigmine or glycolates such as 3-quinuclidinyl benzilate or BZ), hallucinogenic agents (e.g., LSD), marijuana derivatives, and opiates. Riot control agents cause profound ocular or mucus membrane irritation, or vomiting. These are considered briefly at the end of this chapter.

There have been multiple suspected uses of incapacitating agents following their discovery in the late 1950s and 1960s. During the Vietnam War, the United States was accused on at least one occasion of dropping BZ-filled hand grenades on Vietcong guerrillas, killing upward of 300 people. North Vietnamese troops were also suspected of using BZ-like agents in Cambodia. In the 1980s, the Soviet Union was accused of using hallucinogenic chemical warfare agents against Afghan fighters. In July 1995, Bosnian refugees fleeing the conflict described the sudden onset of disorientation, bizarre behavior, and hallucinations, leading some international observers to suspect that the Serbian government had used a glycolate incapacitant. Although the charge was denied and remains unproven, the Federal Republic of Yugoslavia was the sole European state not to join 120 other signatories to the 1993 International Chemical Weapons Convention Treaty banning the use of chemical warfare agents. In 2002, the Russian military used BZ gas in a botched attempt to free hostages held by Chechnyan terrorists in a Moscow theater. The gas disabled 50 Chechnyan terrorists, but killed 115 hostages, due to the military's unfamiliarity with the chemical weapon.

The Moscow theater debacle underscores one of the reasons why these agents are considered too unreliable for use as military weapons. In his 1968 book, *We All Fall Down—The Prospects of Biological and Chemical Warfare*, author Robbing Clarke states:

> One of the major problems is that it is impossible to predict what effects the hallucino-genic drugs will produce. It is quite conceivable that they will increase belligerency and yet at the same time make a man less effective in his duties. The aim of using such a weapon could hardly be to produce a belligerent, maniacal and depressed machine gun operator or, worse, Army commander with nuclear power at his elbow. Further, there is considerable doubt as to how reversible large doses of such drugs might be; certainly if the doses were really high death could result and permanent psychological changes might be expected from slightly smaller doses.

Sources

BZ and similar anticholinergic chemicals are not difficult to synthesize and so may be produced in clandestine laboratories. Anticholinergics are used widely in medicine and are available as prescription or over-the-counter drugs—atropine, oxybutynin, scopolamine, and antihistamines, to name a few. Numerous plant sources for anti-cholinergic compounds exist, many with hallucinogenic effects used for medicinal and religious purposes by indigenous peoples, shamans, and healers. These include members of the plant family Solanaceae that contain atropine, hyoscine, and hyoscyamine in varying proportions. Examples include Jimson weed or thornapple (*Datura stramonium*); belladonna or deadly nightshade (*Atropa belladonna*); black henbane *(Hyoscyamus niger);* woody nightshade *(Solanum dulcamara),* and Jerusalem cherry (*Solanum pseudocapsicum*). Pharmacologic researchers also use BZ (known as QNB for 3-quinuclidinyl benzilate) as a marker for muscarinic receptors.

Route of Exposure

Incapacitating agents are aerosolized readily and so pose an inhalation risk. They may also be placed in solution, or adsorbed to particles, and therefore absorbable through the skin and mucus membranes, or ingested. Solutions of BZ and other anticholinergic agents may be the preferred mechanism of delivery for terrorists, in part because incorporating the chemical into a solvent significantly enhances dermal absorption and thereby augments toxicity. Aerosolization is a probable route of exposure in a military setting.

Pathophysiology

Glycolate incapacitating agents—of which BZ is the prototype—act by competitive inhibition of muscarinic and nicotinic receptors in the parasympathetic nervous system and CNS. Muscarinic receptors are present on numerous organ systems, including the eye, heart, respiratory system, skin, GI tract, and bladder. Sweat glands, innervated by the sympathetic nervous system, also are modulated by muscarinic receptors.

Properties

BZ is an odorless gas that has significant biopersistence with a half-life in air of 3 to 4 weeks. It remains stable in water, heat, and most solvents. These properties enhance its toxicity and value as a chemical agent. Anticholinergics cross the placenta and can be found in breast milk. This is true of both atropine and hyoscyamine, but similar information on BZ and other chemical incapacitating agents is lacking. The toxicokinetics of BZ are reasonably well worked out. The chemical is metabolized by the liver and the parent compound, or its metabolites are excreted in the urine. The concentration needed to produce the desired effect, incapacitation, is nearly 1,000-fold less than the fatal concentration. Although BZ has a high safety ratio, in the real-life setting, as the Vietnam and Moscow anecdotes suggest, a variety of factors leave less room for safety than laboratory testing might indicate. The concentration needed to incapacitate 50% of an exposed group (LCT_{50}) for BZ is 112 mg-min/m^3, compared to an LCt_{50} (lethal concentration to 50% of exposed group) of approximately 200,000 mg-min/m^3.

Clinical Effects

The physiological impact of BZ is time and dose dependent. Clinical effects of BZ are similar to those caused by atropine's effects on the eccrine and apocrine glands of the body and cause understimulation of the organs where these glands are located. The clinical symptoms are dry skin, decreased sweating, cutaneous vasodilation, and papillary dilation with inability to accommodate. The clinical syndrome is perhaps best characterized as a syndrome of similes: individuals appear "dry as a bone," "mad as a hatter," "hot as a hare," "red as a beet," and "blind as a bat." Additional symptoms indicative of unopposed cholinergic input are dry mouth, blurry vision, and urinary retention. The application of deadly nightshade from the belladonna plant and the resultant pupillary dilation explains the vernacular name of the plant "belladonna" or ("beautiful woman"). An important clinical corollary of the peripheral nervous system effects is that exposed individuals are at risk for heat stroke. CNS effects of BZ are many and reflect nicotinic and anticholinergic overstimulation. Clinical symptoms reflecting CNS effects include muscle weakness, discoordination, ataxia, changes in sensorium, poor judgment, hallucinosis or delusional perceptions, distractibility, and fluctuating activity ranging from restlessness to relative inactivity. Onset of symptoms occurs within four hours of exposure, though impairments in levels of consciousness occur later. Used properly, affected individuals are fully recovered in about four days. Delayed recovery, however, is not uncommon. As the Moscow theater incident demonstrates, if used incorrectly, incapacitating agents may indeed be lethal. Recovery is also affected by the victim's preexisting medical status and time from exposure to initiation of decontamination and medical treatment. Symptoms may appear as late as thirty-six hours after percutaneous exposure, even if the skin is washed within an hour. Delayed development of clinical symptoms is common, often as long as a few hours or even a day following exposure.

Differential Diagnosis

Patients with acute confusional states, delirium, or irrational behavior require consideration of a broad differential (Table 25–9). The absence of anticholinergic peripheral nerve symptoms (dry mouth, dry skin, and mydriasis) and orientation to person, time, and place will make incapacitating agents, specifically BZ and other anticholinergics, unlikely. Intoxication syndromes (alcohol, marijuana, LSD, barbiturates, and opiates) should be considered, along with anxiety or conversion reactions and toxic environmental or occupational exposure to tetraethyl lead, mercury, or bromide. Heat stroke may accompany BZ exposure but is by itself in the differential diagnosis. This poses diagnostic difficulties because first responders, hazmat team members, or soldiers may be wearing protective gear under environmental conditions that predispose to heat stress, heat exhaustion, or heat stroke. Diagnostic considerations include: heavy metal exposure; alcohol intoxication; illicit drug abuse; psychiatric pathology; TIA/CVA; heat stroke.

Personal Protective Equipment

Chemical protective masks with HEPA filters are effective in protecting the face and the respiratory system from airborne BZ. Chemical protective suits prevent dermal exposure to incapacitating agents in the form of solids, dusts, or solutes.

TABLE 25–9   Incapacitating Agents: Signs and Symptoms

Signs and Symptoms	Possible Etiology
Restless, dizziness, or giddiness; failure to obey orders, confusion, erratic behavior; stumbling or staggering; vomiting	Anticholinergics (e.g., BZ), indoles (e.g., LSD), cannabinols (e.g., marijuana), anxiety reaction, other intoxications (e.g., alcohol, bromides, barbiturates, lead)
Dryness of mouth, tachycardia at rest, elevated temperature, flushing of face; blurred vision, pupillary dilation; slurred or nonsensical speech, hallucinatory behavior, disrobing, mumbling and picking behavior, stupor and coma	Anticholinergics
Inappropriate smiling or laughter, irrational fear, distractibility, difficulty expressing self, perceptual distortions, labile increase in pupil size, heart rate, blood pressure; stomach cramps and vomiting may occur	Indoles (schizophrenic psychosis may mimic in some respects)
Euphoric, relaxed, unconcerned daydreaming attitude, easy laughter; hypotension and dizziness on sudden standing	Cannabinols
Tremor, clinging or pleading, crying; clear answers, decrease in disturbance with reassurance; history of nervousness or immaturity; phobias	Anxiety reaction

Decontamination

Soap and copious amounts of water are necessary to achieve decontamination. A mildly alkaline solution of sodium bicarbonate or sodium carbonate may also be used, whereas bleach is contraindicated.

Treatment

Provided the victim has been adequately decontaminated, first response and medical personnel are not at risk for exposure. However, because BZ often presents in a delayed fashion, inadvertent exposure is possible so that appropriate PPE is recommended. General treatment for the glycolate incapacitating agents such as BZ consists of close observation, judicious restraint, and confinement and supportive care. It is important to bear in mind the "red as a beet" simile because there is significant risk of hyperthermia or heat stress in victims. The often bizarre behavioral and psychological effects also warrant careful attention to objects that might be used as potential weapons or to inflict self-injury in victims.

TABLE 25-10 Comparison of Symptoms, Signs for Bioterrorism Agents with Neurologic Syndromes

Agent (Onset)	Nerve Agents (min to hr)	BZ (hr)	Botulinum (hr-days)	Fentanyl Opiates (min to hr)
GI	N/V/D	Decreased motility	N/V/D	Decreased motility
CNS	Depressed sensorium and respiration	Slow onset of behavioral changes, later delirium	MS normal	Depressed sensorium and respiration
Muscle	Weakness Fasciculations	Weakness	CN palsies Weakness	
Fever	Afebrile	Febrile	Afebrile	None
Pupils	Miosis	Mydriasis	Mydriasis	Miosis
PNS	Cholinergic Hyper secretions	Cholinergic No secretions	Cholinergic No secretions	No secretions

Adapted with permission, Stephan Kales, AIM; ACP, Annual Session.

Specific pharmacologic treatment for BZ and other anticholinergic agents is available. The antidote of choice for BZ is physostigmine (eserine, antilirium). Physostigmine works by reversibly inhibiting anticholinesterases in the postsynaptic junction, thereby effectively raising the level of acetylcholine and counteracting the chemical's pharmacologic effects. Structurally, the compound is nonpolar and has a tertiary amine group that enables it to cross the blood–brain barrier where it reverses BZ's muscarinic and nicotinic effects. Repeat dosing is often necessary to achieve clinical remission and maintenance therapy is followed by slow tapering. The duration of treatment varies from hours to weeks, depending on the duration of exposure, absorbed dose, and associated medical conditions. Any co-morbid medical conditions should be treated in standard fashion.

A 1-mg test dose of physostigmine is sometimes useful if the diagnosis of an anticholinergic agent is uncertain. If the patient's clinical status improves, full dosing can begin. Intravenous administration of physostigmine is the best treatment option as it affords the opportunity for rapid adjustment of dose. Slow intravenous administration of 30 μg/kg (or 1 mg/min) is recommended. If given intramuscularly, the recommended dose is 45 μg/kg. Individuals should have serial mental status examinations to monitor treatment benefit and determine when to begin tapering off the drug. If response is unsatisfactory, a second identical dose may be administered. Pediatric dosing is 20 mg/kg intravenously. In the setting of mass casualties where field administration of antidotes are being considered, either intramuscular or oral dosing is possible. Oral administration (60 μg/kg) requires masking or diluting the drug due to its intensely bitter taste.

TABLE 25-11	Quick Reference Guide: Nerve Agents
Diagnostic Keys	Multiple cases, multiple organ systems; triad of pupillary constriction, hypersecretion, and fasciculations
Risk Factors	None
Transmittable	Yes, if liquid and no decontamination of patient
	No, if vapor
Transmission	Only via direct contact with liquid (on patient skin, clothes, etc.)
Management	Atropine, PAM
Prevention	
Preexposure	None
Postexposure	None
Containment	Patient: none, if decontaminated
	Secondary contacts: none, if decontaminated
Populations at Heightened Risk	Asthmatics, children
Long-term Issues	Neurocognitive and psychological impairment
Public Health Concerns	One case means chemical attack has occurred
Contacts	Local, state DPH, CDC, FBI

Overdose of physostigmine is characterized by diaphoresis, nausea, vomiting, diarrhea, cramping, fasciculations, weakness, and other cholinergic features. The antidote's half-life is roughly thirty minutes so such features should subside quickly, usually without requiring intervention. Treatment of BZ may be paused in such a case but not necessarily stopped as anticholinergic symptoms may return as the physostigmine is cleared.

Patients with mild symptoms may be watched without treatment. All patients who are ambulatory and symptom free after eight hours may be released safely.

Field Testing

At present, chemical detection devices for incapacitating agents such as BZ are not available. Confirmatory testing requires analysis of environmental samples.

☣ Irritating or Vomiting Agents

A number of irritating or vomiting agents have been developed primarily as crowd-control agents by the military and law enforcement communities. Some irritants caused profound tearing. These agents, referred to as "lacrimators" or "tear gas," cause exposed individuals to tear up and/or develop blepharospasm, temporarily limiting their vision. These agents have high LCt_{50} and a low effective Ct_{50}, and it is this high safety ratio that makes them useful as riot-control agents. The pain, discomfort, or difficulty with vision caused by these agents usually subsides within a few minutes, although examples of acute bronchospasm, anaphylaxis, and persistent ocular injury have been reported, probably an individual or dose-dependent side effect. These agents may be dispersed as fine particulate smoke (aerosols) or in solutions as droplet aerosols. Vomiting agents cause intense irritation of the eyes and upper respiratory tract with resultant sneezing, coughing, nausea, and vomiting. Treatment consists of supportive measures and decontamination—flushing of eyes and washing of skin. For a more detailed discussion of irritating and vomiting agents, see chapter references at the end of *Bioterrorism Source book*.

CHAPTER 26

Blood Agents

Clinical Vignette

One of your patients is a newly minted volunteer firefighter in a rural community in Colorado. That morning he was called to a metal finishing plant where a chemical accident, involving hydrochloric acid and cyanide salts occurred. Fearing that the HCl might interact with the CN^-, the area surrounding the plant was evacuated. The local hazmat team entered the area wearing SCBA gear and chemically protective clothing.

Your patient did not enter the building, but complained of feeling tired, dizzy, and short of breath in the building's parking lot and was told to seek medical attention. In your office, he denies any headache, coughing, or GI complaints. No other firefighters complained of any symptoms. His physical exam was unremarkable: specifically no tachycardia, tachypnea, or flushing. Your differential diagnosis includes mild cyanide toxicity or possible anxiety reaction. What tests, if any, would you order?

$$H — C \equiv N$$
Hydrogen Cyanide

Background

Cyanide has been used for thousands of years in intentional poisonings, though it wasn't until the end of the 18th century that the actual compound was identified.

FIGURE 26–1   The bodies of Iraqi Kurds after attacks with, among other things, cyanide gas in the late 1980s.
Courtesy of the Kurdistan Democratic Party.

Its role in infamy is unquestionable with such uses by Nero to murder his family and "friends," by Reverend Jim Jones in Guyana for a mass cult suicide of over 900 people in 1978, and in the 1982 Tylenol contamination episode in which seven people died. More recently, Iraqi military used cyanide gas as part of the Iraqi chemical attack on Kurdish citizens, killing thousands in the late 1980s (Fig. 26–1).

Hydrogen cyanide (HCN) is a byproduct of the combustion of molecules containing carbon and nitrogen, including most plastics. HCN is felt to be a major cause of inhalational injury from residential and commercial fires. Worldwide, probably the single biggest source of cyanide exposure comes through cigarette smoking.

Cyanide (CN^-) itself is omnipresent in living things. In fact, cyanide is naturally occurring in relatively low levels in many plants as well as in plant foods; such as corn, spinach, lima beans, cherries, soy, tapioca, peaches, bitter almonds, and cassava beans (Fig. 26–2). It is even produced by certain species of bacteria, fungi, and algae and is an important and vital part of many metabolic processes such as in the making of vitamin B_{12}. Cyanide becomes toxic beyond a certain threshold of exposure. Because of the biological ubiquity of cyanide, there are intrinsic metabolic pathways for its removal. However, exposure levels at sufficiently high levels can readily overwhelm normal clearance mechanisms.

FIGURE 26-2 **Numerous foods, such as corn and cherries, are natural sources of cyanide although at relatively low levels.**
Courtesy of USDA.

Cyanide is also a byproduct of a number of industrial and commercial processes including metallurgy and mining, organic chemical production, and photographic development. It is found in vehicle exhaust, in some pesticides, and commonly in combustion products. The result of all of this is that there is a presence of cyanide in the natural and industrial environments, which means some degree of exposure is probably more common than we realize.

As a chemical weapon, cyanide was first used during WWI by the French. The Nazis used a form of cyanide gas, known as Zyklon B, in the concentration camps systematically killing millions of Jews and other "undesirables" (Fig. 26–3). In the 1980s, cyanide gas is believed to have been used by Saddam Hussein both against the Kurds in Northern Iraq and against Iran during the Iran–Iraq War. It is still used as a method of capital punishment in some places.

FIGURE 26-3 Canisters of Zyklon from a Nazi death camp storage facility.
Courtesy of the United States Holocaust Memorial Museum.

The term *blood agent* came out of WWI as a way of differentiating the effects of cyanide gas from other gases at the time, which all had specific target organs: skin, eyes, or lungs, whereas cyanide, it was thought, affected primarily the blood.

As potential weapons, the forms of cyanide that are of concern are cyanide chloride and hydrogen cyanide.

Sources

Cyanide is a naturally occurring chemical that is present in a variety of food sources, soil, and water. Industrial exposures occur through metallurgical, iron and steel mills, electroplating, gold refining, photochemical manufacturing processes, plastics, fumigants, and mining. It is used as an industrial intermediate and as a byproduct in chemical industries. Environmental sources of CN^- include cigarette smoke, vehicle exhaust, and biomass burning. Cyanide is commonly present as waste, and many hazardous waste sites have cyanide in both the soil and in water runoff. Firefighters are at particular risk given the ubiquity of cyanide in building materials.

Physical Properties

Cyanide is likely to be disseminated via bomb or aerosolization. It is stored as a colorless liquid but is extremely volatile so that, once released, it almost entirely

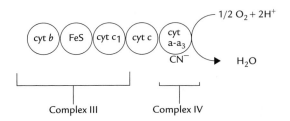

FIGURE 26–4 Cyanide inhibition of cytochrome oxidase.
Adapted from Marc Tischler, Ph.D.

vaporizes. Because it has such low persistence, cyanide vapor rather than liquid is the major threat. Although there is a characteristic description of cyanide smelling like "bitter almonds," this is only so for the hydrogen cyanide form. Even then it is not a reliable means of identifying the gas; only 50% of people can detect it even under ideal environmental conditions.

Pathophysiology

Cyanide's most important toxicity results from inhibition of normal cellular energy metabolism. Specifically, CN^- binds to the iron moiety on cytochrome oxidase, a key enzyme in mitochondrial electron transport (Fig. 26–4). This inhibition disrupts oxidative phosphorylation, shifting the cell from aerobic to anaerobic metabolism. By interrupting normal cellular oxygen utilization, adenosine triphosphate (ATP) production comes to a halt and the ATP/ADP ratio declines, choking off all energy-dependent cellular metabolism. The resultant histotoxic hypoxia shifts the oxygenhemoglobin dissociation curve and diminishes tissue oxygen uptake.

The CNS is especially vulnerable to CN^-, particularly the respiratory drive center. This is a major cause of fatalities in cyanide exposure. Neuronal vulnerability results not only from histotoxic hypoxia but also from increased intracellular calcium and peroxide formation. Rising intracellular calcium in the myocardium causes arrhythmias and myocyte necrosis. Cyanogen chloride, a dimer of cyanide, has an additional toxicity due to the presence of chloride. As noted in the pulmonary agent chapter, chloride is highly irritating to the respiratory epithelium, activating inflammatory pathways and causing pulmonary congestion, interstitial edema, and bronchospasm. Last, CN^- is a potent nucleophile, but its potential genotoxic effect is eclipsed by histotoxic effects that occur at extremely low concentrations.

Toxicity

Cyanide is ubiquitous in nature, and not surprisingly, the human body has evolved mechanisms to clear the toxin from the body. The existence of built-in protection specific to cyanide differentiates this class of agent class from other chemical agents. In the same vein, however, we are equipped similarly to live in a world where we

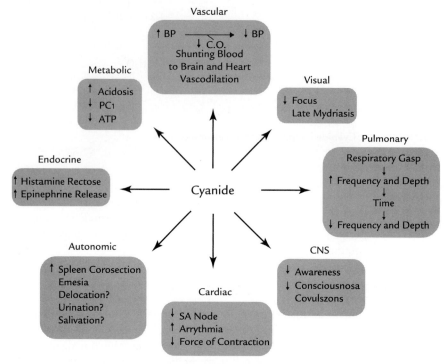

FIGURE 26–5 Signs and symptoms of cyanide poisoning.
Reprinted from the Office of the Surgeon General, Department of the Army from The Textbook of Military Medicine: Medical Aspects of Chemical and Biological Warfare.

are constantly bombarded by cosmic and earth-born ionizing radiation. A consequence of this distinction is that the usual means of measuring toxicity—LD_{50} and LCt_{50}—become less useful due to potential upregulation of metabolic clearance pathways. Put differently, lethal doses are higher with chronic exposure and lower for more acute exposures.

Means of Absorption

In the context of a chemical agent attack, the key port of entry for cyanide is the respiratory tract. Other means of exposure include ingestion, or absorption through skin (particularly wounds or other lesions), or the eyes. Cyanide has both local and systemic affects (see following). Cyanide passes easily through epithelial tissue regardless of the means route exposure.

Signs and Symptoms

As with all the chemical agents, clinical features depend on the severity and means of exposure as well as on the general health of the individual. The onset of symptoms is dose and route dependent varying from seconds with severe exposures to half an hour for mild exposure or ingestion. Cyanide has effects on virtually all systems in the body (Fig. 26–5). The danger of cyanide exposure is primarily in the resultant hypoxic state, especially CNS and myocardium because of the vital role of oxygen for their function. Onset occurs shortly after exposure, and the outcome is either death or timely resolution. If symptoms have not occurred by the time the patient is seen and examined in a safe and clean setting, they are unlikely to occur.

The use of cyanide agents as chemical weapons will present with the acute clinical features, whereas chronic effects are seen more commonly in the occupational setting and are addressed in a subsequent section.

Mild Exposures

The signs and symptoms of cyanide are nonspecific and often transient, typically resolving in mild exposures, but persisting or progressing in more severe cases. These include fatigue, dizziness or vertigo, tachypnea, tachycardia, hyperpnea, agitation, headache, anxiety, nausea, vomiting, peripheral vasodilation, hyperemia, sweating, tinnitus, and behavioral changes. Very low-level or transient exposures may cause coughing, upper respiratory and nasal irritant symptoms, and dysosmia. Mild exposure, and therefore milder illness, occurs with serum levels less than 1.0 mg/mL.

Severe Exposures

More severe illnesses are seen when serum cyanide levels are greater than 1.0 μg/mL, and death occurs at levels greater than 3.0 μg/mL. The severe exposures present within seconds, beginning with those features described in mild exposures, notably hyperpnea, but then progress to serious cardiovascular and CNS toxicity, including loss of consciousness; seizures; coma; and dilated, fixed pupils; and hemodynamic compromise, including an initial hypertension followed by hypotension; slowed respirations and apnea; and arrhythmias and cardiac arrest. Convulsions, tremor, muscular rigidity, and nausea and vomiting occur commonly. Progression to death often occur within as few as five minutes. If the patient survives, CNS symptoms may persist for long periods, or permanently, even after clearance of the poison (see Chronic Effects).

A variable physical finding worth mentioning is the reddening of the skin to a cherry color, similar to that seen with carbon monoxide poisoning. This results from increased O_2 levels in the venous blood due to the inability of tissues to absorb oxygen.

Cyanogen chloride has additional features related to respiratory tract damage caused by the chlorine moiety, including pronounced dyspnea, lacrimation, cough, rhinorrhea, increased secretions, and pulmonary edema.

Laboratory Tests

The most prominent feature associated with cyanide exposure is anion gap acidosis with an accompanying high lactate level—secondary to the anaerobic respiration that occurs. Cyanide is often detected by testing for its primary metabolite, thiocyanate, in serum, urine, or saliva. Because it does not persist in the body, delayed testing may miss it. Serum cyanide levels should be checked both to confirm exposure and to help assess the severity of exposure. Serum cyanide levels of less than 1.0 μg/ml are generally considered mild. Exposure risk of death increases substantially as serum levels approach 3.0 μg/ml. A third lab finding that may be useful is a venous O_2 level, an elevation of which is consistent with cyanide exposure. The standard means of monitoring methemoglobin levels are of less value with cyanide poisoning often because they give falsely low values.

Differential Diagnosis

As always, the clinical setting is a key factor in diagnosing cyanide poisoning. An identified attack reduces the differential diagnosis to the various agents. Multiple patients with similar signs and symptoms can also be suggestive of the release of agents. Nerve gas and cyanide may be distinguished by skin color—nerve agent exposure is associated with cyanosis, whereas cyanide is associated with skin that has a cherry red color. In an isolated patient, with an anion gap metabolic acidosis, the differential includes diabetic ketoacidosis, ethanol, renal failure, iron overdose, starvation, seizures, sepsis, salicylism, rhabdomyolysis, and poisoning from various alcohols, such as methanol or ethylene glycol. The clinician must also consider sepsis, carbon monoxide poisoning, and other causes of tissue hypoxia such as myocardial infarction, asphyxiation, or nitroprusside toxicity.

Chronic Effects

Cyanide causes long-term endocrine, neurologic, and possibly behavioral effects. By interfering with the uptake of iodine, cyanide inhibits thyroid hormone production. Occupational studies have demonstrated excess prevalence of chest pain and palpitations, increases in lymphocyte and hemoglobin levels, headaches, paresthesias, and a range of neurasthenic symptoms, such as chronic fatigue, sleep disorders, memory deficits, and dizziness. Neurotoxic sequelae, including tremors, incoordination, and restless legs, have been described. Developmental, reproductive, or carcinogenic effects in humans have not been demonstrated, although lab animal testing indicates possible developmental effects. Konzo, a nonprogressive demyelinating disorder presenting with ataxia and upper motor neuron signs, is found in individuals who consume cassava beans. No teratogenic, carcinogenic, or reproductive developmental effects have been associated with H_2S exposures. Cyanide may be capable of crossing the blood–brain barrier; however, the health implications are uncertain.

TABLE 26–1	Making the Diagnosis: Blood Agents
Pattern of Presentation	Multiple, concurrent cases of flushing, weakness, loss of consciousness, reports of bitter almond odor prior to symptoms
Features	Mild—flushing, dizziness, headache, nausea, and vomiting Severe—hemodynamic compromise, arrhythmias, seizures, and coma
Findings	Lab—metabolic acidosis with elevated lactate and elevated venous O_2 levels, as well as presence of CN^- in the serum

Vulnerable Groups

Cyanide's toxicity knows no distinctions based on age, race, gender, pregnancy status, or co-morbid conditions. However, individuals with protein and vitamin deficiencies, particularly B12 and riboflavin, are more sensitive to the neurotoxic effects of cyanide. Cyanide's arrythmogenic effect represents a potential risk at lower concentrations for individuals with preexisting cardiac disease. Children's greater body surface area-to-volume ratios and increased baseline ventilatory rate experience greater exposure for a given concentration compared to adults.

Diagnosis

The presentation of a dyspneic, stricken individual with rose-color skin and a faint almond odor on the breath works well for the classic murder mysteries, but cannot be relied on in many instances. A deliberate attack with cyanide will be diagnosable based on a clustering of patients presenting with the constellation of symptoms described previously (Table 26–1). The *sine qua non* of cyanide exposure is the rapid onset of dyspnea, tachypnea, and tachyarrythmias, followed by hypotension, apnea, bradycardia, and death. If recognized early, treatment is life saving.

Decontamination

Victims should be removed from exposure. Assuming you are seeing the patient in a stable and safe setting, there is an extremely low risk of secondary exposure as cyanide vapor dissipates readily. If by some chance there is liquid cyanide still present on the skin, it can be decontaminated with soap and water. Contaminated clothes should be removed by those wearing appropriate safety equipment. The clothes then may be double bagged or stored in rubber containers. First responders should use SCBA and chemically protective clothing. As noted, once decontamination has occurred, risk to treating HCWs is minimal.

TABLE 26–2	Critical First Steps: Blood Agents
Medical Treatment	100% O₂, amyl nitrate, sodium thiosulfate, supportive
Infection Control	None
Public Health	Contact local or state health department

Treatment

Management of cyanide toxicity varies according to the severity of the exposure; however, supportive care should be provided (Table 26–2). Patients with severe respiratory difficulty may require intubation or assisted ventilation. Delivery of 100% oxygen is recommended, as it has been shown in animal studies to be beneficial for cyanide-poisoned patients regardless of severity of exposure and despite the inhibition of aerobic metabolism by cyanide. Hyperbaric O₂ has not, however, been shown to be of real benefit.

Medical management includes antidotes that act to clear CN^- from the blood, primarily using intravenous sodium thiosulfate, methemoglobin-forming agents, (such as amyl nitrate, inhaled from crushed perles), or sodium nitrite given intravenously (see Appendix). The methemoglobin-forming agents are given first—amyl nitrate initially, and if no improvement is seen then a sodium

TABLE 26–3	Quick Reference Guide: Blood Agents
Diagnostic Keys	Mild: dizziness, headache, and vomiting
	Severe: arrhythmias, seizures
Risk Factors	Industrial workers
Transmittable	No
Transmission	None
Management	100% O₂, amyl nitrate, sodium thiosulfate, supportive
Prevention	
Preexposure	None
Postexposure	None
Containment	Patient: none
	Secondary contacts: none
Populations at Heightened Risk	Children, pregnant women, elderly
Long-term Issues	CNS impairment
Public Health Concerns	
Contacts	Local, state DPH, CDC, FBI

nitrite infusion is indicated. This should be followed by administration of sodium thiosulfate. Sodium thiosulfate works as a sulfur donor, converting cyanide to thiocyanate, a non-toxic, readily cleared compound. The methemoglobin-forming agents work by causing the iron portion of hemoglobin to oxidize, resulting in the formation of methemoglobin for which CN^- has much higher affinity than it does for cytochrome oxidase. Once unbound, cytochrome oxidase resumes aerobic metabolism. Methemoglobin serves as a site for binding extracellular CN^-, and once bound, the CN^- is excreted readily from the body. The production of too high a level of methemoglobinemia impairs O_2 delivery in and of itself, an obvious concern in patients who are already hypoxic. However, mild cyanosis in a patient who has received nitrite therapy is evidence of a clinically beneficial methemoglobinemia. Men and women should receive standard adult dosing schedules, whereas children should receive standard pediatric dosing where applicable. Administration of these nitrites should be done slowly to minimize adverse effects. Since they cause vasodilation, hypotension occurs commonly with nitrite administration. This is not usually a significant problem in those with severe exposures, as they are in a prone position and should remain so.

Other treatments used internationally include Kelocyanor (cobalt edetate), which is used in the United Kingdom and France, and dimethylaminophenol (DMAP) has been used in Germany. Trials are under way in the United States investigating the use of hydroxycobalamin.

Summary

Blood agents are metabolic poisons that disrupt cellular oxygen utilization and cause tissue hypoxia. Their primary sites of action are the CNS and cardiovascular systems. If severe, patients develop dyspnea, convulsions, arrythmias, and death by histotoxic asphyxiation. These effects occur whether the route of exposure is respiratory, oral, or dermal. Chronic effects involving the CNS, thyroid, and cardiovascular are documented. Carcinogenic, developmental, and genotoxic effects have not been described in humans. Antidotes combined with supportive therapy are effective treatments.

SECTION V

Nuclear
and
Radiation
Syndromes

CHAPTER 27

A Brief History of Nuclear Weapons and the Atomic Age

"I am become death, the destroyer of worlds."
> Quote from Bhagavita attributed to J. Robert Oppenheimer
> "Father of the Atomic Bomb"

The history of the nuclear age is a series of steps—many quite small, others of medium stride, and others still veritable leaps. Discoveries such as the presence of electromagnetic forces noted by Maxwell to Roentgen's stumbling onto the presence of x-rays, to Becquerel finding naturally occurring x-rays from uranium, to the Curies discovering the key components of are just a few examples of such steps (Fig. 27–1). From the late 19th century and into the 20th century, several important larger discoveries were made: Rutherford discovered alpha and beta particles, and Planck articulated the quantum theory of physics. During that time fundamental principles of radioactivity were being sorted out; such as isotopes, radioactive decay, and the properties of the types of radiation.

In 1905, the history of science and of the world was altered irrevocably when Albert Einstein, working full time as a patent clerk in Bern, Germany, conceived of and derived what is perhaps the most recognizable equation in history: $E = mc^2$. Einstein's theory of relativity had profound implications, a worthy discussion of which is beyond the scope of this text. For our purposes, the most important was its inference that enormous energy is contained in even a single atom. It was not long before the hunt to access that energy began (Fig. 27–3).

The critical breakthroughs to harnessing atomic energy came in the late 1930s. When nuclei were bombarded with slow-moving neutrons, they became unstable and

FIGURE 27-1 **Marie Curie (pictured here) and her husband Pierre's work led them to identify radioactivity and thus win them the Nobel Prize.**
Courtesy of the Library of Congress, Prints and Photographs Division.

split into smaller particles. However, these smaller particles added up to a weight less than that of the original nuclei. The missing mass had become energy: a vast amount of energy. The implications of this research, particularly its military potential was realized quickly in several Western nations, including Germany, England, and the United States. As the 1930s progressed, tensions between Hitler's Germany and other European powers erupted into World War II and the work of atomic physicists worldwide became a critical resource for the war effort.

The story of the development of nuclear weapons is a fascinating and pivotal one that altered the history of the 20th-century world. During WWII, both the Nazis and the Americans raced to develop viable nuclear weapons (Fig. 27–2). Although Germans had a head start, when the allied forces forced the unconditional surrender of German forces on May 7, 1945, Nazis scientists had yet to complete their work. In the United States, the responsibility for developing an atomic weapon began with the so-called Manhattan Project. In fact, some effort to build a bomb began in late 1939, but the work has not coordinated, poorly funded, and generally lethargic. Full-fledged research began after the U.S. government heeded repeated warnings by the British government and escaped German scientists, including the

FIGURE 27–2 Enrico Fermi performed the first controlled chain reaction of uranium.
Courtesy of Ernest Orlando Lawrence Berkeley National Laboratory.

world's most famous scientist, Albert Einstein. Once the Manhattan Project was fully funded, (ultimately receiving $2 billion) and had full presidential and military backing, research progressed rapidly, facilitated by advancements made at universities and government laboratories throughout the country. Considering its late start, the Manhattan Project went quite quickly, and bomb tests were first conducted in New Mexico on July 16, 1945.

President Truman's decision to use the atom bomb is one of the most controversial decisions in U.S. history. Personal letters and journal entries reflect the

FIGURE 27-3 Aspects of Einstein's work formed the theoretical basis for the development of the atomic bomb.
Courtesy of the Library of Congress, Prints and Photographs Division.

personal struggle he went through in making the decision. The argument to use the bomb was driven by the number of lives, both American and Japanese, that would be saved if a direct invasion of Japan could be avoided. A second often understated concern was the emerging antagonism between the West and the USSR. Fear that the Soviet involvement in the Pacific theatre—and a postwar presence in the region—underlaid the great reluctance to "share" information about the bomb with an ally who most believed would soon be an adversary. Still, Truman gave Japan a last opportunity for surrender. Three weeks before Hiroshima "The Potsdam Declaration" was issued to the Japanese government warning that failure to surrender would lead to Japan's "prompt destruction." The Japanese refused to surrender and as promised, the Enola Gay dropped the first atomic bomb, dubbed "Little Boy," on Hiroshima in August 1945. When the Imperial Japanese government still did not surrender, a second bomb, dubbed. "Fat Man," was dropped on Nagasaki three days later (Fig. 27–4). Over 100,000 people died and nearly twice that were injured.

The arms race that developed during the Cold War is both sufficiently well known and well documented. Nuclear tensions waxed and waned throughout the Cold War. In the 1960s, three other nations joined the United States and the USSR as nuclear powers: France, Britain, and China. The 1970s saw the first attempts at limiting the arms race with Strategic Arms Limitation Talks (SALT) being signed

FIGURE 27–4 **Mushroom cloud from the detonation of the atomic bomb dropped on Nagasaki, Japan.**
Courtesy of the U.S. National Archives and Records Administration.

between the United States and the USSR. The 1980s began with a marked inten-sification of the Cold War (Fig. 27–5), but ended with the dramatic and symbolic destruction of the Berlin wall. Fortunately, no nuclear weapon was ever used as a strategic weapon, even though testing by the world's nuclear powers was common from the 1950s to mid-1970.

However, other events highlight the risks associated with the many nonmilitary uses of radioactive elements in the Atomic Age. In 1979, a core meltdown in a re-actor at the Three Mile Island facility in Pennsylvania (Fig. 27–6) led to the release of a small amount of radioactivity to the surrounding area. No injuries or deaths resulted, but the accident was a warning of what would later be seen in Chernobyl. In 1987, an abandoned medical source of cesium-137 exposed 249 Brazilians to ionizing radiation, resulting in 52 hospitalizations, 4 deaths, and thousands of postevent medical visits, most of which were "worried well" (see Of Note: Goiania, Brazil). Similarly, in 2001 discarded strontium-90 canisters from nuclear batteries in the Republic of Georgia led to acute radiation sickness in three exposed individuals.

FIGURE 27-5   A test launch of a Minuteman I intercontinental ballistic missile (ICBM): one of several ICBMs developed during the Cold War.
Courtesy of the United States Air Force.

One of the largest and worst industrial accidents in the history of humanity occurred April 26, 1986, near Belarus, Ukraine, at the Chernobyl Nuclear facility (Fig. 27–7). Human and mechanical errors combined with a reactor that had major design flaws resulted in two explosions of the reactor core and the release of large quantities of radioactive material into the environment. The Soviets had tried to keep the disaster a secret, but when technicians at a Swedish nuclear plant were found to have contaminants on their clothes, the truth quickly came to light. In the aftermath of the accident, some 100,000 people were evacuated and a geographically large and agriculturally productive area was contaminated with radioactive fallout. Thirty workers at the plant died from acute radiation sickness, and thousands more died over the ensuing months. Psychological fallout was significant, with a large number of individuals committing suicide fearing the consequences of radiation exposure.

Deemed a "global and ecological disaster" the consequences of Chernobyl are difficult to assess with full confidence. However, some details are unquestioned. In total, over forty radionuclides were released into the environment, most significantly

FIGURE 27-6 Three Mile Island.
Courtesy of the CDC.

FIGURE 27-7 The Chernobyl nuclear plant following the reactor meltdown.
Courtesy of Environmentalists for Nuclear Energy.

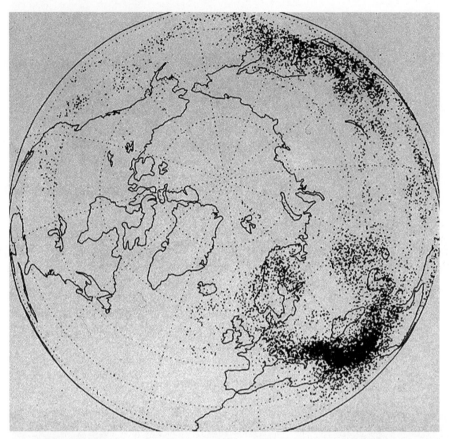

FIGURE 27-8   High winds carried radioactive particles from Chernobyl with world-wide contamination.
Courtesy of the Lawrence Livermore National Laboratory and the U.S. Department of Energy.

iodine-131 (half-life 8 days), cesium-137 (half-life 30 years), strontium-90 (half-life 29 years), and numerous plutonium isotopes (half-life 24,000 years). Approximately 8,000 people died in the immediate aftermath. Nearly the entire Northern hemisphere experienced at least some environmental effects of the radiation, although northern Europe was impacted most heavily (Fig. 27–8). All residents within a 10-km radius of the site were resettled permanently. In the decade following the accident, researchers observed a ten-fold increase in thyroid cancer in the pediatric population. The United Nations acknowledges that the full extent of human and environmental damage will not be known until approximately 2016 when the radiation will be reduced sufficiently to allow a better assessment of the long-term consequences and genetic damage. It is of little solace that Chernobyl provide improved understanding of acute radiation sickness, and prompted international calls to implement better safety protocols and design of nuclear plants in Russia and elsewhere.

FIGURE 27-9   Following the reactor meltdown, the Chernobyl nuclear plant was
covered in a large concrete shelter.

Courtesy of Greenpeace/Clive Shirley.

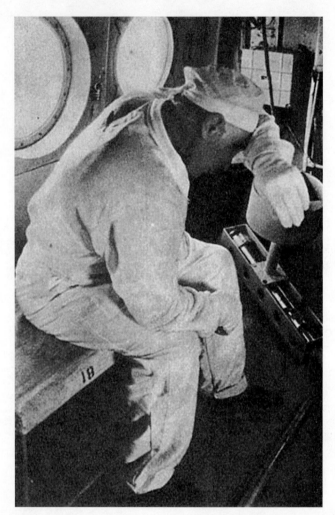

FIGURE 27-10   Clean-up worker from Chernobyl accident.
Courtesy of Progetto Humus.

The Chernobyl incident provides some insight into what might be expected with a terrorist attack on a nuclear facility. In the immediate aftermath of the explosion and fire, 187 people fell ill from acute radiation sickness; 31 of these died (Fig. 27–9 and Fig. 27–10). Most of these early casualties were firefighters who combated the blaze (Fig. 27–11). Nearly 400,000 workers were brought in over the subsequent years as "liquidators" and were tasked with burying radioactive waste and a concrete "sarcophagus" around the burned out reactor core. Nearly 30,000 of these workers became sick, and 5,000 were permanently incapacitated for work. Accurate figures are difficult to come by, but somewhere between 100,000 and a

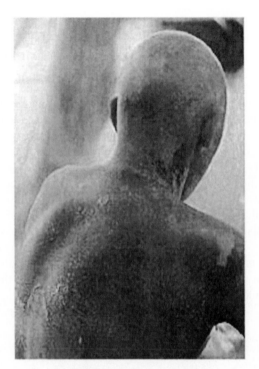

FIGURE 27-11   Radiation burns on a Chernobyl firefighter.

quarter of a million people were evacuated, most of them within a 30-km radius of the plant. Over a half-million individuals participated in the plant cleanup and remediation. Subsequent radiation-related deaths, including cancers, are estimated at approximately 2,500 people. Excesses in thyroid cancer in children were seen, although anticipated excesses in hematologic malignancies were not seen. The total amount of radioactivity released will never be known, but the official Soviet figure of 90 million curies represents a minimum estimate. Other estimates indicate that total radioactivity might have been several times higher. In terms of the amount of radioactive fallout, Chernobyl was comparable to the fallout expected with a medium-size nuclear strike.

The destroyed reactor liberated hundreds of times more radiation than that produced by the atomic bombings of Hiroshima and Nagasaki. The intensity of gamma radiation on the site of the power plant reached more than 100 roentgens an hour. This level produces in an hour doses hundreds of times the maximum dose the International Commission on Radiological Protection recommends for members of the public a year.

Chernobyl remains one of the most radioactive places on earth, containing tens of thousands of metric tons of nuclear fuel. Radiation levels within the reactor core itself remain lethal to all forms of life, and radioactive waste buried in pits throughout a 30-km circumference hot zone will ensure that the region remains uninhabitable for centuries.

Unfortunately, the story of nuclear power as a danger does not end with the Chernobyl disaster. In the post-September 11 world, nuclear materials in any form are now understood by all sides to be potential weapons. The various means that can potentially be used will be discussed more in depth, but the possibility of this threat should not be ignored. Large quantities of nuclear weapons left over from the Soviet arsenal are unaccounted for. Based on acknowledged concern with existing safety systems, the G7 nations decided to give Russia $20 billion to protect and dismantle its nuclear arsenal in 2002. Despite such international efforts, multiple efforts to sell, purchase, smuggle, or steal nuclear material worldwide have been reported in the past 10 years. There have been 14 separate arrests related to illegal nuclear trade in the 8-year period between 1992 and 2000 in Europe or Russia. Radiological threats in the United States exist as well. A *Washington Post* cover story reported that federal agencies led an emergency search for dirty bombs in five U.S. cities on New Years Eve in 2003. The irony is that though the Cold War has long since warmed and the nuclear arms race is over, the possibility that some kind of nuclear device could be used may be higher than ever.

CHAPTER 28

An Overview of Radiation Physics and Injury

Life on earth evolved in a shower of ionizing radiation. Indeed, evolution as we currently understand it is singularly dependent on ionizing radiation's capacity to alter genomic sequences and change the structure and function of intracellular proteins. The human body experiences some 10^{12} collisions daily from ionizing radiation from a wide range of cosmic and earth-bound sources (Fig. 28–1). For obvious reasons, therefore, humans have evolved repair mechanisms and genetic redundancies that enable us to withstand these exposures without obvious effect. Much of the information used to determine the health effects of radiation on humans derives from the atomic bombing of Hiroshima and Nagasaki, as well as industrial, research, and environmental accidents. To better understand the health effects of ionizing radiation, and specifically, the issue of nuclear terror, this section provides a short introduction to radiation terminology and radiation physics.

Radiation is the transfer of energy through space as electromagnetic waves or particulate matter. The energy contained in radiation is inversely proportional to the wavelength. There are two types of radiation: nonionizing and ionizing. Ionizing radiation is of short electromagnetic length and high energy and has the capability to break covalent nuclear bonds; thereby altering genomic sequences, destroying the structure and function of proteins, and generating free radicals that are directly cytotoxic. Higher-frequency wavelengths (e.g., x- or gamma rays and alpha particles) have higher energy and are more destructive than low-frequency radiation (e.g., radiowaves or microwaves). Nonionizing radiation (e.g., microwaves and radiowaves) is not considered potential weapon of terror and is not discussed further (Table 28–1).

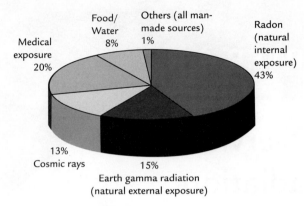

FIGURE 28–1   Sources of radiation exposure.
Courtesy of WHO.

Multiple manmade and natural sources of radiation exist (Fig. 28–1). Background radiation from natural and artificial sources averages about 360 millirem per year per person in the United States. This is an order of magnitude below the current permissible exposure limit and below the LD_{50} for radiation of 450,000 millirem. A natural source of radiation is cosmic rays. These include charged and neutral particles as well as electromagnetic waves that originate from space. Another natural source of radiation is found within the earth itself. One example is radon gas, a familiar environmental concern to homeowners. Radon represents the largest single source of ionizing radiation exposure for Americans. Medical diagnostic testing, cigarette smoke, and many other sources of radiation exist as well.

Routes of exposure to ionizing radiation are three-fold: dermal, inhalation, and ingestion. Various types of ionizing radiation exist, including alpha, beta, neutron, x-ray, and gamma rays. These forms of radiation are differentiated by their physical properties and the kinetics of their energy, which in turn, determine their potential for adverse health effects. Low linear energy transfer (LET) particles—such as x-rays and gamma rays—pass through most tissues and transfer little of their energy to surrounding tissues. In contrast, high LET particles—such as electrons, neutrons, protons, and alpha particles—expend most of their energy over a short distance and are more destructive to human tissues.

There are three types of ionizing radiation particles: alpha, beta, and neutrons. Alpha particles (Fig. 28–2) are the most massive of the radioactive particles and the weakest energetically. They are comprised of two protons plus two neutrons. By composition, they are helium nuclei missing their electrons, a fact that leaves them intensely electrophilic. Alpha particles cannot travel far and are stopped by virtually any physical barrier: a sheet of paper, clothing, even the epidermis. Alpha particles pose a hazard mainly if internalized through ingestion or inhalation.

FIGURE 28-2 An example of alpha particle decay.
Courtesy of Lawrence Berkeley National Laboratory.

Beta particles (Fig. 28–3) are high-energy electrons or positrons (positively charged electrons) emitted from a nucleus with a variable amount of energy. They travel about a meter through the air, a millimeter into lead, and many millimeters into tissue. If accumulated on the skin in large numbers, they produce damage to the basal stratum of the skin resulting in a "beta burn." Internalized beta particles cause extensive damage to tissues and cells in their path. However, beta particles cannot penetrate through glass or even aluminum foil.

Neutrons (Fig. 28–4) are electrically neutral, but once emitted from a nucleus, they cause damage in two ways. First, neutrons can change nonionizing isotopes into ionizing isotopes, a process known as "neutron capture." In humans, for example, neutron capture of sodium atoms changes them from Na^{23} to Na^{24}, turning the

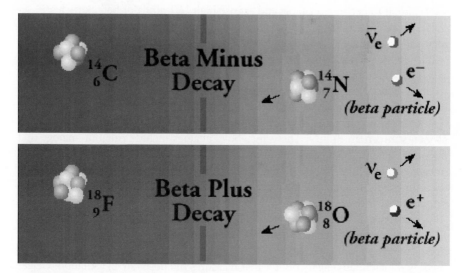

FIGURE 28-3 The two forms of beta decay with either an electron or a positron resulting.
Courtesy of Lawrence Berkeley National Laboratory.

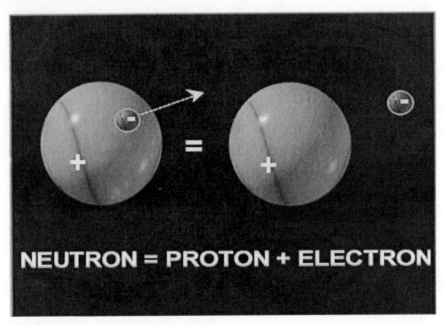

FIGURE   28–4   Neutron decay.
Courtesy of OSHA.

latter into an internal radiation source. Neutrons directly impacting another nucleus may cause it to become unstable and release high LET alpha or beta particles.

Gamma rays are similar to x-rays in that they are short wavelength, high energy, and deeply penetrating (Fig. 28–5). Gamma rays result from energy transitions within the nucleus and travel deeply into tissues. Gamma rays are released as part of a nuclear detonation, as well as in radioactive fallout. X-rays have a longer wavelength and therefore less energy than gamma rays. X-rays are produced as the result of high-energy electrons passing through a positive nucleus.

Radioactive elements may be incorporated preferentially along with normal elements in certain tissues. These metabolic analogs include calcium analogs that have their effects in bone (e.g., radium and strontium), water analogs (e.g., tritium), or radioactive iodine will be taken up by the thyroid. These metabolic analogs are used for medical diagnostic and treatment purposes because of their ability to be interchanged with necessary cellular elements.

A variety of terms are used when talking about ionizing radiation (Table 28–2). In general, these terms refer to dose, dose rate, and biological system effects. Radioactivity refers to the rate at which a material emits radiation, and it is measured by the familiar Geiger counter that measures the dose rate of radiation. Geiger counters count the number of disintegrations per unit time (minutes, seconds, or hours) in units of Curies or Becquerels. A rad, or radiation-absorbed dose, is the amount of energy deposited per unit tissue by ionizing radiation. This term is

FIGURE 28–5   Gamma and x-rays are emitted from excited electron orbitals.
Courtesy of OSHA.

antiquated and clinicians should be familiar with the term *Gray*. One Gray equals 100 rads. A roentgen is a measure of exposure amount of electromagnetic radiation (gamma and x-rays) and measures the ionizations of the molecules in a mass of air. The rem, or roentgen equivalent man, is a term derived for the purposes of defining the biological effects of radiation. Rem is a dated term as well, superseded by the Sievert (1 Sv = 100 rem). Both rem and Sievert define the biological system's response to absorbed radiation because ionizing radiation's biological effects may differ even when the absorbed dose is equivalent. This difference is due to the type of radiation exposure.

TABLE 28–1   Characteristics of the Electromagnetic Spectrum

Region	Wavelength (Cm)	Frequency (Hz)	Energy (Joules)
Radio	>10	$<3 \times 10^9$	$<10^{-24}$
Microwave	$10 - 0.01$	$3 \times 10^9 - 3 \times 10^{12}$	10^{-24}
Infrared	$0.01 - 7 \times 10^{-5}$	$3 \times 10^{12} - 4.3 \times 10^{14}$	$10^{-23} - 10^{-19}$
Visible	$7 \times 10^{-5} - 4 \times 10^{-5}$	$4.3 \times 10^{14} - 7.5 \times 10^{14}$	10^{-19}
Ultraviolet	$4 \times 10^{-5} - 10^{-7}$	$7.5 \times 10^{14} - 3 \times 10^{17}$	$10^{-19} - 10^{-17}$
X-rays	$10^{-7} - 10^{-9}$	$3 \times 10^{17} - 3 \times 10^{19}$	$10^{-17} - 10^{-14}$
Gamma	$<10^{-9}$	$>3 \times 10^{19}$	$10^{-14} - 10^{-10}$

TABLE 28-2   Radiation Measurements and Units

Characteristic Measured	SI Unit	Old Unit	Conversion
Exposure	None	Roentgen (R) = 2×10^9 ions per cm^3 of air	—
Dose absorbed	Gray (Gy)	Radiation-absorbed dose (rad) =100 ergs energy/ 1 g mass	1 Gy = 100 rad 1 cGy = 1 rad
Dose equivalent	Sievert (Sv)	Roentgen equivalent unit (rem) 1 rem (α) = 20 $\times$ rad 1 rem (neutron) = 2-11 $\times$ rad	1 Sv = 100 rem 1 cSv = 1 rem
Atomic transformations (disintegrations) per unit time	Becquerel (Bq) = 1 transformation per second	Curie (Ci) = 3.7×10^{10} transformations per second	1 Bq = 2.7 $\times$ 10^{-11} Ci

Radiation injuries result from a variety of exposures and can occur rapidly or with varying degrees of latency depending on the dose. The types of exposures are:

- *Irradiation.* Defined as the process of exposure to radiation from an external source without physical contact with the body, that is, without direct contamination. Irradiation occurs with x-rays, gamma rays, neutrons, and alpha or beta particles. Depending on the energy level, the radiation may be non-penetrating (affecting only the skin) or penetrating (affecting areas deeper than the skin). Because decontamination is not necessary with irradiation, medical care should be the only concern in an irradiated patient. Similarly, it follows that irradiated food poses no radiation threat to consumers.
- *Internal contamination.* Defined as radionuclides entering the body via inhalation, ingestion, or wounds and potentially injured internal organs and tissues. No risk is posed to others from internal contamination.
- *External contamination.* Defined as the presence of radionuclides on exposed body surfaces, often in the form of particles or liquid, and that may further contaminate the patient via absorption and inhalation. These particles are transferrable, mandating adequate decontamination to avoid contamination of health care providers and emergency response workers.

CHAPTER 29

Introduction to Nuclear and Radiation Weapons

"It's not a matter of if; it's a matter of when."

> General Eugene Habiger, head of American strategic weapons until 1998, referring to the likelihood of a terrorist act in the United States involving a radiologic device.

Humans evolved in a universe where ionizing radiation is a fact of life. Yet for many it remains a mysterious force, deadly, unseen, and impossible to detect with our senses. Atomic energy and the atomic bomb loom large in a public psyche shaped by post-WWII geopolitics. For these reasons, public concern with atomic energy, nuclear waste, and the risk of nuclear terrorism probably exceeds that of any other agents. The world is filled with a variety of sources of ionizing radiation, and consequently, many experts consider the likelihood of a terrorist act involving nuclear material to be simply a matter of time. For decades, life on earth was threatened by the Cold War. In the post-September 11 world, the nuclear threat is more limited and ironically more unpredictable. This chapter provides a brief synopsis of the possible scenarios in which terrorists could use nuclear weapons or nuclear material (Table 29–1).

Detonation of Nuclear Warheads

The destructive potential of a nuclear weapon is enormous, encompassing explosive force, radiation, and extreme heat. The latter two are discussed in terms of medical management. An attack on a nuclear reactor would be more an issue of radioactive fallout rather than one of thermonuclear explosion. The detonation of a

TABLE 29–1   Classification of Radiological Terror

Radiation Source	Example
Point source radiation	Placing a gamma source in densely populated area
Radiologic dispersal device (a.k.a. dirty bomb)	Spreading radiation via detonation of a conventional explosive
Nuclear facility attack/sabotage	Disrupting nuclear reactor safety/protective features
Improvised nuclear device	Detonating a modified or homemade nuclear device in a city
Nuclear weapon	Detonating nuclear weapon that was built by a country for its nuclear arsenal

Reproduced, with permission, from Walter FG [editor]: Advanced Hazmat Life Support Provider Manual, 3rd ed. Arizona Board of Regents, 2003.

nuclear bomb is the most complicated event, both in scale of damage and in mechanisms of injury.

In the event of a nuclear bomb detonation, the energetics of impact are accounted for in the following ways: the blast itself accounts for roughly 50% of the released energy and occurs in the form of pressure waves. Heat (thermal radiation) and light account for 35% of the energy. This includes lower energy radiation such as infrared, visible, and ultraviolet light. Temperatures at sites closest to the blast reach over 20 million degrees Fahrenheit. It is the heat and light that give the classic feature of the flash followed by the mushrooming fireball. Ionizing radiation, in contrast, accounts for only 5% of the energy released by a nuclear weapon, primarily in the form of gamma rays and neutrons. Radioactive fallout accounts for another 10% of the released energy. Fallout includes hundreds of radioactive products from the blast. These particles primarily give off beta and gamma radiation. Lighter particles are carried high into the atmosphere and travel around the globe. Heavier particles represent a more serious health risk as they settle in the area surrounding the detonation site. Finally, an electromagnetic blast accounts for 1% of the energy released by a nuclear bomb. Although this does not pose a direct medical threat, it disrupts electrical equipment, including vehicles, hospital equipment, and communications. Direct injuries from the force of the blast and from the thermal energy released are responsible for most of the immediate mortality and morbidity following a nuclear attack.

Several countries—including North Korea, Pakistan, and Iran—are known or believed to have the technological capability to develop and fire a nuclear warhead. Concern centers around these nations' willingness, aided by established nuclear powers (China or France) to assist other nations or terrorist organizations in weapons development by sharing technology, information, and resources. Arms experts fear such nations are less constrained than established nuclear powers due to political instability, aggressive histories, or ideological or religious zealousness. Since the fall of the Soviet Union, the security and tracking of its nuclear stockpile has been a major international concern. A significant number of nuclear weapons are unaccounted for, perhaps as many as 200. Included among the missing are dozens of Soviet-built nuclear weapons that fit into small suitcases. A major concern is that these devices could end up in the hands of extremist groups, potentially causing death, destruction, and devastation on an unimaginable scale. Thankfully, many experts believe that the likelihood of a terrorist act involving a stolen warhead is quite small, as the built-in antitheft and security measures are too sophisticated. Such a weapon may be disassembled, the radioactive material extracted, and then used in any number of ways to be summarized in the following sections.

☣Improvised Nuclear Devices

A second nuclear scenario, and a more likely one, involves the use of an improvised nuclear device (IND). Although logistically difficult, individuals or extremist groups may acquire the means and materials needed to construct a nuclear device capable of a fission reaction; alternatively they could modify an acquired nuclear weapon. Such a device would be comparable to the nuclear weapons used in military settings, capable of killing thousands of people. As with conventional devices, detonation of an IND would result in large-scale radiation and thermal injuries.

The capacity to build an IND hinges on access to either plutonium or highly enriched uranium (HEU), which are the essential raw materials needed to build a nuclear bomb. Between 5 and 15 kg of plutonium is all that's needed to make a nuclear device. Fortunately, there are other considerations that arbitrate this concern. For example, plutonium is not as readily available as HEU and possesses considerable and immediate health risks to those handling it. There are currently about 500 tons of plutonium available—half is in the hands of the military (Fig. 29–1).

HEU is a different story. Securities experts express concern because of the abundance and relative safety in the handling of HEU. HEU can be used to make a bomb, and it further poses no risk to the handlers (Fig. 29–2). Roughly 1,000 tons of HEU exists in the former Soviet states alone, enough to make 10,000 bombs. Since September 11, there has been a more organized effort to reduce the availability of HEU. Whether this effort has mitigated the risk posed by HEU is uncertain.

☣Attacks on Nuclear Facilities

The third scenario involves a coordinated attack or sabotage on nuclear power plants. There are approximately one hundred nuclear reactors in the United States in thirty-one states, and over four hundred reactors worldwide. Other sites involved in U.S. nuclear weapons production exist as well. Each site is a potential target, although assessing the risk to workers, communities, and environments is difficult due to variability in the facilities' structural ability to withstand attack, onsite security, and the nature of the attack.

FIGURE 29-1   One of the many U.S. government efforts to reduce the availability of plutonium; here, removing soil contaminated with it.
Courtesy of the U.S. Environmental Protection Agency.

Prior to September 11, 2001, the nuclear industry's security systems were grossly inadequate. Testing of security at nuclear facilities against sabotage and theft found a sobering 50% failure rate. Even more astonishing, the National Research Council website provided access to design and layout details, security weaknesses, and structural flaws for the nation's nuclear facilities up until September 11. Although security systems and information access have been tightened dramatically over the last several years, how well these efforts are working is unclear.

Experts disagree on the likelihood of such a scenario, but there is no doubt that it figures prominently in the public eye. Some facts are pretty clear, and to a smaller degree, reassuring. First, an attack on a nuclear facility will not result in a thermonuclear explosion. In a worst-case scenario, a reactor breach and radiation release as occurred at Chernobyl would result.

Second, the safety design of reactors reduces greatly the likelihood of a successful attack as the amount of explosives required to cause a breach is prohibitive. Four-foot-thick steel and reinforced concrete and one-foot steel-encased reactors make this scenario improbable as a terrorist option. Indeed, crashing a jumbo jet into a reactor would not necessarily cause a significant nuclear event. However, not all nuclear power or security experts agree with this assessment, including the International Physicians for the Prevention of Nuclear War. Should such an attack be successful, the risk to workers, communities, and the environment is sufficiently great that steps are required to prepare for and prevent such an eventuality. They argue that a large

FIGURE 29-2   Low enriched uranium powder—a form that cannot be weaponized.
Courtesy of Department of Energy/National Nuclear Security Administration.

enough explosion might cause radioactive coolant to leak; and that the contamination from this alone is a serious threat to plant workers and those responsible for cleanup (Fig. 29–3). Further, such an event will have grave psychological effects on the surrounding communities locally and at other communities in which nuclear plants exist.

Radiologic Dispersal Device

The most often discussed nuclear scenario is the so-called radiologic dispersal device, commonly referred to as a dirty bomb or RDD. This involves using radioactive materials within a conventional explosive device that, when detonated, disperses radioactive material. Dirty bombs are relatively easy to make, but pose a much smaller danger in terms of severity of injuries and the number of people affected caused by its detonation. The radioactive material could include plutonium, HEU, or medical isotopes, such as cobalt-60 (a gas sterilant), cesium-137 (tracer), iridium-192 (medical imaging equipment), or Americium-241 (smoke detectors). There are many other isotopes that are used in a wide range of medical, laboratory, and research facilities. Greater health risks are posed by more potent radiation—such as gamma emitters or plutonium—these materials are harder to obtain because they are subject to more stringent tracking requirements. They also pose significant risks to those transporting them and require advanced protective gear that is difficult to disguise. Unfortunately, there are over 10 million sealed nuclear sources in over fifty countries worldwide. Indeed, more than 1,200 of these sealed nuclear containers have been stolen since the collapse of the Soviet Union, of which nearly half have yet to be recovered. Lost

FIGURE 29-3   Three Mile Island, the site of the worst nuclear accident in U.S. history.
Courtesy of the U.S. Environmental Protection Agency.

nuclear material in Pakistan and India are frequent occurrences. In the past several years, smugglers have been caught transporting plutonium, uranium, and cesium. In May 2002, Jose Padilla (aka Abdullah-al-Muhajir) was arrested in Chicago and charged with intent to detonate an RDD with radioactive material gathered from university laboratories. Clearly, tracking and securing these multiple radiation sources pose substantial difficulties for those international agencies responsible for doing so.

Another consideration with RDD is caused by its detonation and environmental contamination. For example, small pieces of radioactive material, such as dust or even small shards, may spread to other areas from contaminated clothing, shoes, or even skin. The greatest health and safety risk from the detonation of RDD devices is the conventional explosion itself. Radiation exposure would be minimal and unlikely to cause immediate or long-term health effects. The psychological impact would likely affect far more people than any direct physical or medical effects.

☢Point Source Radiation

The simplest nuclear scenario involves the simple placement of radioactive materials in a location where individuals passing are unknowingly exposed. This is the most likely scenario because radioactive material is plentiful and placement in public places requires no nuclear expertise.

Still, not all radiation material is equally hazardous to human health. The risk to humans depends on the type of material used. The most dangerous radioemitters pose the greatest risk in terms of storage, transport, and delivery and are more readily detectable by monitoring devices that are employed for security purposes. It follows that the simpler methods and materials are more likely, but pose less risk of injury. For example, a suitcase containing a gamma radiation source would very likely cause acute radiation syndrome (ARS) in the terrorists planting the suitcase. Aircraft dispersal is also plausible, although misdirection by wind and other environmental conditions make this scenario a more risky one in terms of successful ground contamination. The worst-case scenario involves placing highly radioactive material in proximity to a densely populated or highly traveled area. Ventilation systems, mail boxes, and trash bins at transit areas such as train stations or airports, all offer hidden sites for the deposition of radioactive material. Known nuclear accidents again provide an example of how such a low-technology scenario could play out. One of the most infamous cases of point source radiation (PSR) exposure occurred in 1987. In Goiania, Brazil, two men stole radioactive cesium from a hospital to sell for scrap metal. Contaminated pieces ended up all over the city. Over one hundred thousand people required medical evacuation, and over one hundred people received significant external and internal radiation exposure.

Attack on the Food or Water Supply

An often overlooked means of exposing the public to radiation is contaminating crops or livestock with radioactive substrates, or agroterrorism. Radioactive nuclides aerosolized over grazing areas if eaten by livestock could ultimately find their way into consumer meats. Forms of radiation that are metabolic substitutes may incorporate radioactivity into various organs of the body that use the substituted analog or metabolite. For example, the thyroid would take up radioactive iodine. Similarly, cesium and strontium are potassium and calcium analogs, and radioactive isotopes—such as iodine or tritium—are candidates for dissemination through water sources. Natural bodies of water are less likely scenarios compared with closed sources, such as water-processing plants, small aquifers, or water towers. Scenarios such as these would likely yield low numbers of affected individuals, but the psychological effects would be expansive (see Chapters 4 and 7).

Although any of these scenarios is possible, one factor is constant: the consequences of radiation exposure vary greatly in terms of the public health response and the nature of the health threat. Relevant variables include the method of exposure (aerosol and point source) and the type of radiation (e.g., alpha vs. gamma radiation).

TABLE 29–2 Radioactive Isotopes and Target Organs of Deposition[a]

Radionuclide	$t_{1/2}$	Emission	Shielding	Target Organs	Absorption Route	Sources[a] l	i	m	n
Americium: ^{241,243}Am	140 years	α	Sheet of paper	Bone, liver	Respiratory				
Californium: ^{252}Cf	3 years	α	Sheet of paper	Liver, bone, lung	Respiratory				
Cerium: ^{144}Ce	285 days	β	Aluminum foil, thick plastic	Liver, bone, lung, intestine	Respiratory			√	√
Cesium: ^{137}Cs	30 years	β, γ	Thick concrete	Any tissue	Respiratory, gastrointestinal			√	√
Curium: ^{242,244,245}Cu	0.5–9000 years	α	Sheet of paper	Liver, bone	Gastrointestinal				
Iodine: ^{125,131}I	8 days	β, γ	Thick concrete	Thyroid	Gastrointestinal, respiratory	√	√	√	
Phosphorus: ^{32}P	14 days	β	Aluminum foil, thick plastic	Genomic structures	Gastrointestinal	√		√	
Plutonium: ^{239}Pu	82×10^6 years	α	Sheet of paper	Lung, liver, bone	Respiratory				√
Radium: ^{226}Ra	1622 years	α, β, γ	Thick concrete	Bone	Gastrointestinal	√	√	√	
Strontium: ^{90}Sr	28 years	β	Aluminum foil, thick plastic	Bone	Gastrointestinal, respiratory	√			√
Tritium: ^{3}H	12 years	β	Aluminum foil, thick plastic	Any organ with water	Gastrointestinal				√
Uranium: ^{234,235,238}Ur	2.5×10^6 to 4.5×10^9 years	α, γ	Thick concrete	Bone, lung	Respiratory	√	√		√

[a] l, laboratory; i, industry; m, medicine; n, nuclear plant.

Modified from Walter FG [editor]: Advanced Hazmat Life Support Provider Manual, 3rd ed. Arizona Board of Regents, 2003.

It is axiomatic and yet still too often underappreciated that even when the health threat is minimal—for example, a low-level exposure to a low-energy medical isotope using a stationary PSR—the psychological aftermath of any attack involving radiation is likely to be amplified. In addition to exposure, contamination is a concern with PSR as well. Metal sources pose no contamination risk, whereas loose material such as powders could potentially be tracked by foot, spread by direct contact, and possibly inhaled or ingested.

If there is a silver lining in the ominous cloud of nuclear terrorism, it may be that because nuclear materials are so prevalent in medicine and science, there are a considerable number of people available trained in safety and emergency strategies in regard to radioactive materials. These individuals form a critical resource in terms of preparedness efforts and emergency response.

CHAPTER 30

Radiologic Terrorism

"I am become death, the destroyer of worlds."
From Bhagavita attributed to J. Robert Oppenheimer, *Father of the Atomic Bomb*

Clinical Vignette

You are the part-time local health director in a small community. You are called with the news that an unknown person or group has introduced radiologic material into the town's water processing plant. Radioactive iodine is found in the water. What do you do? What do you tell your patients?

Background

As described in the previous chapter, terrorists could use nuclear material in one of a half dozen scenarios: detonation of a nuclear weapon, detonation of an IND, sabotage of a nuclear power facility, or more probably, RPSs or RDDs. This chapter summarizes the known health effects from exposure to nuclear materials.

Properties

Although neither the detonation of a thermonuclear device, an atomic bomb, nor an attack on a nuclear reactor would go unnoticed, the use of RPSs and RDDs may not readily be detected as radioactivity and cannot be detected by our senses. Exposure from an RPS or from an RDD would only be known once symptoms of radiation-related illnesses began or by detection using sensing devices such as a Geiger counter.

Minimizing the health effects of radiation exposure is consequence of three factors: time, distance, and barrier. There is a dose-dependent relationship for the time of exposure and the amount of radiation absorbed that directly impacts health effects. Similarly, exposure decreases exponentially and proportionally to the distance from a radiation source. Physical barriers, even if not leaded, provide an effective shield. The denser the shielding, of course, the more protection afforded.

Sources

Radiation exposure comes from a wide range of natural and manmade sources. Most naturally occurring atoms (e.g., thorium, uranium, radium) that exist in the earth exist at thermodynamically low-energy levels and are quite stable. The few that are unstable, (e.g., uranium) attempt to become stable by releasing energy from their nuclei in the form of radiation. There are internal body deposits of radiation in living cells, including potassium-40, carbon-14, radium, and others. External cosmic rays originate in outer space and exposure to them increase twofold in mountainous regions, and even more at jet aircraft altitudes. Human exposure to natural radiation is greater in that area of the earth's crust rich in radioactive elements. X-rays from medical diagnostics and treatment are the largest source of artificial ionizing radiation exposure. Additional exposures occur from cigarettes, radioactive material in building materials or bedrock, radon gas, and phosphate fertilizers. Production, regulatory, and construction workers in the nuclear industry are exposed, although stringent monitoring requirements are in place. Medical, emergency response, and first-responder personnel may inadvertently be exposed to radiation before they are aware that a nuclear terror event has transpired.

Pathophysiology

Radiation injuries result from three basic methods of exposure: irradiation, external contamination, and internal contamination. The location and severity of injury are influenced by these routes of exposure. Irradiation is radiation exposure from an external source without physical contact with the body. Irradiation from x-rays, gamma rays, neutrons, or alpha or beta particles may be non-penetrating (affecting only the skin) or penetrating (affecting areas deeper than the skin). The depth of penetration is a function of the energy contained within the electromagnetic wave or radiation particle. Decontamination is unnecessary with irradiation so HCWs may focus their attention on acute medical issues. External contamination occurs when radionuclides come into contact with external body surfaces. Adequate decontamination is needed with external contamination to avoid exposure to care providers. These particles are transferable to hands, clothing, and other items, and can inadvertently add to the patient's body burden if allowed to be inhaled, ingested, or entered into open wounds. Finally, internal contamination results from

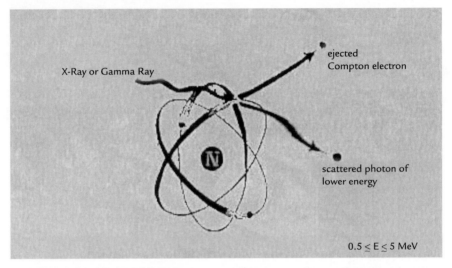

FIGURE 30-1   Ionizing radiation.
Courtesy of OSHA.

the inhalation, ingestion, injection, or intrusion through wounds of radionuclides. Once inside the body, they emit damaging radiation to local tissues and organs.

Three properties govern how tissues respond to radiation. First, radiation sensitivity varies directly with the rate of cell proliferation and with the number of potential future cell divisions. Simply put, rapidly growing cells (e.g., gastrointestinal epithelium) or cell lines that are generative (e.g., spermatogonia or hematopoeitic stem cells) are more sensitive to radiation than are other cells. Radiosensitivity varies inversely with the degree of morphologic and functional differentiation; that is, more mature and differentiated cells are generally less susceptible to radiation injury than are immature or dedifferentiated cells.

Ionizing radiation (Fig. 30–1) breaks the covalent bonds of nuclear material and alters genomic sequences, irreversibly destroys the structure and function of formed proteins in tissues, and generates free radicals from water that may damage tissue. Linear energy transfer (LET), which determines the risk of injury, occurs maximally at a distance equal to the width of a DNA molecule, an inspiring fact of nature that explains in part how nature's genetic abundance, evolution, and ionizing radiation are intertwined. Damage to any key cellular compounds with important structural or metabolic functions can result in cell damage or death. The extent of the damage determines whether the consequences occur immediately or with delay. DNA may experience hundreds of damaged sites after irradiation exposure. Types of damage that can occur vary from point mutations, partial and complete gene deletions, misrepair, to translocations. Ionizing radiation may cause sufficient DNA damage as to fracture the double helix itself. Indirect damage occurs via water molecules that have been ionized into free radical form or electrophilic hydroxyl ions.

radiation

$$H_2O \xrightarrow{\hspace{1.2cm}} H_2O^+ + e^- \qquad H_2O^+ + H_2O \longrightarrow H_3O^+ + OH^-$$

Certain cell types and sites are classically affected by radiation exposure. The most radiosensitive cells are those that are most actively dividing or dedifferentiated, or stem cells, the latter due to the fact that they undergo repeated cell division. Relative biologic effectiveness (RBE) is a term used to describe the quality of injury caused by LET in cells. Those cells that have the most need of frequent sublethal repair turn out to be most radiosensitive. These include stem cells and gastrointestinal tract epithelium. Radiation may disrupt normal cell growth and replication, impair cell motility, and damage membranes. These effects may occur below the threshold for cellular apoptosis in cell death.

The high rate of cell turnover and large surface area makes the gastrointestinal tract among the most radiosensitive organs—with changes occurring within hours of exposure. Ingested radionuclides are absorbed or pass through, depending on physicochemical properties, such as solubility. Whatever the biological path of radioisotopes, they emit radiation into surrounding structures, tissues, cells, and cell organelles. The gut immune system components can also be affected, particularly intramucosal lymphocytes.

Stem cells within the hematopoetic system are also highly radiation sensitive, including the multipotent stems cells in the erythropoietic, thrombopoietic, and myelopoietic cell lines. There is a group of stem cells that, because of their lower frequency of dividing, tend to show greater survival with exposures of less than 6 Gy, and thereby play an essential role in restoration of cell lines. Lymphocytes are said to be the most vulnerable to radiation injury of all the blood elements.

Nervous tissue, being so well differentiated, can be damaged if exposed to high enough levels of radiation. Axonal demyelination, microvascular injury and hemorrhage, and apoptosis may all occur, resulting clinically in sensorimotor and autonomic dysregulation.

Pulmonary injury may occur if radionuclide particles are inhaled. Their effects depend on deposition site, which in turn is a function of particle size. Particles less than 5 microns reach the alveoli, whereas larger ones deposit higher up the respiratory tree where they cause local tissue, or if cleared by mucociliary escalator, find their way to the gastrointestinal tract (Table 30–1).

Signs and Symptoms

Acute Radiation Syndrome

ARS refers to the multiorgan system injury resulting from exposure to ionizing radiation. It requires a high dose of penetrating radiation (greater than 1 Gy) to all or much of the body (referred to as "whole body exposure") over the course of several minutes to develop. ARS involves four organ systems: hematopoietic, gastrointestinal, pulmonary, and central nervous system (CNS). The severity of ARS depends on a combination of factors relating to exposure, such as duration and dose of radiation, how deep the radiation penetrates, and total amount of tissue exposed to radiation. Host factors, such as age and co-morbidities and are also pertinent.

The clinical presentation of ARS is independent of the type and the source of ionizing radiation. The Atomic Age is rife with examples where individuals developed ARS from therapeutic, industrial, accidental, or deliberate exposure to radiation.

ARS occurs as four clinically distinct stages: prodromal, latent, illness, and resolution. The severity of signs and symptoms worsens, and the time of each stage lessens as radiation exposure increases.

1. Prodrome. This stage occurs within minutes to hours after exposure, depending on the severity of the exposure. It is characterized by several days of nausea, vomiting (possibly bloody), diarrhea, malaise, and anorexia. Higher exposures (>5 Gy) may prolong and broaden the symptoms seen in the prodromal stage, including dypsnea, fever, fatigue, erythema, conjunctivitis, and irritability. The ARS prodrome typically resolves in several days.

2. Latency. This is a 2- to 6-week period during which the individual may feel relatively well or even completely asymptomatic. However, subclinical damage to stem cells, epithelial cells, and hematopoetic cell lines are underway, accompanied by functional immunodeficiencies and risk of opportunistic infections. The duration of the latent phase is inversely related to the severity of the exposure and may not occur at all with severe exposure.

3. Illness. A wide range of signs and symptoms can arise, including those seen during the prodromal phase, as well as bleeding, bruising, and fevers. Secondary clinical effects—immunodeficiency, sepsis, and thrombocytopenia—may develop. The sloughing of injured gastrointestinal epithelial cells and luminal stem cells causes malabsorption, hemorrhage, ulceration, and hypersecretion of fluid into the lumen. Dramatic fluid losses, hypotension, paralytic ileus, and microbial invasion may result in shock and infection. With severe radiation exposures (greater than 30 Gy), the CNS suffers diffuse microvascular damage, causing vomiting, mental status changes, fever, seizures, increased intracranical pressure, coma, and death. It follows that mortality is high and occurs quickly following the onset of CNS symptoms, typically within seventy-two hours.

4. Resolution. This stage results either in death or in clinical resolution of the ARS. Recovery may take months, but residual risk for malignancy remains.

Generally, the prognosis in ARS is proportional to the severity and duration of signs and symptoms. Using the aspects of concept of "biodosimetry," signs and symptoms can help give an indication of exposure levels (Table 30–3). Whole body radiation below 2 Gy will resolve with little more than transient nausea and minimal or no extreme permanent injury, excepting the potential for carcinogenic transformation. Exposure to between 2 and 8 Gy will produce the ARS prodrome, but survival is still likely. In

TABLE 30–3 Acute Radiation Symptoms: Variability with Exposure Level[a]

Signs and Symptoms	Mild (1–2 Gy)	Moderate (2–4 Gy)	Severe (4–6 Gy)	Lethal (>8 Gy)
Vomiting				
Onset	≥2 hr	1–2 hr	<1 hr	<10 min
Incidence (%)	10–50	70–90	100	100
Diarrhea	None	None	Mild	Heavy
Onset	—	—	3–8 hr	<1 hr
Incidence (%)	—	—	<10	~100
Temperature	Normal	<38.5°C	Fever	High fever
Onset	—	1–3 hr	≥38.5°C	<1 hr
Incidence (%)	—	10–80	1–2 hr 80–100	100
Headache	Slight	Mild	Moderate	Severe
Onset	—	—	4–24 hr	1–2 hr
Incidence (%)	—	—	50	80–90
Level of consciousness	Normal	Normal	Normal	Unconscious for sec-min
Onset	—	—	—	Sec-min
Incidence (%)	—	—	—	100 at ≥50 Gy

[a] Acute radiation syndrome.

Reproduced, with permission, from Walter FG [editor]: Advanced Hazmat Life Support Provider Manual, 3rd ed. Arizona Board of Regents, 2003.

part this is due to the heterogeneity of exposure (e.g., due to partial shielding), so that some stem cells and other radiosensitive cells escape exposure and are then capable of aiding recovery once the acute, often infectious, risks pass. Whole body radiation doses higher than 8 Gy are almost uniformly fatal, with severe prodromal symptoms, CNS involvement, and rapid death.

The absolute lymphocyte count is a useful predictor of mortality. A >50% decline in absolute lymphocyte count over twenty-four hours indicates substantial radiation exposure and risk of injury or death.

Physical exam should include frequent vital signs, especially temperature, because early treatment with antibiotics significantly impacts mortality. Poor prognostic signs include fever, hypotension, or orthostasis. The latter two are a measure of fluid losses and possibly sepsis. A careful skin exam to locate injuries requiring treatment, or retained potentially radioactive fragments from detonation devices is needed. Cardiovascular and pulmonary examinations to identify congestive failure, abdominal swelling and pain, indicative of GI damage, and neurologic and hematologic examinations, looking for petechia or ecchymoses is warranted. If there is minimal contamination and minor injuries, individuals may be evaluated in the emergency department and fully decontaminated later.

Skin

Focal, or partial, body radiation results in superficial radiation burns. Burns can result from any type of radiation if severe enough. Beta radiation exposure is classically associated with thermal-type burns to the skin of exposed areas, so-called "beta burns." Generally, a dose of 3 Gy or greater is needed to induce burns that present the same as thermal burn injuries. Pain and erythema can occur within hours, followed by a second wave of erythema developing some 2 to 4 weeks later. Erythema that sets in immediately is more likely to be thermal or chemical in origin. More severe exposures (up to 6 Gy) may affect deeper layers of the skin and result in desquamation, depilation, and even necrosis. This effect occurs within 2 to 4 weeks but may reoccur months or even years later. Blistering or bullae may be seen as well, appearing over a 2- to 8-week period. As dose increases, dermal fibrosis and vascular insufficiency can ensue, resulting in chronic ulceration and skin necrosis.

Pulmonary

Pulmonary effects of radiation occur in a delayed fashion. A subacute inflammatory fibrogenic process may result from any substantial lung insult with progressive respiratory embarrassment and cor pulmonale.

CNS

Neurovascular symptoms and signs include fever, headache, nausea and vomiting, and hypotension. An electroencephalogram may demonstrate paroxysmal spiking and wave discharges. Head magnetic resonance imaging (MRI) or CT will demonstrate edema and later dystrophy and calcification.

Psychological Effects

The psychological impact of a nuclear event is enormous. Acute and chronic stress responses, altered sleep patterns, as well as depression, anxiety, PTSD, and social alienation have all been described in survivors of nuclear trauma, as well as in unexposed victims. Individuals with psychological sequelae may require referral as well to mental health professionals skilled in addressing PTSD, anxiety, disturbed sleep, and other psychological symptoms. It is estimated that roughly three-fourths of a population suffer psychological sequelae in the aftermath of a nuclear weapon attack (see Chapter 7).

Chronic Effects

Survivors need to be aware of the long-term health consequences and cancer risk associated with radiation exposure. Serial medical surveillance to identify early chronic effects is justifiable given the risks involved. Longitudinal epidemiologic and clinical surveillance—such as the case with the World Trade Center survivors and rescuers—is justifiable. Patients and their families need to be informed about the cancer risk incurred by their exposure and what if any oncologic and general health surveillance is warranted.

The chronic sequelae of ionizing radiation exposure are far better known than nearly all the biological or chemical agents discussed previously. Exposure to ionizing radiation has substantial long-term medical consequences ranging from chronic tissue scarring and loss of function to carcinogenesis, mutagenesis, and reproductive effects. Much of what is known derives from the Japanese atomic bomb cohort or Chernobyl survivors, although unlike in Japan, epidemiologic studies in the latter instance fell well short of what should have been the case. In utero exposure to radiation is highly injurious to the developing embryo and fetus. Spontaneous abortions, birth defects, microcephaly, mental retardation, and impaired intellectual development have all been shown in survivors. Decreased life span, cataract formation, chronic radiodermatitis, decreased fertility, and progeny with genetic mutations occur as well. Incidence of cancers of the thyroid gland and hematopoeitic cell lines—childhood leukemia, solid tumors, papillary thyroid carcinoma—are elevated. Women and children are at particularly high risk for thyroid cancers. Chronic radiation syndrome (CRS) may also result from less intense but longer-term exposure to ionizing radiation (>3 years of 1 Gy or less). CRS affects the bone marrow with alterations in normal cell growth and immunodeficiency. Along these lines, a syndrome referred to as "Chernobyl AIDS" has been described in the liquidators brought into to clean up the ruined reactor. Those workers complained of chronic fatigue and identified immunodeficiencies, especially a loss of natural killer cells, the lymphocytes responsible for killing virus-infected and mutated tumor cells. Increased rates of respiratory tract infection, cardiac disease, leukemia, and other malignant tumors have been identified as well. Chronic thyroiditis and a growing incidence of thyroid cancer have been found in Russian children who inhaled iodine-131 generated by Chernobyl's radioactive plume.

Finally, long-term psychological symptoms occur in response to the acute event and often become chronic. Findings include PTSD, depression, emotional lability, anxiety, and anhedonia. Such patients should be monitored carefully and referred to mental health professionals as warranted (see Chapter 7).

Vulnerable Groups

Pregnant women, developing fetuses, children, the elderly, and those with immuno-compromised conditions are less capable of surviving the acute and chronic sequelae of ionizing radiation exposure.

Decontamination

If radiation exposure is expected, experts should be consulted as soon as available (Table 30–4). The goal of decontamination is radiation levels at background, though double the background level is possibly acceptable from a health risk assessment perspective. HCWs and first responders are at risk for exposure to radiation particles

TABLE 30–4 Comparison of Contamination and Exposure

Incident	Radiation Desposition	Source Type	Physical State	Patient Decontam-ination	Risk to HCWs/ First Responders
Exposure	External	Rays (energy)	None	No	No
External contam-ination	External (skin surface)	Particles (matter)	Solid or liquid	Yes	Yes
Internal	Internal	Particulates	Solid	Yes	Yes
			Gas	No	No

Modified, with permission, from Walter FG [editor]: Advanced Hazmat Life Support Provider Manual, 3rd ed. Arizona Board of Regents, 2003.

depending on the source of radiation. In contrast, electromagnetic radiation exposures (gamma ray or x-ray exposure) pass through tissue (where they do cause damage), but they do not persist, posing little contamination risk to others.

Decontamination measures should be taken for protection of HCWs and the individuals in any case where radiation exposure is suspected. Removing clothing (and disposal into polyethylene bags or containers for later analysis by public health officials or radiation safety specialists) will eliminate 90% of any contaminants present. Washing the individual down three times, following removal of clothing, removes another 90% of the residual, making transport safer for all involved.

PPE is mandatory for those coming into contact with a potential radiation victim. This includes double gloving, gowns, masks, overshoes, and face shields. Taping seams to form a functional space suit is recommended as an added barrier to any radiation particles that may be on the person. Dosimeters should be worn by those in contact with the patient. Contamination of skin, hair, or clothing is not a medical emergency. Prevention and control of further contamination are warranted, but stabilizing the patient medically takes priority. Once removed from the radiation source, contaminated patients pose little risk to those wearing PPE properly.

Skin should be decontaminated serially with soap and water, beginning at outer regions moving inward (thereby reducing spread of the contaminants beyond the site). Gauze or cotton may be used to help clean the site; being mindful not to further irritate or overly debride the area because this may facilitate absorption of radionuclides through nonintact skin. To help gently ensure containment of the site, sterile, waterproof drapes should be taped around the area. Decontamination need not be perfect because any skin contamination that is missed will slough off as the stratum corneum repletes within ten to fourteen days. Contaminated wound sites should be isolated from other sites with waterproof drapes, and irrigation should be done with normal saline saving both debris and washings for later identification. Excision of contaminated tissue is a viable option. Surgical consultation may be needed according to the nature of the wound. Noncontaminated wound sites

TABLE 30-5   Lymphocyte and Prognosis with Varying Dose

Lymphocyte Level	Dose	Expected Treatment
1,500/mm³	Insignificant exposure	Unlikely to require treatment
1,000–1,500/mm³	Mild exposure	May require treatment in 3 weeks for moderate neutropenia and thrombocytopenia
500–1,000/mm³	Severe radiation injury	Symptoms in 2–3 weeks, hospitalization required
>500/mm³	Possible fatal dose	Hospitalization required, inevitable pancytopenia and related complications
Undetectable	Superlethal dose	Death likely in 2 weeks

ᵃ The lymphocyte level may help in estimating the dose of radiation received and possibly help establish prognosis.

Modified from Mettler, FA Jr, Voelz GL. Major radiation exposure—what to expect and how to respond. N Engl J Med. 2002, 346:1554–61.

should be dressed securely to prevent contamination. Hair is a commonly contaminated area and can be decontaminated by two to three washings with shampoo. This should be done prior to showering and washing of the body with soap and water.

Laboratory Findings

For identification of the extent and type of exposure, a nasal swab of each nostril is to be taken marking the time postincident. Throat swabs, sputum, urine, and fecal samples should all be obtained to help determine if radionuclides had internally contaminated the patient. The fluid used to irrigate wounds or for decontamination should also be saved. Serial monitoring for hematologic effects and cytogenetic analysis are recommended. A rapidly dropping absolute lymphocyte count is a bad prognostic sign, indicating significant radiation exposure. Bone marrow biopsies are not useful clinically or diagnostically (Table 30–5).

Diagnosis

Making the diagnosis will be difficult with index cases for RDD, RPS, or agroterrorism (Table 30–6). Nuclear devices will be obvious events and present more of an acute management issue due to the higher likelihood of ARS. Keys to making a diagnosis for index cases include evidence of skin burns with or without any trauma or thermal contact noted. Whole body radiation will likely require diagnosis during the prodromal stage, which can be fairly nonspecific, but evidence of multisystem involvement is suggestive.

TABLE 30-6	Making the Diagnosis: Radiation Exposure
Pattern of Presentation	Sentinel events, such as detonations or multiple cases consistent with ARS
Features	Unexplained skin burns, multiorgan system findings: ARS prodrome, including nausea, vomiting, fatigue, fever
Findings	Radiographic studies are not helpful Positive swabs of body fluids Possible decline of hematopoetic cell lines

Treatment

The first priorities in treating life-threatening conditions (e.g., extensive radiation burns) are isolating the contaminated individual, restricting access to the contaminated zone, and contacting local radiation physicists or radiation safety specialists to assist with controlling the contamination zone. Depending on the nature of the incident, life-threatening injuries and management take precedence over radiation management decisions. Once the acute triage issues are addressed, decontamination can proceed, along with setting up a perimeter barrier to limit inadvertent tracking of radiation to other areas.

Delayed clinical effects complicate initial assessment of radiation-related injuries. Uncertainty regarding the type of radiation and intensity of exposure is probable in the setting of a subtle nuclear terror attack, unlike in the occupational setting where the source and type of radiation are known with reasonable confidence. As mentioned earlier biodosimetry may allow exposure levels to be estimated.

Management of radiation injury is largely supportive though complex (Table 30–7). Careful monitoring of volume status is needed in the event of extensive radiation skin burns or GI fluid losses, not unlike the care of a burn patient. Similarly, risk of infection requires careful clinical assessment, monitoring for declining hematologic parameters, and a low treatment threshold for antibiotic administration. Prophylactic antibiotic use is not recommended until severe neutropenia develops (ANC $< 100/\mu$) at which time quinolones are the drugs of choice, although penicillins still have a role to play. Antivirals, such as acyclovir, should be administered if there is any evidence of HIV infection, and antifungals, such as fluconazole, have a role if clinical suspicion suggests fungal infection. Use of granulocyte colony-stimulating factor or erythropoietin are justified, depending on estimated dose or clinical parameters indicating hematopoeitic compromise. Neutropenic precautions should be instituted when standard criteria are met.

Skin sites, particularly areas not covered by clothing, should be carefully examined repeatedly for the presence of erythema, blisters, or bullae. Burns, once decontaminated, should be managed similarly to conventional thermal burns with intravenous crystalloid fluids, pain control, and cleaning of sites with dressing for partial and full thickness burns. Patients with open wounds are at particularly high

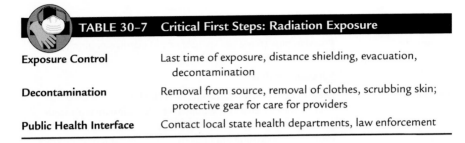

TABLE 30-7	Critical First Steps: Radiation Exposure
Exposure Control	Last time of exposure, distance shielding, evacuation, decontamination
Decontamination	Removal from source, removal of clothes, scrubbing skin; protective gear for care for providers
Public Health Interface	Contact local state health departments, law enforcement

risk for infection, especially should immune function wane. Wound care—which includes cleansing, decontamination, debridement and protecting the wound—is critical.

Once hematologic and GI signs and symptoms are present, instrumentation should be avoided if at all possible. This includes surgery. If it must be done, it is best to do so within 24 hours to limit the risk for infectious and hematologic complications and poor tissue repair.

The use of antiemetics such as granisetron or ondansetron may be warranted and are known to be effective in symptomatic relief. Electrolytes should be carefully monitored and replaced intravenously as needed. Total parenteral nutrition should be administered to maintain adequate caloric and nutritional status. Transfusion of blood products should be initiated as clinically warranted. As with other scenarios for transfusion, leukoreduced products may be appropriate when immunocompetency is questioned. With mass casualty situations, severe shortages of blood products is very likely.

The use of colony-stimulating factors, such as erythropoietin, pegfilgrastim, and filgrastim, may be used for stimulation of specific cell lines, depending on the extent of stem cell injury and subsequent clinical status. Case reports regarding allogeneic stem cell transplants in radiation victims reveal poor outcomes.

Pain control and psychological support are essential to the management of these patients. Over 75% of radiation-exposed individuals and their families will experience clinically significant psychologic reactions.

Management of radionuclides taken up into the body is predicated on four basic principles: complexing the substance with a chelator (Table 30–8), binding the radionuclide with a resin, minimizing GI absorption, minimizing bowel transit time, and diuresis. If information regarding the ingestion or absorption of heavy metal radionuclides is known, appropriate detoxification using chelating agents should be performed.

Aggressive life support and antibiotic use have decreased mortality from radiation exposure. For example, the LD50/60 (the lethal dose of radiation, where 50% die at 60 days) is 3 to 4 Gy without antibiotics, but nearly doubles with early antibiotic use. Growth factors have shown great promise in lab animal trials to the

TABLE 30–8   Radiation Antidotes and Chelating Agents

Cesium-137	Prussian blue (ferric hexacyanoferrate) adsorbs cesium in GI tract; may enhance elimination
Iodine-131	Potassium iodide
Plutonium-239	Diethylenetriaminepentaacetic acid (DTPA) can be used as a chelator as well as for wounds; ethylenediaminetetraacetic acid (EDTA) also suitable chelator Aluminum hydroxide antacids may bind plutonium in GI tract
Radium-226	Immediate lavage with 10% magnesium sulfate, followed by saline solution and magnesium purgatives Ammonium chloride may increase fecal elimination
Strontium-90	Aluminum hydroxide antacids may bind strontium in GT tract Aluminium phosphate can decrease absorption by 85% Ammonium chloride can acidify urine and enhance excretion Barium sulfate may reduce strontium absorption Stable strontium can competitively inhibit metabolism and increase excretion of radiostrontium
Tritium	Oral fluids, 2–4 liters/day will reduce biologic half-life from 12 to approximately 6 days Caution: Do not overhydrate patients
Uranium	Sodium bicarbonate renders uranyl ion less nephrotoxic Diuretics

Modified, with permission, from Walter FG [editor]: Advanced Hazmat Life Support Provider Manual, 3rd ed. Arizona Board of Regents, 2003.

point that the FDA has approved their use in radiation exposures. In fact, the National Pharmaceutical Stockpile now includes cytokines in their push packs.

☣The Role of Potassium Iodide Tablets

Many in the general public have come to think of potassium iodide (PI) as providing protection to the effects of radiation. This is not the case for many types of radiation. By saturating the thyroid with iodide, PI tablets offer protection against uptake of radioactive iodide and are protective against the development of thyroid cancer. This benefit works only if radioactive iodide is present, such as would occur in a nuclear blast or in a nuclear reactor accident. KI is of no benefit in the event of a RDD or with RPS where radioactive iodide is not present. Further, KI must be taken prior to exposure or as soon as possible up to four hours after exposure ends. Potassium iodide offers no protection against other forms of cancer or radiation injury. The number one public health priority in the event of any nuclear terror event is evacuation, and second, protection of water and food sources. Allowing individuals to remain in a contaminated zone with or without PI prophylaxis is unacceptable.

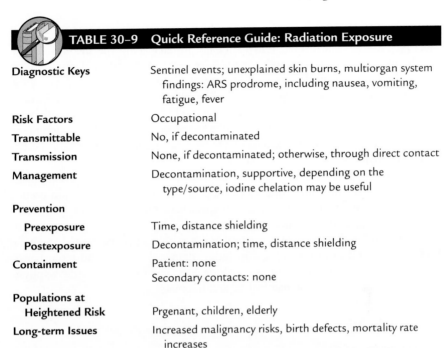

TABLE 30–9   Quick Reference Guide: Radiation Exposure

Diagnostic Keys	Sentinel events; unexplained skin burns, multiorgan system findings: ARS prodrome, including nausea, vomiting, fatigue, fever
Risk Factors	Occupational
Transmittable	No, if decontaminated
Transmission	None, if decontaminated; otherwise, through direct contact
Management	Decontamination, supportive, depending on the type/source, iodine chelation may be useful
Prevention	
Preexposure	Time, distance shielding
Postexposure	Decontamination; time, distance shielding
Containment	Patient: none Secondary contacts: none
Populations at **Heightened Risk**	Prgenant, children, elderly
Long-term Issues	Increased malignancy risks, birth defects, mortality rate increases
Public Health Concerns	Evacuation, decontamination
Contacts	Local, state DPH, CDC, FBI

Public Health Considerations

A patient without a therapeutic or occupational exposure who has evidence of radiation-related injuries represents a medical and a public health emergency. Without delaying necessary acute treatment or triage, notification of appropriate public health and law enforcement officials is necessary (see Decontamination Section, as well as Chapter 3 and Appendix for Radiation Resources).

Aerosols from an RPS that enter a building's ventilation system can be prevented by systems designed to filter out radiation particulates. Serial HEPA filters decrease the risk to building inhabitants but buildings are expensive to retrofit and require regular maintenance. Any contamination necessitates subsequent decontamination of the entire heating, ventilating, and HVAC system.

In the event of a nuclear terrorist attack, there are fairly well-delineated crisis and consequence management teams in place in most states. The initial response will come from local first-response systems and hazmat teams as well as from the ICS manager who will contact the state health department and federal agencies. Locally, the incident commander is usually the local fire chief. The local agencies or state health departments have trained teams who can respond quickly to any emergency and conduct onsite environmental monitoring to determine what, if any, radiation hazard exists. Regional response teams, comprised of state radiation experts and Department of Energy officials have also been created to assist local and state officials. FEMA is the overall lead nationally for managing the medical, public health, and logistical aspects following a BCN event. A presidential directive has named the Department of Homeland Security the lead agency for coordinating the crisis and consequence management for all BCN events. The Homeland Security supervises the FBI in its criminal investigations and FEMA in its public health management. These systems have yet to be fully tested in a true crisis, although many table-top exercises and field exercises have been accomplished as part of the nation's preparedness efforts (see Chapter 3).

Hospital response teams exist in most of the nation's hospitals. Disaster planning occurs frequently, including mock disasters. In the case of radiation events, hospital teams are comprised of radiation specialists, nuclear medicine physicians and technicians, radiation oncologists, emergency and intensive care department physicians and staff, and a host of medical and surgical subspecialities to handle the breadth of potential radiation effects, including dermatologists and burn specialists, maternal–fetal medicine specialists, occupational and environmental medicine physicians, gastroenterologists, hematologists, and traumatologists.

Patient Education

As the Homeland Security Advisory System changes its colors and as the geopolitical climate evolves, patients are likely to have questions in anticipation of terrorist attacks. In regard to nuclear attacks, such questions are likely to include the need and use of PI tablets. As discussed earlier, clinicians should be clear on the benefit and indications for their use and educate patients accordingly.

Clinicians will also be looked to for answers regarding preventive measures and safety practices in the event of an attack. Current recommendations are that people stay indoors for up to 48 hours if a dose of 1 rem is likely. Evacuation from an area is necessary for up to one week if the dose exceeds 5 rem. Areas are inhospitable to life when the estimated lifetime cumulated dose is more than 100 rem. PI prophylaxis and food and water restrictions depend on the nature, method, and intensity of radiation exposure. Clinicians should stress the three key factors that can limit dose: time, distance, and shielding.

☣Summary

Ionizing radiation is found throughout nature, and a wide variety of radioactive elements are used in military, industrial, medical, commercial, and consumer applications. Health effects from radiation depend on the type of radiation (alpha, beta, and gamma), the route of exposure (irradiation, external, or internal contamination), and the dose and duration of exposure. Acute health effects range from skin burns to acute radiation syndrome to death. The absolute lymphocyte count and the time of onset for nausea and vomiting are directly proportional to dose of radiation.

These are among the most radiosensitive cells in the body and provide prognostic data. For example, someone presenting with nausea and vomiting within thirty minutes of a radiation exposure has received a lethal dose. In a mass casualty situation, where resources are going to be quickly consumed, recognizing the futility of treatment and the necessity of compassionate care is essential. Long-term neurologic, developmental, carcinogenic, and psychological sequelae are seen. Active medical surveillance for the wide range of chronic effects is warranted and is best done in conjunction with a formal longitudinal epidemiologic study.

Conclusion

Until recently, few considered the medical and public health implications of biological, chemical, and nuclear terrorism subjects relevant to the professional lives of the vast majority of medical and public health practitioners. Diseases like anthrax, tularemia, plague, and smallpox were presented largely as curiosities of medical history or geography, or in the case of smallpox as the paramount example of the stunning potential that medical science and international public health cooperation offered the world. We might recall a factoid from an infectious disease lecture about anthrax as an occupational hazard for farmers or veterinarians, but who could have imagined a time when postal workers, government employees, politicians, or media personalities would be at risk of contracting the disease?

This view changed in the months following the aerial attacks on the Twin Towers and the Pentagon. In rapid succession, death and disruption caused by anthrax-contaminated mail, a national smallpox vaccination campaign, and the initial uncertainties regarding the capability of the Iraqi army to use biological weapons in the lead up to the Second Gulf War highlighted the critical need for a reinvigorated and proactive public health workforce and an informed health care community.

Much has been accomplished in the past several years, but preparing the nation's public health workforce and medical community to respond to the immediate and long term consequences of biological, chemical, and nuclear events remains an enormous challenge. One reason for our present dilemma is that for too long, the public health systems have been under-funded and, many have argued, antiquated. The reasons for this state are complex. For one, the nation's discretionary health care dollars have flowed increasingly toward curative medicine as socioeconomic, technological and pharmacological advances—such as better nutrition, hygiene, housing, and vaccinations—caused many of the major health issues of the early and mid 20th century to fade.

The deterioration of the nation's public health infrastructure did not happen overnight, however, and remedying this situation is no short term task. To be sure, federal and state funding for public health has increased meaningfully in the past several years, much of it earmarked specifically for bioterrorism. However, many in the public health community honestly and fairly question whether the focus on bioterrorism is justified in light of equally compelling public health priorities, such as AIDS, environmental pollution, gun violence, or tuberculosis. Increasingly, one hears a compelling argument in favor of the concept of "dual use." Put simply, supporters of the dual use concept argue that pragmatism and prudence dictates that we use the opportunity afforded by the influx of bioterrorism funding to make longer term investments in the nation's public health infrastructure. Put more concretely, improvements in disease surveillance, communication systems, and public health preparedness are as applicable to other contemporary public health threats as they are to bioterrorism. The international SARS epidemic is a relevant case in point. As noted in Chapter 3, a more rapid and efficient mobilization of the nation's medical and public health community in order to contain the threat of SARS was almost certainly facilitated by ongoing bioterrorism-related preparedness efforts on the part of federal, state, and municipal governments.

Preparing physicians and the health care community to recognize and respond to bioterrorism is a challenge easily on par with the need to update, expand, and maintain the nation's public health system. According to one survey, 75% of primary care physicians feel ill-prepared to recognize the signs and symptoms of biological, chemical, or nuclear agent exposures, and a similar percentage acknowledge that they would be at a loss in knowing what to do "as a doctor" in the event of a bioterrorist attack. Diagnosis, management, and uncertainty regarding professional responsibility lie at the core of what it means to be a health care professional. That so many physicians lack the knowledge and confidence to assume roles that will be thrust upon them in such circumstances indicate how much more needs to be done.

When we first conceived this book, efforts to focus preparedness efforts beyond hospitals and the emergency response system were nascent. Much has changed since then. Local and state bioterrorism preparedness efforts are beginning to address the educational needs of health care workers and physicians in our communities. The necessity of expanding these efforts into the community are manifest: in any public health emergency, the public typically turns to those individuals—often their primary care physicians—that they know and trust best. Further, terrorist events of any kind leave long term scars among those affected. Delayed medical and psychological effects require long term medical, emotional, and socioeconomic interventions for which clinicians must be as prepared as they are for diagnosing the acute diseases or syndromes themselves, arguably more so. While local, state, and national individual state's preparedness efforts are more well-established than they were a short time ago—and increasingly include the primary care, mental health and local public health communities—these efforts require continued emphasis. Bioterrorism is unlikely to return to the backburner of the nation's public health agenda anytime soon.

A decade from now bioterrorism as a significant medical and public health concern will still be with us. It is probable that we will have developed and (unfortunately) even field-tested clinical algorithms and crisis management systems so that current uncertainties will lessen. Ten years from now, bioterrorism will be perceived less as an episodic event or threat and more as a condition of modern society requiring the cooperation and expertise of the nation's medical and public health communities.

The historical schism between public health and clinical medicine has exerted a significant and troublesome influence on the medical and public health community's ability to confront significant medical and public health issues in our society at the individual and the community level. This schism is no longer acceptable. Now, perhaps more than at any other time in recent memory, community-based practitioners are a critical element of the nation's public health workforce. *The Bioterrorism Sourcebook*—emphasizing as it does the necessity of bridging clinical medicine and public health—promotes our view that community practitioners bring a unique set of skills to the public health and clinical tasks that they need to fulfill their professional obligations in the post-September 11th world. These skills may be underrecognized by those responsible for training, planning, and responding to public health issues, whether immediate, emergent or threatened. For example, clinicians are trained to examine subjective, objective, and epidemiological evidence within the context of their patient's lives to determine what is happening and how to best proceed. They are reminded daily that individuals perceive risk in ways that are often intensely personal, experience-based, and predicated infrequently on the scientific principles dear to health professionals. This clinical perspective is not routine in most public health training or practice. Such individuals are highly versed in epidemiology, biostatistics, as well as risk assessment and risk communication, but statistics, generalities and probabilities too often fall short of what citizens and communities expect and need in circumstances that are volatile and uncertain. This is one domain in which practicing clinicians can offer much to our public health colleagues. Doctor–patient communication occurs typically in situations of uncertainty, but sharing information openly, providing guidance, and offering reassurance are almost always needed.

The irony should not be missed that the plagues and pestilences of the past are being revisited upon the world in the guise of biological terror. Deliberate or accidental exposure to biological, chemical, or nuclear agents represent a plausible concern and for that reason it is no longer outside the scope of the professional lives of physicians, health care and public health workers the world over. While the threat of biological, chemical, and nuclear terrorism is not as great as some alarmists might suggest, neither is it zero. By providing a balanced view of an important evolving public health issue, we hope that *The Bioterrorism Sourcebook* offers those at the forefront of our nation's medical and public health defenses the information, insight, and reassurance that our patients, our peers, and our communities deserve.

Definitions/ Abbreviations

Aerosols: Solid and liquid airborne particles, typically ranging in size from 0.001 to 100 μm. Mists, fumes, and dusts are all examples of aerosols.

Alpha Particles: The largest and weakest radioactive particles and comprised of two protons plus two neutrons. Alpha particles do not travel far and are stopped by virtually any physical barrier. Internalized alpha particles can cause significant cellular damage.

BCN: Acronym for biological, chemical, nuclear.

Beta Particles: High-energy electron or positron (a positively charged electron) emitted from a nucleus with a variable amount of energy usually as a result of radioactive fallout and capable of traveling a meter through air and many millimeters into tissue. Internalized beta particles cause extensive damage to the tissues.

Bioterrorism: The intentional use of any microorganism, virus, infectious substance, or biological product to cause death, disease, or other biological malfunction in a human, an animal, a plant, or another living organism. Commonly used, as in the case with this book, to also encompass chemical and radiological terrorism as well.

Contact Precautions: Contact precautions include standard precautions plus placing patients in a private or semiprivate room; wearing gloves and gown if contact with the patient is anticipated or if the patient has diarrhea; a colostomy or drainage of a wound not covered by a dressing; limiting the movement or transport of the patient from the room; and ensuring bedside equipment, frequently touched surfaces and other patient-care items are being cleaned daily. Contact precautions are recommended for patients with severe gastrointestinal, dermatologic, or wound infections that may be transmitted easily by touching the patient or by handling objects the patient has touched, especially if the infection is caused by a multidrug-resistant organism.

Diffusion: The movement of fluid particles from an area of high concentration to an area of low concentration.

Dose–Response: A gradient of risk or effect that is associated with the "dose" or degree of exposure. The therapeutic (or toxic) effect of material absorbed over a given period (the dose) varies according to the circumstances of exposure.

Epidemic: The occurrence in a community or region of a group of similar conditions of public health importance in excess of normal expectancy and derived from a common source.

Gamma Rays: Deeply penetrating energy waves similar to x-rays but higher in energy (therefore of a shorter wavelength), capable of traveling many centimeters into tissues and causing ionization. Emitted from the nucleus of a radioactive atom.

Gas: Formless fluids that tend to occupy an entire space uniformly under ordinary temperatures.

Half-Life: The time taken for the activity of a radionuclide to lose half its value by decay.

HAN (Health Alert Network): Restricted website designed for state and local health directors to securely view posted documents, submit/collect data, obtain town/district specific aggregate data, enter planned absences, e-mail, view bulletin board. The overall goal of the HAN is to securely facilitate communication of critical health, epidemiological and bioterrorism-related information on a 24/7 basis to local health departments, health organizations, and other partners.

Health Care Provider: Any person or facility that provides health care services including, but not limited to, hospitals, medical clinics and offices, special care facilities, medical laboratories, physicians, pharmacists, dentists, physician assistants, registered and licensed practical/vocational nurses, paramedics, emergency medical or laboratory technicians, community health workers, and ambulance and emergency medical workers.

Health Insurer: An entity subject to the insurance laws and regulations of a state, or subject to the jurisdiction of a state insurance commission, that contracts or offers to contract to provide, deliver, arrange for, pay for, or reimburse any of the costs of health care services. This may include insurance companies, health maintenance organizations, hospitals, or any other entity providing a plan of health insurance or health benefits.

ICt$_{50}$: The product of concentration (C) and time (t) of exposure to a vapor or aerosol found in air or water needed to incapacitate 50% of a specific population.

Ionizing Radiation: The energy released through either electromagnetic waves or particulate matter that deposits into a bystander atom, exciting an electron in the bystander atom to where it can overcome the energy binding it to that nucleus.

Isolation: The physical separation and confinement of an individual or group who are infected or reasonably believed to be infected with a contagious or possibly contagious disease to prevent or limit the transmission of the disease to nonisolated individuals.

LC_{50}: The concentration of a substance in air which is lethal to 50% of the animals under test.

LCt_{50}: The product of concentration (C) and time (t) of exposure found in air or water needed to cause death in 50% of the animals under test.

LD_{50}: The identification of a dose of a substance, given at one time, lethal to 50% of the animals used in the test.

MEDSTAT: A highly reliable telephone connection to be used in response to a public health emergency or in the event of a localized or catastrophic failure of the conventional telephone network, linking public health authorities to key emergency response partners.

Neutron Radiation: Though electrically neutral, once emitted from a nucleus, neutrons cause damage in two ways. First, they cause damage via "neutron capture." In humans, neutron capture is most likely to occur with sodium atoms, thus changing sodium-23 into -24, which is a radioactive isotope and becomes an internal source of radiation. Second, neutrons can cause damage through direct impact with another nucleus, thereby causing the struck nucleus to become unstable and release alpha or beta particles.

Particulate Radiation (includes alpha and beta particles): Results from instability of the atomic nucleus such that the energy level overcomes the "binding energy" of the nucleus (comprised of the strong and weak nuclear forces). Their presence on skin, clothes, or objects represents a risk of contamination to others including health care givers. *The distinctions of radiation types should not be construed as occurring separately. All, some, or one form may be present concurrently depending on the nature of the radiation source.*

Persistence: The tendency of a compound to remain once present; persistence is inversely related to volatility.

Personal Protective Equipment (PPE): Devices (e.g., respirators, gloves, hearing protection) worn by workers to protect against environmental hazards.

Physicochemical Properties: Physical and chemical characteristics of sorbents (pore size, shape, surface area, affinities, etc.).

Private Sector Partner: Nongovernmental entities, including community organizations, contractors, education institutions, health care facilities, health care providers, health insurers, private businesses, media, nonprofit organizations, and volunteers that provide essential public health services and function or work to improve public health outcomes in collaboration with a state or local public health agency.

Public Health Emergency: An occurrence or imminent threat of an illness or health condition that is caused by bioterrorism, a novel, previously controlled, or eradicated infectious agent or biotoxin, natural disaster, a chemical attack or accidental release, or a nuclear attack or accident and poses a high probability of death, serious or long-term disabilities, a significant risk of substantial future harm.

Public Health Official: The head officer or official of a state or local public health agency responsible for the operation of the agency and authorized to manage and supervise the agency's activities.

Quarantine: The physical separation and confinement of well individuals or groups of individuals who were or may have been exposed to a contagious disease to prevent or limit the transmission of the disease to nonquarantined individuals.

Rad (radiation absorbed dose): The amount of energy deposited per unit tissue by ionizing radiation. Superseded by the Gray. 1 gray = 100 rads.

Radiation: Energy that is released from the nuclei of unstable atoms. The forms of energy released can be particles, such as neutrons, protons, or electrons, or it can be in the form of waves, such as gamma or x-rays. The greater the energy released, the greater the potential health risks. Cellular damage occurs when the energy level is high enough to cause ionization of cellular atomic structure.

Radioactivity: Spontaneous nuclear energy resulting in the release of particles and electromagnetic radiation.

Rem (roentgen equivalent man): The extent of the absorbed dose in human tissue in relation to biological damage; a quantity called an equivalent dose. One rem equals the number of rads times n, a conversion factor. The rem has been superseded by the Sievert.

Roentgen: A measure of exposure amount of electromagnetic radiation (gamma and x-rays). It measures the ionizations of the molecules in a mass of air.

Standard Precautions: Practices designed to reduce the risk of transmission of pathogens from moist body substances including handwashing, gloves, face shield (with eye protection), and gown. Standard precautions are intended to be used with all patients even in the absence of known infection.

Threshold: The point on an exposure continuum where a toxic effect is first seen; of particular importance in determining risks associated with exposure to chemical carcinogens.

Toxicokinetics: The study of the dynamic relationship between the concentration of a chemical in body fluids and tissues and its biological effects. Factors that affect the toxicokinetics of a particular substance are rate of absorption, distribution, metabolism, and excretion.

Vapor: A gaseous form of substances that are normally solid or liquid at ambient temperatures; sometimes used interchangeably with gas. Vaporized substances may return to liquid form and thus can cause both inhalational and topical effects.

Vapor Pressure: The pressure exerted by a liquid or solid that maintains equilibrium.

Volatility: The tendency of a liquid to evaporate and form a vaporous form. The more volatile an agent, the quicker it evaporates and disperses.

WANS (Wide Area Notification System): A telephonic system with autodialing and voice messaging capabilities. It allows 24/7 notification of key components of the emergency response, law enforcement, medical and public health community via landline phones, cell phone, pages, fax, and e-mail.

WMD (weapons of mass destruction): Weapons that have the potential to injure or kill large numbers of people through the release, dissemination, or impact of microbial radioactive or chemical agents.

X-Rays: A form of high energy, ionizing electromagnetic radiation released by high energy electrons.

Dosing Regimens

Nearly all of the regimens detailed below lack FDA approval as there is too little in the way clinical data to justify such approval. These regimens below are a composite of the recommendations by civilian, military and government experts based on extrapolation, previous results, and clinical information.

Biological Weapons

Anthrax

CDC-Recommendations for Postexposure Prophylaxis for Prevention of Inhalational Anthrax After Intentional Exposure to *Bacillus Anthracis*

Category	Initial Therapy	Duration
Adults (including pregnant women and immunocompromised persons)	Ciprofloxacin 500 mg po BID or Doxycycline 100 mg po BID	60 days

Children	Ciprofloxacin 10–15 mg/kg po every 12 hrs*
	or
	Doxycycline:
	  >8 yrs and >45 kg: 100 mg po BID
	  >8 yrs and ≤45 kg: 2.2 mg/kg po BID
	  ≤8 yrs: 2.2 mg/kg po BID
		60 days

*Ciprofloxacin dose should not exceed 1 gram per day in children.
Courtesy of the CDC.

CDC Recommendations for Management of Inhalational and Gastrointestinal Anthrax

Category	Initial Therapy (Intravenous)[b,c]	Duration

Adults	Ciprofloxacin 400 mg every 12 hrs
	or
	Doxycycline 100 mg every 12 hrs[e]
	and
	One or two additional antimicrobials[c]
		IV treatment initially.[d] Switch to oral antimicrobial therapy when clinically appropriate:
		  Ciprofloxacin 500 mg po BID
		or
		  Doxycycline 100 mg po BID
		Continue for 60 days (IV and po combined)[f]

Children	Ciprofloxacin 10–15 mg/kg every 12 hrs[g,h]
	or
	Doxycycline:[e,i]
	  >8 yrs and >45 kg: 100 mg every 12 hrs
	  >8 yrs and ≤45 kg: 2.2 mg/kg every 12 hrs
	  ≤8 yrs: 2.2 mg/kg every 12 hrs
	and
	One two additional antimicrobials[c]
		IV treatment initially.[d] Switch to oral antimicrobial therapy when clinically appropriate:
		  Ciprofloxacin 10–15 mg/kg po every 12 hrs[h]
		or
		Doxycycline:[i]
		  >8 yrs and >45 kg: 100 mg po BID
		  >8 yrs and ≤45 kg: 2.2 mg/kg po BID
		  ≤8 yrs: 2.2 mg/kg po BID
		Continue for 60 days (IV and po combined)[f]

Pregnant women[j]	Same for nonpregnant adults (the high death rate from the infection outweighs the risk posed by the antimicrobial agent)	IV treatment initially. Switch to oral antimicrobial therapy when clinically appropriate.[a] Oral therapy regimens same for nonpregnant adults
Immunocompromised persons	Same for nonimmunocompromised persons and children	Same for nonimmunocompromised persons and children

[a] Ciprofloxacin or doxycycline should be considered an essential part of first-line therapy for inhalational anthrax.

[b] Steroids may be considered as an adjunct therapy for patients with severe edema and for meningitis based on experience with bacterial meningitis of other etiologies.

[c] Other agents with *in vitro* activity include rifampin, vancomycin, penicillin, ampicillin, chloramphenicol, imipenem, clindamycin, and clarithromycin. Because of concerns of constitutive and inducible beta-lactamases in *Bacillus anthracis,* penicillin and ampicillin should not be used alone. Consultation with an infectious disease specialist is advised.

[d] Initial therapy may be altered based on clinical course of the patient; one or two antimicrobial agents (e.g., ciprofloxacin or doxycycline) may be adequate as the patient improves.

[e] If meningitis suspected, doxycycline may be less optimal because of poor central nervous system penetration.

[f] Because of the potential persistence of spores after an aerosol exposure, antimicrobial therapy should be continued for 60 days.

[g] If intravenous ciprofloxacin is not available, oral ciprofloxacin may be acceptable because it is rapidly and well absorbed from the gastrointestinal tract with no substantial loss by first-pass metabolism. Maximum serum concentrations are attained 1–2 hours after oral dosing but may not be achieved if vomiting or ileus are present.

[h] In children, ciprofloxacin dosage should not exceed 1 g/day.

[i] The American Academy of Pediatrics recommends treatment of young children with tetracyclines for serious infections (e.g., Rocky Mountain spotted fever).

[j] Although tetracyclines are not recommended during pregnancy, their use may be indicated for life-threatening illness. Adverse effects on developing teeth and bones are dose related; therefore, doxycycline might be used for a short time (7–14 days) before 6 months of gestation.

CDC Recommedations for Management of Cutaneous Anthrax

Adults[a]	Ciprofloxacin 500 mg BID or Doxycycline 100 mg BID	60 days[c]
Children[a]	Ciprofloxacin 10–15 mg/kg every 12 hrs (not to exceed 1 g/day)[b] or Doxycycline[d] >8 yrs and >45 kg: 100 mg every 12 hrs >8 yrs and ≤45 kg: 2.2 mg/kg every 12 hrs ≤8 yrs: 2.2 mg/kg every 12 hrs	60 days[c]
Pregnant women[e]	Ciprofloxacin 500 mg BID or Doxycycline 100 mg BID	60 days[c]
Immunocompromised persons[d]	Same for nonimmunocompromised persons and children	60 days[c]

[a] Cutaneous anthrax with signs of systemic involvement, extensive edema, or lesions on the head or neck require intravenous therapy, and a multidrug approach is recommended. (See previous table)

[b] Ciprofloxacin or doxycycline should be considered first-line therapy. Amoxicillin 500 mg po TID for adults or 50 mg/kg/day divided every 8 hours for children is an option for completion of therapy after clinical improvement. Oral amoxicillin dose is based on the need to achieve appropriate minimum inhibitory concentration levels.

[c] Previous guidelines have suggested treating cutaneous anthrax for 7–10 days, but 60 days is recommended in the setting of this attack, given the likelihood of exposure to aerosolized B. anthracis (6).

[d] The American Academy of Pediatrics recommended treatment of young children with tetracyclines for serious infections (e.g., Rocky Mountain epotted fever).

[e] Although tetracyclines or ciprofloxacin are not recommended during pregnancy, their use may be indicated for life-threatening illiness. Adverse effects on developing teeth and bones are dose related; therefore, doxycycline might be used for a short time (7–10 days) before 6 months of gestation.

Plague

Pneumonic Form

Category	Initial Therapy	Duration
Adults	Streptomycin: 15 mg/kg lean body mass IM every 12 hrs or Gentamicin 5 mg/kg IV 24 hrs or Ciprofloxacin 400 mg IV every 12 hrs[a] or Doxycycline 200 mg IV loading dose followed by 100 mg IV every 12 hrs[b]	10–14 days
Children	Streptomycin, 15 mg/kg IM every 12 hrs (maximum daily dose 2 g) or Gentamicin, 2.5 mg/kg IM or IV every 8 hrs	10–14 days
Pregnant women	Gentamicin, 5 mg/kg IM or IV every 24 hrs or 2 mg/kg loading dose followed by 1.7 mg/kg IM or IV every 8 hrs or Doxycycline, 100 mg IV every 12 hrs or 200 mg IV every day	10–14 days

[a] Can switch to oral form at 750 mg q 12 h when clinically improved.
[b] Can switch to oral form at 100 mg q 12 h after clinically improved.

Adapted from Plague as a Biological Weapon: Medical and Public Health Management: Inglesby TV, Dennis DT, Henderson DA, et al. JAMA, May 3, 2000; vol. 283, no. 17: 2281–2290 and, Treatment of Biological Warfare Agent Casualties Headquarters Departments of The Army, The Navy and The Air Force and Commandant, Marine Corps http://www.vnh.org/FM8284/index.html

Plague Meningitis

Adults	Chloramphenicol 25 mg/kg IV loading dose, followed by 15 mg/kg IV qid[a]	10–14 days

[a] Oral therapy may be given after the patient is clinically improved.

From Treatment of Biological Warfare Agent Casualties Headquarters Departments of The Army, The Navy and The Air Force And Commandant, Marine Corps. http://www.vnh.org/FM8284/index.html

Widespread Outbreak and/or Post-Exposure Prophylaxis

Category	Initial Therapy	Duration
Adults	Ciprofloxacin, 500 mg po every 12 hrs Doxycycline, 100 mg po every 12 hrs	7 days

Adapted from Plague as a Biological Weapon: Medical and Public Health Management: Abstracted from: Inglesby TV, Dennis DT, Henderson DA, et al. JAMA, May 3, 2000; vol. 283, no. 17: 2281–2290.

Botulinum

Dosing for Botulinum Antitoxin for A, B, E

Category	Initial Therapy	Duration
Adults	A single 10-mL vial is used per patient, diluted to 1:10 in normal saline administered IV over 10 min infusion.	1 dose

Tularemia

Tularemia: Known Infection[a]

Category	Initial Therapy	Duration
Adults, including pregnant women	Streptomycin, 1 g every 12 hrs or Gentamicin, 5 mg/kg IM or IV every day or Ciprofloxacin, 400 mg IV every 12 hrs[a]	14 days
Children	Streptomycin: 15 mg/kg lean body mass IV every 12 hrs or Gentamicin, 2.5 mg/kg lean body mass IV every 24 hrs, then, 1.75 mg/kg lean body mass IV every 8 hrs	14 days

[a] Can switch to oral form at 500 mg every 12 hrs when clinically improved.

Adapted from "Consensus Statement: Tularemia as a Biological Weapon: Medical and Public Health Management." Dennis DT, Inglesby TV, Henderson DA, et al. JAMA, 2001; 285(21):2763–2773.

Mass Infection and/or Prophylaxis

Category	Initial Therapy	Duration
Adults, including pregnant women	Doxycycline, 100 mg po every 12 hrs or Ciprofloxacin, 500 mg po every 12 hrs	14 days
Children	Ciprofloxacin, 15 mg/kg po every 12 hrs[a]	14 days

[a] Not to exceed 1 g/day.

Adapted from "Consensus Statement: Tularemia as a Biological Weapon: Medical and Public Health Management." Dennis DT, Inglesby TV, Henderson DA, et al. JAMA, June 6, 2001; vol. 285, no. 21: 2763–2773.

VHF

Dosing of Ribavirin[a]

Category	Initial Therapy	Duration
Adult, pregnant women, and children	Ribavirin, loading dose—30 mg/kg IV (max of 2 g) × 1 dose then	Single dose
	16 mg/kg IV (max of 1 g per dose) every 6 hrs	For 4 days then
	8 mg/kg IV (max of 500 mg per dose) every 8 hrs	For 6 days Total: 7 day course

[a] Patients with VHF with unknown pathogen, or pathogen known to be Arenavirus or Bunyavirus.
Adapted from Borio L, Inglesby T, Peters CJ, et al. Hemorrhagic fever viruses as biological weapons: Medical and Public Health Management JAMA, 2002; 287; 2391–2405.

Ribavirin for Mass Infection and/or Prophylaxis

Category	Initial Therapy	Duration
Adults, pregnant women	Loading dose—2 g po × 1, then If >75 kg, 600 mg po bid If <75 kg, 400 every am and 600 mg every pm	10 days

Adapted from Borio L, Inglesby T, Peters CJ, et al. Hemorrhagic fever viruses as biological Weapons: Medical and Public Health Management. JAMA 2002;287:2391–2405.

Animal-borne Pathogens

Psittacosis

Category	Initial Therapy	Duration
Adults[a]	Doxycycline 100 mg po every 12 hrs	Until defervescence, then take an additional 14–21 days

[a] For critically ill patients, give intravenous treatment of 4.4 mg/kg every 12 hrs.

Q Fever

Fever: Acute Illness[a]

Category	Initial Therapy	Duration
Adults	Doxycycline 100 mg po every 12 hrs	14–21 days

[a] Chronic form requires double antibiotic combination therapy for a minimum of 2 years.

Mass Infection and/or Prophylaxis

Category	Initial Therapy	Duration
Adults	Doxycycline 100 mg po every 12 hrs	5–7 days

Brucellosis

Category	Initial Therapy	Duration
Adults	Doxycycline 200 mg po every day and Rifampin 600 mg po every day	6 weeks[a]

[a] If osteoarticular involvement, continue treatment for 3 months; if CNS involvement, up to 9 months.

Glanders/Melioidosis

Category	Initial Therapy	Duration
Adults	Amoxicillin/clavulanate, 20 mg/kg/day every 8 hrs or Tetracycline, 40 mg/kg/day in 3 divided oral doses or Trimethoprim/sulfa (TMP, 4 mg/kg every day; sulfa, 20 mg/kg every day) po every 12 hrs	2–5 months[a]

[a] If the clinical picture seems more severe, oral treatment should include 2 out of the 3 regimens for 1 month followed by monotherapy for an additional 2–5 months. For extrapulmonary suppurative disease, an additional 6–12 months is needed if, with the pulmonic form, abscesses are present with drainage according to the clinical setting.

Adapted from Treatment of Biological Warfare Agent Casualties, Headquarters Departments of The Army, The Navy and The Air Force and Commandant, Marine Corps.

Food-borne Pathogens

Salmonella Species,[b] Escherichia Coli O157:H7, Shigella Species

Category	Initial Therapy	Duration
Adults	Ciprofloxacin 500 mg po or IV every 12 hrs	7–10 days

[a] Use of antibiotics should be clinically warranted.

[b] Chronic carriers of *Salmonella typhi* are to be treated for four weeks.

Water-borne Pathogens

Cholera[a]

Category	Initial Therapy	Duration
Adults	Doxycycline, 300 mg po, single dose	Single dose
	or Ciprofloxacin 1 g po as a single dose	Single dose
	or 250 mg every 12 hrs	3 days

[a] ORS to replete volume loss is the first priority.

Chemical Weapons

Nerve Agents

Category	Initial Therapy	Duration
Adults	Atropine 2 mg IM every 5–10 min[a] and 1–2 g IV in 250 mL of normal saline or D_5W for adults given over 30 mins[b] and	Until clinical improvement
	Diazepam 5–10 mg IV every 10–20 min, (not to exceed 30 mg per 8 hrs); repeat in 2–4 hrs, PRN	Until seizures well controlled
Children	Atropine, 0.05 mg/kg IM every 5–10 min and PAM 15 mg/kg over 30 min and	Until resolution of dyspnea
	Diazepam, 0.2 to 0.5 mg IV every 2–4 hrs PRN for seizures	Until seizures well controlled

[a] With more severe cases, dosages may need to be increased and given more often (e.g., 5 mg in adult, or 0.05 mg/kg in children every 2 to 5 min).

[b] If IV access is not possible, IM administration an option.

Adapted from ATSDR.

Incapacitating Agents

BZ

Category	Initial Therapy	Duration
Adults	Physostigmine[a], 45 mcg/kg IV or 30 mcg/kg IM or 60 mcg/kg po every 2 hrs (bitter taste, dilute in juice)	Until clinical improvement
Children	Physostigmine, 20 mg/kg IV every 2–4 hrs	Until clinical improvement

[a] A 1 mg test dose of physostigmine is sometimes useful if the diagnosis of an anticholinergic agent is uncertain. If the patient's clinical status improves, full dosing can begin.

Blood Agents

Category	Initial Therapy	Duration
Adults	Amyl nitrate: broken onto a gauze pad and held under the nose, over the ambuvalve intake, or placed under the lip of the face mask. Inhale for 30 seconds every min and use a new perle every 3 min if sodium nitrite infusions will be delayed or Sodium nitrite: 10 mL of a 3% solution (300 mg) infused over no less than 5 min then Sodium thiosulfate, 50 mL of the 250 mg/mL (12.5 g), given intravenously over 10 min. A second treatment with half of the initial dose may be given	Until clinical improvement
Children	Sodium nitrite, 0.12 to 0.33 mL/kg body wieght up to 10 mL infused as above then 1.65 mL/kg of the standard 25% solution, IV. Second treatments with each of the two antidotes may be given at up to half the original dose if needed	Until clinical improvement

Adapted from ATSDR.

☣Radiological Weapons

Prophylaxis for Exposure to Radioactive Iodine

Category	Initial Therapy	Duration
Adults	Potassium iodide, 130 mg po, single dose	Single dose
Children	Potassium iodide, 65 mg po, single dose	Single dose
Infants	Potassium iodide, 32.5 mg po, single dose	Single dose

Selected Bibliography

In writing this book, we have been able to call on the truly remarkable work accomplished by several organizations, much of which is in the public domain. The CDC has extensive information relating to bioterrorism. Similarly, governmental websites such as those of USAMRIID, Homeland Security, the FBI, and the DHHS all have dedicated web pages on this issue. Countless nongovernmental websites were also useful resources, in particular the Center for Biosecurity, the American College of Physicians (ACP), the American College of Occupational and Environmental Medicine (ACOEM), and the Agency for Toxic Substances and Disease Registry (ATSDR). What follows is a partial list of source material for this book that would serve anyone interested in exploring their subject more deeply.

Chapter 1: Introduction to Terrorism and Bioterrorism

Top Officials (TOPOFF) 2000 Exercise Observation Report Volume 2: State of Colorado and Denver Metropolitan Area. Washington, DC: Office for State and Local Domestic Preparedness Support, Office of Justice Programs, Dept of Justice, and Readiness Division, Preparedness Training, and Exercises Directorate, Federal Emergency Management Agency; December 2000.

T. V. Inglesby. "Lessons from TOPOFF." Presented at: Second National Symposium on Medical and Public Health Response to Bioterrorism; November 28, 2000; Washington, DC.

H. Garrison. *How the World Changed: A History of the Development of Terrorism*, Presented at the Delaware Criminal Justice Council Annual Retreat October 28–29, 2001.

B. Lewis. *The Crisis of Islam: Holy War and Unholy Terror*, Random House, 2003.

Dark Winter Exercise
www.homelandsecurity.org/darkwinter/index.cfm

Chapter 2: A Clinical Approach to Biological, Chemical, and Nuclear Terrorism

E. Croddy. Chemical and Biological Warfare: A Comprehensive Survey for the Concerned Citizen. New York, NY: Copernicus Books, 2002.

J. Tucker, ed. Toxic Terror: Assessing the Use of Chemical and Biological Weapons. Cambridge, MA: MIT Press, 2001.

CT Train Site: TRAIN programs are national state-based collaborations between State Health Departments, the CDC, and regional academic centers. Both face to face and online courses are available: See https://ct.train.org/DesktopShell.aspx

Incident Command Systems: see http://www.osha.gov/SLTC/etools/ics/

USAMRIID Medical Management of Chemical and Biological Hazards course: http://www.usamriid.army.mil/education/index.htm

"Preparing for and Responding to Bioterrorism: Information for Primary Care Clinicians" by Jennifer Brennan Brady and Jeffrey S. Duchin. Northwest Center for Public Health Practice at the University of Washington School of Public Health and Community Medicine. Updates available at: http://healthlinks.washington.edu/nwcphp/bttrain/

"Worker Preparedness and Response to Bioterrorism (2003 CD)" by Edward W. Cetaruk, M.D. Available from the Association of Occupational and Environmental Clinics (AOEC) Washington, DC. Available at www.aoec.org

Chapter 3: Clinicians' Role in Public Health Preparedness and Emergency Response

R. Grunow, E. J. Finke. A procedure for differentiating between the intentional release of biological warfare agents and natural outbreaks of disease: Its use in analyzing the tularemia outbreak in Kosovo in 1999 and 2000. *Clin Microbiol Infect,* 2002;(8):510–521.

D. L. Noah, et al. Biological warfare training: Infectious disease outbreak differentiation criteria. *Annals NY Acad Sci,* 1999;894:37–43.

Chapter 4: Environmental Terrorism

Terrorism: Are America's Water Resources and Environment at Risk?

C. Copeland, B. Cody. Terrorism and Security Issues Facing the Water Infrastructure Sector. Congressional Research Service, 2003.

Risk Assessment for Food Terrorism and Other Food Safety Concerns

A. Kohnen. "Responding to Threats of Agroterrorism."

Agricultural Biosecurity: Center for Infectious Disease Research & Policy

P. Chalk. US Agriculture and Terrorism Terrorism, Infrastructure Protection, and the U.S. Food and Agricultural Sector. RAND Publications, October 2001.

The House of Representatives Subcommittee on Water Resources and Environment.
http://www.house.gov/transportation/water/10-10-01/10-10-01memo.html

National Food Safety Programs, USFDA
http://vm.cfsan.fda.gov/~dms/rabtact.html

USDA Homeland Security Efforts
http://www.usda.gov/homelandsecurity/factsheet0504.pdf

Belfer Center for Science and International Affairs (BCSIA)
http://bcsia.ksg.harvard.edu/BCSIA_content/documents/Responding_to_the_Threat_of_Agroterrorism.pdf

Academic Health Center—University of Minnesota
http://www.cidrap.umn.edu/cidrap/content/biosecurity/agbiosec/biofacts/agbiooview.html

Chapter 5: Systems in Bioterrorism Surveillance

D. M. Bravata, K. M. McDonald, W. M. Smith, et al. *Annals of Intern Med,* 2004;140(11):910–922.

Centers for Disease Control. Framework for evaluating public health surveillance systems for early detection of outbreaks: Recommendations of the CDC working group. *MMWR,* 2004;53:1–11.

Annals of Emergency Medicine, 2004;44(3):247–252.

R. Benjamin, K. Mandel. *Annals of Emerg Med,* 2004;44:235–241.

R. Heffernan, F. Mostashari, D. Das, et al. *Emerg Infect Dis,* 2004;10(5): 858–864.

Chapter 6: Vulnerable Groups: A Summary of Relevant Concerns

S. R. White, F. M. Henretig, and R. G. Dukes, "Medical Management of Vulnerable Populations and Co-morbid Conditions of Victims of Bioterrorism," *Emergency Medicine Clinics of North America,* 2002;20(2):365–392.

D. C. James. "Terrorism and the pregnant woman," *J Perinat Neonatal Nurs.* 2005; Jul–Sep;19(3):226–237.

Chapter 7: Psychological Effects of the BCN Threat

W. E. Schlenger, J. M. Caddell, L. Ebert, et al. Psychological Reactions to Terrorist Attacks: Findings from the National Study of Americans' Reactions to September 11. *JAMA,* 2002;288:581–588.

R. C. Silver, E. A. Holman, D. N. McIntosh, et al. Nationwide Longitudinal Study of Psychological Responses to September 11. *JAMA,* 2002;288:1235–1244.

A. L. Hassett, L. H. Sigal. Unforeseen consequences of terrorism: Medically unexplained symptoms in a time of fear. *Arch Intern Med,* 2002;162:1809–1813.

American Psychological Association
www.APA.org/monitor/apr99/doc.html

America's Health Together
Facing Fear Together; Blue Print report
http://www.healthtogether.org/healthtogether/ facingFear/bluePrint.html

National PTSD Center: A National Center for PTSD Fact Sheet
http://www.ncptsd.org/facts/disasters/fs_self_care_disaster.html

Chapter 8: Toxicology and Bioterrorism

Casarett and Doull's Toxicology, 5th ed., McGraw-Hill; 1998.

Patty's Industrial Hygiene and Toxicology, 5th ed., Wiley; New York, 2000.

Goldfrank Toxicological Emergencies (7th ed.), McGraw-Hill; 2002.

C. Maltoni, I. J. Selikoff. Living in a chemical world: Occupational and environmental carcinogens. *NY Acad Sci,* 1988;534.I.

Chapter 9: Legal and Ethical Issues

G. J. Annas. Bioterrorism, Public Health, and Civil Liberties. *N Engl J Med,* 2002;346(17):1337–1442.

D. Fuller. The Malevolent Use of Microbes and the Rule of Law: Legal challenges presented by bioterrorism. *Clin Infect Dis,* 2001;33(5):686–689.

L. O. Gostin, J. W. Sapsin, S. P. Teret, et al. The Model State Emergency Health Powers Act: Planning for and response to bioterrorism and naturally occurring infectious diseases. *JAMA,* 2002;288(5):622–628.

D. G. Joseph. Uses of Jacobson v. Massachusetts in the Age of Bioterror. *JAMA,* 2003;290(17):2331.

S. W. Marmagas, L. R. King, M. G. Chuk. Public health's response to a changed world: September 11, biological terrorism, and the development of an environmental health tracking network. *Am J Public Health,* 2003;93(8):1226–1230.

Turning Point: Collaborating for a New Century in Public Health. Model State Public Health Act. A tool for assessing public health laws. September 2003. www.hss.state.ak.us/dph/improving/turningpoint/ www.publichealthlaw.net/

Chapter 10: A Brief History of Biological Weapons

E. Geissler, ed. *Biological and Toxin Weapons Today.* Oxford, England: Oxford University Press; 1986.

P. Williams, D. Wallace. *Unit 731: Japan's Secret Biological Warfare in WWII.* New York: Free Press; 1989.

S. H. Harris. *Factories of Death.* New York: Routledge; 1994.

J. Miller, S. Engelberg, W. Broad, *Germs: Biological Weapons and America's Secret War.* New York: Simon & Schuster; 2001;31–88.

T. Mangold. *Plague Wars: The Terrifying Reality of Biological Warfare.* New York: St. Martin's Press; 1999;81–85, 379–391.

Regis, ed. *The Biology of Doom.* New York: Owl Press; 219–233.

A. Mayor. *Greek Fire: Poison Arrows and Scorpion Bombs.* New York: Overlook Press; 2003.

E. Lesho, D. Dorsey, D. Bunner Feces. Dead horses, and fleas: Evolution of the hostile use of biological agents. *Western J Med,* 1998;168(6):512–516.

A. G. Robertson, L. J. Robertson. From asps to allegations: Biological warfare in history. *Mil Med,* 1995(8);160:369–373.

G. W. Christopher, T. J. Cieslak, J. A. Pavlin, E. M. Eitzen. Biological warfare: a historical perspective. *JAMA,* 1997;278(5):412–417.

The Discovery Channel Series. Bioterror Through Time. http://dsc.discovery.com/anthology/spotlight/bioterror/history/history.html.

D. Lisa Rotz, S. Ali Khan, Public health assessment of potential biological terrorism agents. *Emerg Infect Dis J,* 2002;8(2). http://www.cdc.gov/ncidod/EID/vol8no2/01-0164.htm#appendix

Chapter 11: Introduction to Biological Agents

Center for Disease Control—Emergency Preparedness & Response Site Bioterrorism Agents/Diseases http://www.bt.cdc.gov/agent/agentlist-category.asp

Chapter 12: Smallpox

J. M. Neff. Variola (Smallpox) and Monkeypox Viruses. In: G. L. Mandell, J. E. Bennett, R. Dolin eds. *Mandell's Principles and Practice of Infectious Diseases.* 5th ed. Philadelphia; Pa: Saunders; 2000:1555–1556.

D. A. Henderson, T. V. Inglesby, J. G. Bartlett, et al. Smallpox as a biological weapon: medical and public health management. Working group on civilian biodefense. *JAMA*, 1999;281(22):2127–2137.

J. G. Breman, D. A. Henderson. Diagnosis and management of smallpox. *N Engl J Med*, 2002;346(17):1300–1308.

S. R. Kimmel, M. C. Mahoney, R. K. Zimmerman. Vaccines and bioterrorism: Smallpox and anthrax. *J Fam Pract*, 2003;52(Suppl 1):S56–61.

J. M. Lane, J. Goldstein. Evaluation of 21st-century risks of smallpox vaccination and policy options. *Ann Intern Med*, 2003;138(6):488–493.

J. M. Neff, J. M. Lane, V. A. Fulginiti, D. A. Henderson. Contact vaccinia— transmission of vaccinia from smallpox vaccination. *JAMA*, 2002;288 (15):1901–1915.

U.S. Army Medical Research Institute of Infectious Diseases, Medical Management of Biological Casualties Handbook: Viral Agents.
http://www.vnh.org/BIOCASU/12.html

Office of the Surgeon General, Department of the Army, Textbook of Military Medicine: Medical Management of Biological Warfare.
http://www.vnh.org/MedAspChemBioWar/smallpox

Armed Forces Medical Services, Treatment of Biological Warfare Agent Casualties: Smallpox
http://www.vnh.org/FM8284/index.html

Center for Biosecurity, Clinical Smallpox: Primer for Physicians.
http://www.upmc-biosecurity.org/pages/agents/smallpox.html

Center for Disease Control—Emergency Preparedness and Response
http://www.bt.cdc.gov.

Executive Summary. Smallpox Response Plan and Guidelines.
http://www.bt.cdc.gov/agent/smallpox/response-plan/index.asp.

CDC. Protecting Americans: Smallpox Vaccination Program.
http://www.bt.cdc.gov/agent/smallpox/vaccination/vaccination-program-statement.asp.

Chapter 13: Viral Hemorrhagic Fevers

E. I. Ryabchikova, L. V. Kolesnikova, S. V. Luchko. An analysis of features of pathogenesis in 2 animal models of Ebola virus infection. *J Infect Dis*, 1999;179 (Suppl 1):S199–S202.

L. Borio, T. Inglesby, C. J. Peters, A. L. Schmaljohn, J. M. Hughes, P. B. Jahrling, et al. Hemorrhagic fever viruses as biological weapons: medical and public health management. *JAMA*, 2002;287(18):2391–2405.

U.S. Army Medical Research Institute of Infectious Diseases (USAMRIID), Medical Management of Biological Casualties Handbook: VHF. http://www.vnh.org/BIOCASU/6.html

Office of the Surgeon General, Department of the Army, Textbook of Military Medicine: Medical Management of Virtual Naval Hospital. http://www.vnh.org/MedAspChemBioWar/ VHF

Departments of the Army, the Navy, and the Air Force, and Commandant, Marine Corps, Treatment of Biological Warfare: VHF. http://www.vnh.org/FM8285/cover.html

World Health Organization. http://www.who.int/csr/delibepidemics/en/

Center for Biosecurity, Clinical Anthrax: Primer for Physicians. http://www.upmc-biosecurity.org/pages/agents/anthrax.html

J. K. Brennan, J. S. Duchin. Preparing for and responding to bioterrorism. Bioterrorism training manual. Northwest Center for Public Health Practice and Public Health. http://www.metrokc.gov/health/bioterrorism/education

Chapter 14: Anthrax

T. V. Inglesby, T. O'Toole, D. A. Henderson, J. G. Bartlett, M. S. Ascher, E. Eitzen, A. M. Friedlander, et al. Anthrax as a biological weapon, 2002: Updated recommendations for management. *JAMA,* 2002;287(17):2236–2252.

L. Bush, B. Abrams, A. Beall, C. Johnson. Index case of fatal inhalational anthrax due to bioterrorism in the United States. *N Engl J Med,* 2001;345: 1607–1610.

M. N. Swartz. Recognition and management of anthrax—an update. *N Engl J Med,* 2001;345(22):1621–1626.

J. C. Pile, J. D. Malone, E. M. Eitzen, A. M. Friedlander. Anthrax as a potential biological warfare agent. *Arch Intern Med,* 1998;158(5):429–434.

J. P. Earls, Jr., D. Cerva, E. Berman, J. Rosenthal, N. Fatteh, P. P. Wolfe, et al. Inhalational anthrax after bioterrorism exposure: Spectrum of imaging findings in two surviving patients. *Radiology,* 2002;222:305–312.

J. C. Pile, J. D. Malone, E. M. Eitzen, A. M. Friedlander. Anthrax as a potential biological warfare agent. *Arch Intern Med,* 1998;158(5):429–434.

N. Hupert, et al. Accuracy of screening for inhalational anthrax after a bioterrorist attack. *Ann Intern Med,* 2003;139:337–345.

Center for Biosecurity, Clinical Anthrax: Primer for Physicians. http://www.upmc-biosecurity.org/pages/agents/anthrax.html

U.S. Army Medical Research Institute of Infectious Diseases (USAMRIID), Medical Management of Biological Casualties Handbook: Anthrax. http://www.vnh.org/BIOCASU/6.html

Office of the Surgeon General, Department of the Army, Textbook of Military Medicine: Medical Management of Biological Warfare. Virtual Naval Hospital. http://www.vnh.org/MedAspChemBioWar/anthrax

Armed Forces Medical Services, Treatment of Biological Warfare Agent Casualties: Anthrax. VNH. http://www.vnh.org/FM8284/index.html

Bell DM. Meeting summary: Clinical issues in the prophylaxis, diagnosis, and treatment of anthrax. *Emerg Infect Dis,* 2002;8(2). http://www.cdc.gov/ncidod/ eid/vol8no2/01-0521.htm.

Temte JL. Differential diagnosis of inhalation anthrax: Diagnosis and management based upon initial presentations of eleven cases of bioterrorism-associated inhalation anthrax. University of Wisconsin Department of Family Medicine and the Wisconsin Academy of Family Physicians Web site. http://www.fammed.wisc.edu/research/project/anthrax.html

CDC Anthrax FAQ: Signs and Symptoms. http://www.bt.cdc.gov/agent/anthrax/faq/signs.asp

Food and Drug Administration/Center for Drug Evaluation and Research: Anthrax. http://www.fda.gov/cder/drugprepare/default.htm#Anthrax. Accessed 4/11/03.

Chapter 15: Plague

T. Butler. Yersinia pestis and the Plague. In: G. L. Mandell, J. E. Bennett, R. Dolin eds. *Mandell's Principles and Practice of Infectious Diseases.* 5th ed. Philadelphia. Saunders; 2000:2406–2411.

H. Feldmann, M. Czub, S. Jones, D. Dick, M. Garbutt, A. Grolla, et al. Emerging and re-emerging infectious diseases. *Med Microbiol Immunol (Berl),* 2002;191(2):63–74.

T. V. Inglesby, D. T. Dennis, D. A. Henderson, et al. Plague as a biological weapon: medical and public health management. *JAMA,* 2000;283: 2281–2290.

U.S. Army Medical Research Institute of Infectious Diseases, Medical Management of Biological Casualties Handbook: Viral Agents. http://www.vnh.org/BIOCASU/10.html

Office of the Surgeon General, Department of the Army, Textbook of Military Medicine: Medical Management of Biological Warfare. http://www.vnh.org/MedAspChemBioWar/plague

Armed Forces Medical Services, Treatment of Biological Warfare Agent Casualties: Plague. VNH.
http://www.vnh.org/FM8284/index.html

Center for Biosecurity, Clinical Anthrax: Primer for Physicians.
http://www.upmcbiosecurity.org/pages/agents/plague.html

Chapter 16: Tularemia

J. T. Cross, R. L. Penn. In: G. L. Mandell, J. E. Bennett, R. Dolin eds. *Mandell's Principles and Practice of Infectious Diseases.* 5th ed. Philadelphia: Saunders; 2000;2393–2401.

H. Feldmann, M. Czub, S. Jones, D. Dick, M. Garbutt, A. Grolla, et al. Emerging and re-emerging infectious diseases. *Med Microbiol Immunol (Berl),* 2002;191 (2):63–74.

E. Choi. Tularemia and Q fever. *Med. Clin. North America,* 2002;86(2): 393–416.

D. T. Dennis, T. V. Inglesby, D. A. Henderson, J. G. Bartlett, M. S. Ascher, E. Eitzen, et al. Tularemia as a biological weapon: medical and public health management. *JAMA,* 2001;285(21):2763–2773.

Office of the Surgeon General, Department of the Army, Textbook of Military Medicine: Medical Management of Biological Warfare.
http://www.vnh.org/MedAspChemBioWar/tularemia

Armed Forces Medical Services, Treatment of Biological Warfare Agent Casualties: Tularemia.
http://www.vnh.org/FM8284/index.html

U.S. Army Medical Research, Institute of Infectious Diseases, Medical Management of Biological Casualties Handbook: Tularemia.
http://www.vnh.org/BIOCASU/11.html.

Center for Biosecurity, Clinical Anthrax: Primer for Physicians.
http://www.upmc-biosecurity.org/pages/agents/tularemia.html.

D. Velendzas, Plague. EMedicine.
http://www.emedicine.com/EMERG/topic428.htm#section~differentials.

Medical Aspects of Bioterrorism. ACPonline-Bioterrorism.
http://www.acponline.org/ bioterro/medicalaspects.htm.

Chapter 17: Introduction to Biotoxins

CDC Botulism in the U.S., 1899–1996. Handbook for Epidemiologists, Clinicians, and Laboratory Workers, Atlanta, GA. CDC; 1998.

American Academy of Pediatrics. Clostridial Infection. In: L. K. Pickering, ed. 2000 Red Book: Report of the Committee on Infectious Diseases. 25th ed. Elk Grove Village, IL: American Academy of Pediatrics; 2000;212.

S. S. Arnon, R. Shechet, T. V. Inglesby et al. Botulinum toxin as a biological weapon: Medical and public health management. *JAMA,* 2001;285:8, 1059–1070.

R. L. Shapiro, C. Hathaway, D. L. Swerdlow. Botulism in the U.S.: A clinical and epidemiologic review. *Ann Intern Med,* 1998;129:221–228.

U.S. Army Medical Research Institute of Infectious Diseases (USAMRIID), Medical Management of Biological Casualties Handbook: Botulinum. http://www.vnh.org/BIOCASU/6.html

Office of the Surgeon General, Department of the Army, Textbook of Military Medicine: Medical Management of Virtual Naval Hospital. http://www.vnh.org/MedAspChemBioWar/botulinum

Armed Forces Medical Services, Treatment of Biological Warfare Agent Casualties: botulinum. VNH. http://www.vnh.org/FM8284/index.html

Center for Biosecurity, Clinical Botulinum: Primer for Physicians. http://www.upmc-biosecurity.org/pages/agents/botulism.html

Chapter 18: Botulinum Toxin

CDC Botulism in the U.S., 1899–1996. Handbook for Epidemiologists, Clinicians, and Laboratory Workers, Atlanta, GA. CDC, 1998.

American Academy of Pediatrics. Clostridial Infection. In: L. K. Pickering, ed. 2000 Red Book: Report of the Committee on Infectious Diseases. 25th ed. Elk Grove Village, IL: American Academy of Pediatrics; 2000;212.

S. S. Arnon, R. Shechet, T. V. Inglesby et al. Botulinum Toxin as a Biological Weapon: Medical and Public Health Management. *JAMA,* 2001;285:8, 1059–1070.

R. L. Shapiro, C. Hathaway, D. L. Swerdlow. Botulism in the U.S.: A clinical and epidemiologic review. *Ann Intern Med,* 1998;129:221–228.

U.S. Army Medical Research Institute of Infectious Diseases (USAMRIID), Medical Management of Biological Casualties Handbook: botulinum. http://www.vnh.org/BIOCASU/6.html.

Office of the Surgeon General, Department of the Army, Textbook of Military Medicine: Medical Management of Virtual Naval Hospital. http://www.vnh. org/MedAspChemBioWar/botulinum

Armed Forces Medical Services, Treatment of Biological Warfare Agent Casualties: botulinum. VNH. http://www.vnh.org/FM8284/index.html

Center for Biosecurity, Clinical Botulinum: Primer for Physicians.
http://www.upmc-biosecurity.org/pages/agents/botulism.html

Chapter 19: Ricin

G. J. Moran. Threats in bioterrorism II: CDC category B and C agents. *Emerg Med Clins of North America*, 2002;20:2.

U.S. Army Medical Research Institute of Infectious Diseases (USAMRIID), Medical Management of Biological Casualties Handbook: Ricin.
http://www.vnh.org/BIOCASU/6.html

Office of the Surgeon General, Department of the Army, Textbook of Military Medicine: Medical Management of Virtual Naval Hospital.
http://www.vnh.org/MedAspChemBioWar/ricin

Departments of the Army, the Navy, and the Air Force, and Commandant, Marine Corps Treatment of Chemical Agent Casualties and Conventional Military Chemical Injuries: Ricin.
http://www.vnh.org/FM8285/cover.html

Ricin and the umbrella murder Oct 23, 2003.
http://www.cnn.com/2003/WORLD/europe/01/07/terror.poison.bulgarian/

Nick Allen. Six people held after discovery of deadly poison. 07 January 2003. The Independent.
http://news.independent.co.uk/uk/crime/story.jsp?story=367215

Chapter 20: Category B, C, and Additional Agents

Q fever

Threats in bioterrorism. II: CDC category B and C agents. G. J. Moran *Emerg Med Clin North Am* 01-May-2002;20(2):311–30.
http://www.cdc.gov/ncidod/dvrd/qfever/index.htm

U.S. Army Medical Research Institute of Infectious Diseases (USAMRIID), Medical Management of Biological Casualties Handbook: Q fever.
http://www.vnh.org/BIOCASU/6.html

Office of the Surgeon General, Department of the Army, Textbook of Military Medicine: Medical Management of Virtual Naval Hospital.
http://www.vnh.org/MedAspChemBioWar/Q fever

Armed Forces Medical Services, Treatment of Biological Warfare Agent Casualties: Q fever. VNH.
http://www.vnh.org/FM8284/index.html

Brucellosis

Threats in bioterrorism. II: CDC category B and C agents.

G. J. Moran. *Emerg Med Clin North Am*, 2002;20(2):311–330.

Centers for Disease Control and Prevention, National Center for Infectious Diseases Division of Bacterial and Mycotic Diseases.
http://www.cdc.gov/ncidod/dbmd/diseaseinfo/brucellosis_g.htm

U.S. Army Medical Research Institute of Infectious Diseases (USAMRIID), Medical Management of Biological Casualties Handbook: Q fever.
http://www.vnh.org/BIOCASU/6.html

Office of the Surgeon General, Department of the Army, Textbook of Military Medicine: Medical Management of Virtual Naval Hospital.
http://www.vnh.org/MedAspChemBioWar/Q fever

Armed Forces Medical Services, Treatment of Biological Warfare Agent Casualties: Q fever. VNH.
http://www.vnh.org/FM8284/index.html

Glanders

Threats in bioterrorism. II: CDC category B and C agents. G. J. Moran. *Emerg Med Clin North Am*, 2002;20(2):311–330.

Centers for Disease Control and Prevention, National Center for Infectious Diseases. Division of Bacterial and Mycotic Diseases.
http://www.cdc.gov/ncidod/dbmd/diseaseinfo/glanders_g.htm.

USAMRIID's Medical Management of Biological Casualties Handbook Glanders and Melioidosis Glanders and Melioidosis.
http://www.vnh.org/BIOCASU/8.html

Melioidosis

B. H. Short. Air vice—Marshal. *J Australian Defence Health Service ADF Health,* 2002;3(1):13–21.

Preparedness for Deliberate Epidemics, WHO, Annex 3, Biological and chemical agents (pdf, 83k).
http://www.who.int/csr/delibepidemics/biochemguide/en/index.html

USAMRIID's Medical Management of Biological Casualties Handbook Glanders and Melioidosis U.S. Army Medical Research Institute of Infectious Diseases
www.vnh.org/biocasu/8.html

Centers for Disease Control and Prevention, National Center for Infectious Diseases Division of Bacterial and Mycotic Diseases.
http://www.cdc.gov/ncidod/dbmd/diseaseinfo/melioidosis_g.htm#common

Epsilon Toxin of Clostridium perfringens

S. Baron. *Clostridia: Sporeforming Anaerobic Bacilli Medical Microbiology.* Chapter 18. Philadelphia, PA: Addison-Wesley; 1996.

Threats in Bioterrorism. II: CDC Category B and C Agents. G. J. Moran. *Emerg Med Clin North Am,* 2002;20(2):311–330.

Headquarters Departments of the Army, the Navy, and the Air Force, and Commandant, Marine Corps. Treatment of Biological Warfare Agent Casualties: Chapter 4: Toxins Section III—Clostridium Perfringens Toxins. http://www.vnh.org/FM8284/Chapter4/4-13.html

Public health response to biological and chemical weapons: WHO guidance (2004). Annex 2.Toxins. http://www.who.int/csr/delibepidemics/en/annex2.pdf

Illinois Poison Center Bioterrorism Treatment Guidelines Clostridium perfringens Toxins. http://www.mchc.org/ipc/InfoForProfessionals/HealthProfessionals/Clostridium.pdf

Epsilon Toxin of Clostridium perfringens. Center for Food Security and Public Health Iowa State University College of Veterinary Medicine. http://www.scav.org/Epsilon-toxin%20Fact%20Sheet.htm

Staphylococcal Enterotoxin B

Threats in bioterrorism. II: CDC category B and C Agents.

G. J. Moran. *Emerg Med Clin North Am,* 2002;20(2):311–330.

M. D. J. Williams. FAAEM, ChemBioRadialogicNuclExplosive—Staphylococcal Enterotoxin B. Emedicine.com http://www.emedicine.com/emerg/topic888.htm

Public health response to biological and chemical weapons: WHO guidance (2004). Annex 2. Toxins. http://www.who.int/csr/delibepidemics/en/annex2.pdf

Food-Borne Pathogens

M. Montes, H. L. DuPont. In: *Cohen & Powderly Infect Dis.* 2nd ed. Chapter 24, 93. Elsevier, New York, New York, 2004.

Threats in Bioterrorism. II: CDC Category B and C Agents.

G. J. Moran, *Emerg Med Clin North Am,* 01-May-2002;20(2):311–330.

Foodborne Pathogenic Microorganisms and Natural Toxins Handbook. FDA, Center for Food Safety and Applied Nutrition. http://vm.cfsan.fda.gov/~mow/intro.html

CDC: Food Safety Threats. http://www.cdc.gov/ncidod/dbmd/diseaseinfo/

National Institute of Allergy and Infectious Diseases. National Institutes of Health. Food-Borne Diseases. http://www.niaid.nih.gov/factsheets/foodbornedis.htm

Water-Borne Pathogens

M. M. Herbert, L. DuPont. In: Cohen & Powderly Infectious Diseases. 2nd ed. Chapters 24, 93. Elsevier, New York, New York, 2004.

R. B. Sack, G. Balakrish. *The Lancet Seminar: Cholera,* 2004;363(9404):223–233.

D. E. Katz. Parasitic infections of the gastrointestinal tract. *Gastroenterol Clin North Am,* 2001;30(3):797–815.

World Health Organization
http://www.who.int/mediacentre/factsheets/en/fact107.html

Foodborne Pathogenic Microorganisms and Natural Toxins Handbook. FDA, Center for Food Safety & Applied Nutrition
http://www.cfsan.fda.gov/~mow/chap7.html

Division of Parasitic Diseases, National Center for Infectious Diseases.
http://www.cdc.gov/ncidod/dpd/parasites/cryptosporidiosis/
factsht_cryptosporidiosis.htm

Typhus Fever

G. L. Mandell. *Principles and Practice of Infectious Diseases.* 5th ed. Copyright © 2000; Churchill Livingstone, Inc.: 2050–2052.

M. M. Herbert, L. DuPont. In: Cohen & Powderly Infectious Diseases, 2nd ed. Elsevier, New York, New York, 2004;2221–2230.

F. Azad, S. Abdu. Pathogenic rickettsiae as bioterrorism agents. *Ann NY Acad Sci,* 2003;990:734–738.

World Health Organization
http://www.who.int/csr/delibepidemics/en/annex3.pdf

Psittacosis

M. M. Herbert, L. DuPont. In: Cohen & Powderly Infectious Diseases, 2nd ed., Elsevier, New York, New York, 2004;2331–2335.

G. L. Mandell. *Principles and Practice of Infectious Diseases.* 5th ed. Churchill Livingstone, Inc. 2000;2004–2006.

R. S. Cotran. *Robbins Pathologic Basis of Disease,* 6th ed. Philadelphia, PA. 1999; Saunders, 361.

National Center for Infectious Diseases Division of Bacterial and Mycotic Diseases.
http://www.cdc.gov/ncidod/dbmd/diseaseinfo/psittacosis_t.htm

Alphavirus Encephalitides

G. L. Mandell. *Principles and Practice of Infectious Diseases.* 5th ed. Churchill Livingstone, Inc. 2000; Chap. 140.

G. J. Moran. Threats in bioterrorism. II: CDC category B and C agents. *Emerg Med Clin North Am*, 2002;20(2):311–330.

U.S. Army Medical Research Institute of Infectious Diseases, Medical Management of Biological Casualties Handbook: Viral Agents.
http://www.vnh.org/BIOCASU/14.html

Office of the Surgeon General, Department of the Army, Textbook of Military Medicine: Medical Management of Biological Warfare.
http://www.vnh.org/MedAspChemBioWar/chapters/chapter_28.htm

Armed Forces Medical Services, Treatment of Biological Warfare Agent Casualties: Plague.
http://www.vnh.org/FM8284/Chapter3/3-11.html

Public health response to biological and chemical weapons: WHO guidance (2004). Chapter 4
http://www.who.int/csr/delibepidemics/en/annex3.pdf

Nipah Virus

WHO Weekly Epidemiological Record 27 February 2004, vol. 79, 9 (pp 85–92).

M. M. Herbert, L. DuPont. In: Cohen & Powderly Infectious Diseases, 2nd ed., Elsevier, New York, New York, 2004;974.

Nipah virus—a potential agent of bioterrorism? Sai-Kit Lam *Antiviral Res*, 2003;57(1–2):113–119.

G. J. Moran. Threats in bioterrorism. II: CDC category B and C agents. *Emerg Med Clin North Am*, 2002;20(2):311–330.

K. J. Goh, C. T. Tan, N. K. Chew, et al. Clinical features of Nipah virus encephalitis among pig farmers in Malaysia. *N Engl J Med*, 2000;342: 1229–1235.

Sai Kit Lam, Kaw Bing Chua. Nipah virus encephalitis outbreak in Malaysia. *Clin Infect Dis*, 2002;34:S48–S51.

Hanta Virus

M. M. Herbert, L. DuPont. In: Cohen & Powderly Infectious Diseases, 2nd ed., Elsevier, New York, New York, 2004.

G. L. Mandell. Principles and Practice of Infectious Diseases, 5th ed., Churchill Livingstone, Inc., Philadelphia, Pennsylvania, 2000; Chap. 140.

G. J. Moran. Threats in bioterrorism. II: CDC category B and C agents. *Emerg Med Clin North Am*, 2002;20(2):311–330.

C. J. Peters, S. Ali Khan. Hantavirus pulmonary syndrome: The new American hemorrhagic fever. *Clin Infect Dis*, 2002;34:1224–1231.

R. Riquelme, M. Riquelme, A. Torres. Hantavirus pulmonary syndrome. *Southern Chile Emerging Infectious Dis J,* 2003;9(11).

All About Hantaviruses: Special Pathogens Branch, Division of Viral and Rickettsial Diseases, National Center for Infectious Diseases, CDC.
http://www.cdc.gov/ncidod/diseases/hanta/hps/noframes/phys/technicalinfoindex.htm

Trichothecene (T-2) Mycotoxins

A. Ruth Etzel MD, PhD. Mycotoxins. *JAMA,* 2002;287:425–427.

J. B. Tucker. Conflicting evidence revives "Yellow Rain" Controversy Center for Nonproliferation Studies, Monterey Institute of International Studies.
http://cns.miis.edu/pubs/week/020805.htm

U.S. Army Medical Research Institute of Infectious Diseases (USAMRIID), Medical Management of Biological Casualties Handbook: Trichothecene (T-2) mycotoxins.
http://www.vnh.org/BIOCASU/20.html.

Office of the Surgeon General, Department of the Army, Textbook of Military Medicine: Medical Management of Virtual Naval Hospital.
http://www.vnh.org/MedAspChemBioWar/Trichothecene (T-2) mycotoxins

Armed Forces Medical Services, Treatment of Biological Warfare Agent Casualties: Trichothecene (T-2) mycotoxins.
http://www.vnh.org/FM8284/index.html

Public health response to biological and chemical weapons: WHO guidance (2004) Chapter 4.
http://www.who.int/csr/delibepidemics/en/annex3.pdf

W. S. Augerson. A Review of the Scientific Literature as it Pertains to Gulf War Illnesses. Volume 5: Chemical and Biological Warfare Agents.
http://www.rand.org/publications/MR/MR1018.5/MR1018.5.chap4.html

Aflatoxin

Public health response to biological and chemical weapons: WHO guidance (2004) Chapter 4.
http://www.who.int/csr/delibepidemics/en/annex3.pdf
http://www.rand.org/publications/MR/MR1018.5/MR1018.5.chap4.html

Foodborne Pathogenic Microorganisms and Natural Toxins Handbook. FDA, Center for Food Safety & Applied Nutrition.
http://vm.cfsan.fda.gov/~mow/chap41.html

Public health response to biological and chemical weapons: WHO guidance (2004).
http://www.who.int/csr/delibepidemics/en/annex2.pdf

Saxitoxin

Foodborne Pathogenic Microorganisms and Natural Toxins Handbook.

Treatment of Biological Warfare Agent Casualties: Chapter 4: Toxins.

M. de Carvahlo, J. Jacinto, Paralytic shellfish poisoning. *J Neurol,* 1998;245: 551–554.

Public health response to biological and chemical weapons: WHO guidance (2004).
http://www.who.int/csr/delibepidemics/en/annex2.pdf

Woods Hole Oceanographic Institution. Harmful Algae.
http://www.whoi.edu/science/B/redtide/illness/psp.html

FDA, Center for Food Safety & Applied Nutrition.
http://www.cfsan.fda.gov/~mow/chap37.html

Departments of the Army, the Navy, and the Air Force, and Commandant, Marine Corps.
http://www.vnh.org/FM8284/Chapter4/4–32.html

Chapter 21: Introduction to Chemical Weapons

S. Connor. An Air That Kills: A Familiar History of Poison Gas. Death By Technology conference, Birkbeck College, May, 2003.

K. F. Mckenzie. "An Ecstay of Fumbling." Joint Force Quarterly, Winter '96.

D. Evison. Chemical Weapons. *BMJ,* 2002;324(7333):332–335.

K. B. Olson, A. Shinrikyo. Once and future threat? *Emerg Infect Dis,* 1999;5(4): 513–516.

Center for Disease Control- Emergency Preparedness & Response Site
Chemical Emergencies, Chemical Disasters
http://www.bt.cdc.gov/agent/agentlistchem-category.asp

United States Army Medical Research Institute of Chemical Defense Chemical Casualty Care Division, Medical Management of Chemical Casualties Handbook.
http://www.vnh.org/CHEMCASU/titlepg.html

Departments of the Army, the Navy, and the Air Force, and Commandant, Marine Corps. Treatment of Chemical Agent Casualties and Conventional Military Chemical Injuries.
http://www.vnh.org/FM8285/cover.html

Office of the Surgeon General, Department of the Army, Textbook of Military Medicine: Medical Aspects of Chemical and Biological Warfare.
http://www.vnh.org/MedAspChemBioWar/

Agency for Toxic Substances and Disease Registry, The Agency for Toxic Substances and Disease Registry, Medical Management Guidelines (MMGs) for Acute Chemical Exposures.
http://www.atsdr.cdc.gov/mmg.html

The Center for Nonproliferation Studies (CNS) at the Monterey Institute of International Studies.
http://cns.miis.edu/research/cbw/possess.htm.

Chapter 22: A Brief History of Chemical Weapons

S. Connor. An Air That Kills: A Familiar History of Poison Gas. Death By Technology conference, Birkbeck College, 30 May, 2003.

K. F. Mckenzie. "An Ecstay of Fumbling." Joint Force Quarterly, Winter '96.

D. Evison. Chemical Weapons. BMJ, 2002 Feb9;324(7333):332–335.

United States Army Medical Research Institute of Chemical Defense Chemical Casualty Care Division, Medical Management of Chemical Casualties Handbook
http://www.vnh.org/CHEMCASU/titlepg.html

Departments of the Army, the Navy, and the Air Force, and Commandant, Marine Corps. Treatment of Chemical Agent Casualties and Conventional Military Chemical Injuries
http://www.vnh.org/FM8285/cover.html

Office of the Surgeon General, Department of the Army, Textbook of Military Medicine: Medical Aspects of Chemical and Biological Warfare
http://www.vnh.org/MedAspChemBioWar/

The Agency for Toxic Substances and Disease Registry, Medical Management Guidelines (MMGs) for Acute Chemical Exposures. http://www.atsdr.cdc.gov/mmg.html

The Center for Nonproliferation Studies (CNS) at the Monterey Institute of International Studies. http://cns.miis.edu/research/cbw/ possess.htm

Chapter 23: Pulmonary Agents

R. Shenoi. Chemical warfare agents. Clin Ped Emerg Med, 3:239–247.

Departments of the Army, the Navy, and the Air Force, and Commandant, Marine Corps. Treatment of Chemical Agent Casualties and Conventional Military Chemical Injuries, 1995.
http://www.vnh.org/FM8285/cover.html

ATSDR CDC.
http://www.bt.cdc.gov/agent/agentlistchem-category.asp#choking

United States Army Medical Research Institute of Chemical Defense Chemical Casualty Care Division USAMRIID Medical Management of Chemical Casualties Handbook, 3rd ed.
www.vnh.org/CHEMCASU/02Pulmonaryagents.html

Office of the Surgeon General, Department of the Army. Textbook of Military Medicine: Medical Aspects of Chemical and Biological Warfare Virtual Naval Hospital, 1997.
http://www.vnh.org/MedAspChemBioWar/chapters/chapter_9.htm#Toxic

Treatment of Chemical Agent Casualties and Conventional Military Chemical Injuries: FM8-285: Part 1 Chemical Agent Casualties.
http://www.vnh.org/FM8285/Chapter/chapter5.html

American College of Physicians.
http://www.acponline.org/bioterro/toxic_gas.htm

Chapter 24: Vesicants or Blistering Agents

A. Devereaux, D. E. Amundson. Vesicants and nerve agents in chemical warfare. *Postgrad Med,* 2002;Oct;112(4):90–96.

United States Army Medical Research Institute of Chemical Defense Chemical Casualty Care Division, Medical Management of Chemical Casualties Handbook.
http://www.vnh.org/CHEMCASU/titlepg.html

Departments of the Army, the Navy, and the Air Force, and Commandant, Marine Corps. Treatment of Chemical Agent Casualties and Conventional Military Chemical Injuries.
http://www.vnh.org/FM8285/cover.html

Office of the Surgeon General, Department of the Army, Textbook of Military Medicine: Medical Aspects of Chemical and Biological Warfare.
http://www.vnh.org/MedAspChemBioWar/

Agency for Toxic Substances and Disease Registry, Medical Management Guidelines (MMGs) for Acute Chemical Exposures.
http://www.atsdr.cdc.gov/mmg.html

Chapter 25: Nerve and Incapacitating Agents

Harvey. Pharmacology, 2nd ed. LWW. P43.

Chemical warfare. Nerve agent poisoning. *Crit Care Clin,* 1997;13(4): 923–942.

R. Shenoi. *Chemical Warfare Agents CPEM.* 2002;3(4):239–247.

J. B. Leikin. A review of nerve agent exposure for the critical care physician. *Crit Care Med,* 2002;30(10):2346–2354.

J. S. Rotenberg. Nerve agent attacks on children: diagnosis and management. *Pediatrics,* 2003;112(3 Pt 1):648–658.

Centers for Disease Control.
http://www.bt.cdc.gov/agent/sarin/index.asp

American College of Physicians.
http://www.acponline.org/bioterro/nerve_gas.htm

United States Army Medical Research Institute of Chemical Defense Chemical Casualty Care Division, Medical Management of Chemical Casualties Handbook, 1999.
http://www.vnh.org/CHEMCASU/titlepg.html

Departments of the Army, the Navy, and the Air Force, and Commandant, Marine Corps. Treatment of Chemical Agent Casualties and Conventional Military Chemical Injuries, 1995.
http://www.vnh.org/FM8285/cover.html

Office of the Surgeon General, Department of the Army, Textbook of Military Medicine: Medical Aspects of Chemical and Biological Warfare.
http://www.vnh.org/MedAspChemBioWar/

Chapter 26: Blood Agents

United States Army Medical Research Institute of Chemical Defense Chemical Casualty Care Division, Medical Management of Chemical Casualties Handbook.
http://www.vnh.org/CHEMCASU/titlepg.html

Departments of the Army, the Navy, and the Air Force, and Commandant, Marine Corps. Treatment of Chemical Agent Casualties and Conventional Military Chemical Injuries.
http://www.vnh.org/FM8285/cover.html

Office of the Surgeon General, Department of the Army, Textbook of Military Medicine: Medical Aspects of Chemical and Biological Warfare.
http://www.vnh.org/MedAspChemBioWar/

Agency for Toxic Substances and Disease Registry, Medical Management Guidelines (MMGs) for Acute Chemical Exposures.
http://www.atsdr.cdc.gov/mmg.html

Chapter 27: A Brief History of Nuclear Weapons

Cline B. Lovett. *Men Who Made a New Physics: Physicists and the Quantum Theory.* University of Chicago Press; 1987.

Nuclear Age Peace Foundation
http://www.nuclearfiles.org/index.html

Wikipedia Encyclopedia
http://en.wikipedia.org/wiki/History_of_physics

UCLA Department of Physics.
http://www.physics.ucla.edu/~cwp/articles/EARLYNPC. HTML

Center for History of Physics, American Institute of Physics.
http://www.aip.org/history/newsletter/spring2002/photo-spring2002.htm

Chapters 28, 29, 30: Nuclear Agents

Armed Forces Radiobiology Research Institute. Medical Management of Radiological Casualties.

Guidance for protecting building environments from airborne chemical, biological, or radiological attacks. DHHS (NIOSH) publication no. 2002–139.

F. A. Jr, Mettler, G. L. Voelz. Major radiation exposure—what to expect and how to respond. *N Engl J Med*, 2002;346:1554–1561.

J. B. Leikinl, R. B. McFee, F. G. Walter, K. Edsall. A primer for nuclear terrorism, *Disease-A-Month*, 2003;49(8).

F. Fong, D. Schrader. Disaster medicine: radiation disasters and emergency department preparedness. *Emer Med Clin of North Am*, 1996;14(2): 349–370.

Boutwell. Nuclear terrorism: the danger of highly enriched uranium. September 2002. Pugwash
http://www.pugwash.org/publication/pb/sept2002.pdf

Department of the Navy, Bureau of Medicine and Surgery Initial Management of Irradiated or Radioactively Contaminated Personnel.
http://www.vnh.org/BUMEDINST6470.10A/TOC.html

CDC Radiation Emergencies.
http://www.bt.cdc.gov/radiation/index.asp

K. Edsall. Radiological and nuclear incidents and terrorism. In: F. G. Walter, H. W. Meislin, eds. Advanced Hazmat Life Support Provider Manual. Arizona Board of Regents, Tucson, Ariz.: 2003;361–388. Advanced Hazmat Life Support.
http://www.ahls.org/

Primary Internet Resources

American College of Physicians
www.acponline.org

American College of Occupational/Environmental Medicine
www.acoem.org

American Academy of Pediatrics
www.aap.org/advocacy/releases/cad.htm

Centers for Disease Control
www.bt.cdc.gov

Virtual Naval Hospital
www.vnh.org

U.S. Army Military Research Institute in Infectious Disease (USAMRIID)
www.usamriid.army.mil

American Society for Microbiology
www.usmusa.org

ATSDR
www.atsdr.cdc.gov

Emergency Resources

State Health Departments: (Each state has its own number, and these can be accessed at this CDC site.)
www.cdc.gov/other.htm#states

Another useful site is maintained by the National Governor's Association and the Association of State and Terrotorial Epidemiologists
www.statepublichealth.org/

Local Health Departments
http://www.astho.org/index.php?template=regional_links.php

Reporting An Event

ATSDR—Chemical, radiologic, and natural disasters	(770) 488-7000
CDC Bioterrorism Preparedness and Response Center	(770) 488-7100
USAMRIID Emergency response Line	(888) 872-7443
USPHS Emergency Preparedness Office	(800) USA-NDMS
National Response Center	(800) 424-8802

Index

Page numbers followed by *f* indicate a figure; *t* following a page number indicates tabular material.